Unsteady Turbulent Shear Flows

International Union of Theoretical
and Applied Mechanics

Unsteady Turbulent Shear Flows

Symposium Toulouse, France,
May 5-8, 1981

Editors
R. Michel, J. Cousteix and R. Houdeville

With 283 Figures

Springer-Verlag
Berlin Heidelberg New York 1981

R. MICHEL
J. COUSTEIX
R. HOUDEVILLE

Office National d'Études et de Recherches Aérospatiales
Centre d'Études et de Recherches de Toulouse
2, avenue Edouard Belin
31055 Toulouse – France

ISBN-13:978-3-642-81734-2 e-ISBN-13:978-3-642-81732-8
DOI: 10.1007/978-3-642-81732-8

Library of Congress Cataloging in Publication Data
Main entry under title:
Unsteady turbulent shear flows.
(IUTAM-Symposien)
At head of title: International Union of Theoretical and Applied Mechanics.
Proceedings of the Symposium on Unsteady Turbulent Shear Flows.
Bibliography: p. Includes index.
1. Unsteady flow (Aerodynamics)-Congresses.
2. Turbulence-Congresses.
3. Shear flow-Congresses.
I. Michel, Roger, 1920-.
II. Cousteix, J. (Jean), 1947-.
III. Houdeville, R. (Robert), 1948-.
IV. International Union of Theoretical and Applied Mechanics.
V. Symposium on Unsteady Turbulent Shear Flows
(1981 : Toulouse, France) VI. Series.
TA358.U57 620.1'064 81-18386 AACR2

Softcover reprint of the hardcover 1st edition 1981

2061/3020/ 5 4 3 2 1 0

Scientific Committee

J. Cousteix, France

W.J. Mc Croskey, USA

H. Eckelmann, FRG

R. Michel, France (Chairman)

C.Y. Stepanov, USSR

D.P. Telionis, USA

A.D. Young, UK

Local Organizing Committee

J. Cousteix

R. Houdeville

R. Michel

Symposium Sponsors

International Union of Theoretical and Applied Mechanics (IUTAM)

Centre d'Etudes et de Recherches de Toulouse (ONERA-CERT)

Ecole Nationale Supérieure de l'Aéronautique et de l'Espace (ENSAE)

Direction des Etudes et Recherches Techniques (DRET)

Société Nationale des Industries Aérospatiales (SNIAS)

Preface

It was on a proposal from the "Comité National Français de Mécanique" that the General Assembly of the International Union of Theoretical and Applied Mechanics decided to sponsor a Symposium on Unsteady Turbulent Shear Flows. It was hoped that this might be arranged in Toulouse during the Spring of 1981, and that support might be obtained from the ONERA Toulouse Research Center.

From that time, it was clear that this subject, of considerable importance, had in recent years increasingly attracted the attention of research workers, and it was deemed timely to organize a symposium in order to reveal the current state of knowledge and to define the main areas for future research.

The Scientific Committee decided that the Symposium should concentrate on the behaviour of turbulent shear flows, experiencing various unsteady boundary conditions ; fundamental aspects of turbulent phenomena should be considered, this in view of largely aeronautical applications, but also for other important problems concerned for example with turbomachinery, off-shore installations in rough seas, atmospheric wind effects, etc ...

The program included 3 extended papers of a 60 minute duration and 29 papers of 30 minute duration. The symposium lasted 3 days and a half from Monday morning to Friday afternoon. A free afternoon on Thursday offering a well deserved rest, and the occasion of new personal exchanges between the participants seems to have been appreciated by all.

We can say that the meeting has well fulfilled the goals aimed at by such a symposium by gathering the well known scientific persons from various countries who work on that subject. The lectures have given a very good, clear and sound view of the present knowledge of experimental and theoretical research on the main unsteady shear flows as boundary layers, jets, wakes and vortex shedding .... Very bright discussions leading to coherent suggestions for future research occured after each lecture, and during breaks and lunches.

We do hope that this Symposium on Unsteady Turbulent Shear Flows met the success we hoped it would, but it must be understood that is success was also due to the goodwill and support of many persons and organizations.

We are grateful to the Members of the Scientific Committee for their scientific assistance, which has been unvaluable as well as for their help before and during the symposium.
Let us thank the Session Chairmen, the Authors and all the participants for their active cooperation.
We would also express our thanks to the Director of the "Ecole Nationale Supérieure de l'Aéronautique et de l'Espace", who has so kindly lent us his remarkable installations and means.
A quite generous financial help has been given to us by the "Direction Technique des Etudes et Recherches Techniques" from the French Ministry of Defence, and by the Société Nationale des Industries Aérospatiales.

We would like to thank them very much.
Finally, we did appreciate Springer Verlag's help and efficacy for the publication of the Symposium proceedings.

Toulouse, July 1981

R. Michel, J. Cousteix, R. Houdeville

Participants

Acharya, M.	CH	Fluid Mechanics Group, B.B.C. Research Center, 5405 Baden Dattwill
Barbi, C.	F	Institut de Mécanique des Fluides de Marseille, 1 rue Honnorat, 13003 Marseille
Binder,G.	F	Institut de Mécanique de Grenoble, BP 53X, 38041 Grenoble Cedex
Boisson, H.C.	F	Institut de Mécanique des Fluides de Toulouse, 2 rue Charles Camichel, 31071 Toulouse Cedex
Bouriot, M.	F	Université de Poitiers, Centre d'Etudes Aérodynamiques et Thermiques,86000 Poitiers
Brown, D.T.	UK	London Centre for Marine Technology, Department of Mechanical Engineering, University College, London
Carr, L.W.	USA	U.S. Army Aeromechanics Laboratory, Nasa Ames Research Centre, Moffett Field, California 94035
Cebeci, T.	USA	Mechanical Engineering Department, California State University, Long Beach,California
Cerri, G.	I	Istituto di Macchine e Tecnologie Meccaniche, Universita di Roma, Roma
Charnay, G.	F	Ecole Centrale de Lyon, Laboratoire de Mécanique des Fluides, 36 avenue Guy-de-Collongue, 69130 Ecully
Cervantes, J.G.	MEX	School of Engineering, National University of Mexico, Mexico City
Couston, M.	F	Onera, 29 avenue de la Division Leclerc, 92320 Chatillon
Cousteix, J.	F	Onera/Cert, 2 avenue Edouard Belin, 31055 Toulouse Cedex
De Ruyck, J.	B	Vrije Universiteit Brussels, Department of Mechanics, Pleinlaan 2, 1050 Brussels
Desopper, A.	F	Onera, 29 avenue de la Division Leclerc, 92320 Chatillon
Dyment, M.A.	F	Institut de Mécanique des Fluides de Lille, 5 boulevard Painlevé, 59000 Lille

Favier, D.	F	Institut de Mécanique des Fluides de Marseille, 1 rue Honnorat, 13003 Marseille
Favre-Marinet, M.	F	Institut de Mécanique de Grenoble, BP 53X, 38041 Grenoble Cedex
Fang, L.W.	PRC	Nanjing Aeronautical Institute, Peoples' Republic of China
Fannelop, T.K.	N	Division of Aerodynamics and Gas Dynamics, The Norvegian Institute of Technology, N 7034 Trondheim NTH
Gerber, A.	F	Ecole Centrale de Lyon, Laboratoire de Mécanique des Fluides, 36 avenue Guy-de-Collongue, 69130 Ecully
Ha Minh, H.	F	Institut de Mécanique des Fluides de Toulouse, 2 rue Charles Camichel, 31071 Toulouse Cedex
Hayakawa, M.	JAP	Faculty of Engineering, Hokkaïdo University N-13 W-8 Sapporo 060
Ho, C.M.	USA	Department of Aerospace Engineering, University of Southern California,Los Angeles, California 90007
Houdeville, R.	F	Onera/Cert, 2 avenue Edouard Belin, 31055 Toulouse Cedex
Hussain, A.K.M.F.	USA	Department of Mechanical Engineering, University of Houston, Houston, Texas 77004
Javelle, J.	F	Onera/Cert, 2 avenue Edouard Belin, 31055 Toulouse Cedex
Johnson III, J.A.	USA	Department of Physics, Rutgers University, New Brunswick, New Jersey 08903
Kobashi, Y.	JAP	Faculty of Engineering, Hokkaïdo University, N-13 W-8 Sapporo 060
Kueny, J.L.	F	Institut de Mécanique de Grenoble, BP 53X, 38041 Grenoble Cedex
Lecointe, Y.	F	Ensm, 1 rue La Noé, 44072 Nantes Cedex
Legendre, R.	F	Onera, 29 avenue de la Division Leclerc, 92320 Chatillon
Levi, E.	MEX	Instituto de Ingenierria,Universidad Nacional Autonoma de Mexico, Mexico 20,D.F.
Mc Croskey, W.J.	USA	Usartl, Ames Research Centre, Moffett Field, California 94035

Mainardi, H.	F	Laboratoire d'Energétique et de Mécanique Appliquée, Ecole Supérieure de Transport, d'Energie et de Propulsion, Université d'Orléans, 45046 Orléans Cedex
Maresca, C.	F	Institut de Mécanique des Fluides de Marseille, 1 rue Honnorat, 13003 Marseille
Meier, H.U.	FRG	Deutsche Forschungs und Versuchsanstalt für Luft und Raumfahrt e.v., Institut für Experimentelle Strömungsmechanik, D-3400 Göttingen, Bunsenstrasse 10
Mehta, U.B.	USA	M/S 202A-1, Nasa Ames Research Centre, Moffett Field, California 94035
Michel, R.	F	Onera/Cert, 2 avenue Edouard Belin, 31055 Toulouse Cedex
Orlandi, P.	I	Istituto di Aerodinamica, Universita di Roma, Via Eudossiana, 16
Parikh, P.G.	USA	Department of Mechanical Engineering, Stanford University, Stanford, California 94305
Ramaprian, B.R.	USA	Institute of Hydraulic Research, University of Iowa, Iowa City, Iowa
Reynolds, W.C.	USA	Department of Mechanical Engineering, Stanford University, Stanford, California 94305
Richter, K.	FRG	Drittes Physikalisches Institut, Universitat Göttingen, Bürgerstrasse 42-44, D-3400 Göttingen
Schiestel, R.	F	Institut de Mécanique Statistique de la Turbulence, 12 avenue du Général Leclerc, 13003 Marseille
Shamroth, S.J.	USA	Scientific Research Associates, Inc., PO 498, Glastonbury, Ct 06033
Simpson, R.L.	USA	Department of Civil and Mechanical Engineering, Southern Methodist University, Dallas, Texas 75275
Sirieix, M.	F	Onera, 29 avenue de la Division Leclerc, 92320 Chatillon
Smith, P.D.	UK	R.A.E., Aerodynamics Department, Bedford MK41 6AE

Sutton, E.P.	UK	Cambridge University, Engineering Department, Trumpington street, Cambridge CB2 IPZ
Svehla, K.M.	UK	Cambridge University, Engineering Department, Trumpington street, Cambridge CB2 IPZ
Tartarin, J.	F	Laboratoire de Dynamique des Fluides, 40 avenue du Recteur Pineau, 86022 Poitiers Cedex
Tassa, Y.	USA	Lockheed-Georgia Company, Marietta, Georgia 30063
Van den Berg, B.	NL	N.L.R., Anthony Fokkerweg 2, Amsterdam 1017
Vingut, R.	F	Aérospatiale, BP 13, 13722 Marignane
Weinstein, L.M.	USA	Nasa Langley Research Centre, Hampton, Virginia 23665
Wirz, H.J.	FRG	MTU Motoren und Turbineen Union, Munchen GMBH, Dachauer STR 663, Munchen
Young, A.D.	UK	Department of Aeronautical Engineering, Queen Mary College, University of London, London E1 4NS

Contributors

Acharya, M.	CH	Fluid Mechanics Group, B.B.C. Research Center, 5405 Baden Dattwill
Barbi, C.	F	Institut de Mécanique des Fluides de Marseille, 1 rue Honnorat, 13003 Marseille
Binder, G.	F	Institut de Mécanique de Grenoble, BP 53X, 38041 Grenoble Cedex
Boisson , H.C.	F	Institut de Mécanique des Fluides de Toulouse, 2 rue Charles Camichel 31071 Toulouse Cedex
Bouchard, E.E.	USA	Department of Mechanical Engineering, Stanford University, Stanford, California 94305
Bouriot, M.	F	Université de Poitiers, Centre d'Etudes Aérodynamiques et Thermiques, 86000 Poitiers
Brown, D.T.	UK	London Centre for Marine Technology, Department of Mechanical Engineering, University College, London
Carr, L.W.	USA	U.S. Army Aeromechanics Laboratory, Nasa Ames Research Center, Moffett Field, California
Cebeci, T.	USA	Mechanical Engineering Department, California State University, Long Beach,California
Cenedese, A.	I	Istituto di Aerodinamica, Universita di Roma, Roma
Cerri, G.	I	Istituto di Macchine e Tecnologie Meccaniche, Universita di Roma, Roma
Cervantes, J.G.	MEX	School of Engineering, National University of Mexico, Mexico City
Charnay, G.	F	Ecole Centrale de Lyon, Laboratoire de Mécanique des Fluides, 36 avenue Guy-de-Collongue, 69130 Ecully
Chassaing, P.	F	Institut de Mécanique des Fluides de Toulouse, 2 rue Charles Camichel, 31071 Toulouse Cedex
Chen, S.H.	USA	Department of Aerospace Engineering, University of Southern California, Los Angeles, California 90007

Chew, Y.T.	USA	Department of Civil and Mechanical Engineering, Southern Methodist University, Dallas, Texas 75275
Cousteix, J.	F	Onera/Cert, 2 avenue Edouard Belin, 31055 Toulouse Cedex
De Ruyck, J.	B	Vrije Universiteit Brussels, Department of Fluid Mechanics, Pleinlaan 2, 1050 Brussels
Desopper, A.	F	Onera, 29 avenue de la Division Leclerc, 92320 Chatillon
Dyment, M.A.	F	Institut de Mécanique des Fluides de Lille, 5 boulevard Painlevé, 59000 Lille
Evans, G.P.	UK	Cambridge University, Engineering Department, Trumpington street, Cambridge CB2 IPZ
Fang, L.W.	PRC	Nanjing Aeronautical Institute, Peoples' Republic of China
Favier, D.	F	Institut de Mécanique des Fluides de Marseille, 1 rue Honnorat, 13003 Marseille
Favre-Marinet, M.	F	Institut de Mécanique de Grenoble, BP 53X, 38041 Grenoble Cedex
Gerber, A.	F	Ecole Centrale de Lyon, Laboratoire de Mécanique des Fluides, 36 avenue Guy-de-Collongue, 69130 Ecully
Ha Minh, H.	F	Institut de Mécanique des Fluides de Toulouse, 2 rue Charles Camichel, 31071 Toulouse Cedex
Hayakawa, M.	JAP	Faculty of Engineering, Hokkaïdo University, Sapporo 060
Hirsch, Ch.	B	Vrije Universiteit Brussels, Department of Fluid Mechanics, Pleinlaan 2, 1050 Brussels
Ho, C.M.	USA	Department of Aerospace Engineering, University of Southern California,Los Angeles, California 90007
Houdeville, R.	F	Onera/Cert, 2 avenue Edouard Belin, 31055 Toulouse Cedex
Hussain, A.K.M.F.	USA	Department of Mechanical Engineering, University of Houston, Houston, Texas 77004
Iannetta, S.	I	Istituto di Idraulica, Universita Dell'Aquila
Javelle, J.	F	Onera/Cert, 2 avenue Edouard Belin, 31055 Toulouse Cedex

Jayaraman, R.	USA	Department of Mechanical Engineering, Stanford University, Stanford, California 94305
Jones , G.S.	USA	Virginia Polytechnic Institute and State University, Blacksburg, Virginia 24061
Johnson III, J.A.	USA	Department of Physics, Rutgers Univerity, New Brunswick, New Jersey 08903
Kobashi, Y.	JAP	Faculty of Engineering, Hokkaïdo University, N-13 W-8 Sapporo 060
Kreplin, H.P.	FRG	Deutsche Forschungs und Versuchsanstalt für Luft und Raumfahrt e.v., Institut für Experimentelle Strömungsmechanik, D-3400 Göttingen, Bunsenstrasse 10
Kueny, J.L.	F	Institut de Mécanique de Grenoble, BP 53X, 38041 Grenoble Cedex
Lemonnier, H.	F	Institut de Mécanique de Grenoble, BP 53X, 38041 Grenoble Cedex
Levi, E.	MEX	Instituto de Ingenierria, Universidad Nacional Autonoma de Mexico, Mexico 20,D.F.
Lin, I.	USA	Department of Physics, Rutgers University, New Brunswick, New Jersey 08903
Mc Guiness, M.D.	UK	Cambridge University, Engineering Department, Trumpington street, Cambridge CB2 IPZ
Mainardi, H.	F	Laboratoire d'Energétique et de Mécanique Appliquées, Ecole Supérieure de Transport, d'Energie et de Propulsion, Université d'Orléans, 45046 Orléans Cedex
Maresca, C.	F	Institut de Mécanique des Fluides de Marseille, 1 rue Honnorat, 13003 Marseille
Meier, H.	FRG	Deutsche Forschungs und Versuchsanstalt für Luft und Raumfahrt e.v., Institut für Experimentelle Strömungsmechanik, D-3400 Göttingen, Bunsenstrasse 10
Melinand, J.P.	F	Ecole Centrale de Lyon, Laboratoire de Mécanique des Fluides, 36 avenue Guy-de-Collongue, 69130 Ecully
Orlandi, P.	I	Istituto di Aerodinamica, Universita di Roma, Roma
Panday, P.K.	F	Laboratoire d'Energétique et de Mécanique Appliquées, Ecole Supérieure de Transport, d'Energie et de Propulsion, Université d'Orléans, 45046 Orléans Cedex

Parikh, P.G.	USA	Department of Mechanical Engineering, Stanford University, Stanford, California 94305
Patel, M.H.	UK	London Centre for Marine Technology, Department of Mechanical Engineering, University College, London
Ramaiah, R.	USA	Department of Physics, Rutgers University, New Brunswick, New Jersey
Ramaprian, B.R.	USA	Institute of Hydraulic Research, University of Iowa, Iowa City, Iowa
Reynolds, W.C.	USA	Department of Mechanical Engineering, Stanford University, Stanford, California 94305
Richter, K.	FRG	Drittes Physikalisches Institut, Universität Göttingen,Bürgerstrasse 42-44, D-3400 Göttingen
Ronneberger, D.	FRG	Drittes Physikalisches Institut, Universität Göttingen,Bürgerstrasse 42-44, D-3400 Göttingen
Sankar, N.L.	USA	Lockheed-Georgia Company, Marietta, Georgia
Sevrain, A.	F	Institut de Mécanique des Fluides de Toulouse, 2 rue Charles Camichel, 31071 Toulouse Cedex
Shamroth, J.	USA	Scientific Research Associates, Inc., Glastonbury, CT 06033
Shivaprasad, B.G.	USA	Department of Civil and Mechanical Engineering, Southern Methodist University, Dallas, Texas 75275
Simpson, R.L.	USA	Department of Civil and Mechanical Engineering, Southern Methodist University, Dallas, Texas 75275
Sutton, E.P.	UK	Cambridge University, Engineering Department, Trumpington street, Cambridge CB2 IPZ
Svehla, K.M.	UK	Cambridge University, Engineering Department, Trumpington street, Cambridge CB2 IPZ
Tassa, Y.	USA	Lockheed-Georgia Company, Marietta, Georgia
Telionis, D.P.	USA	Virginia Polytechnic Institute and State University, Blacksburg, Virginia 24061
Tsen, L.F.	F	Université de Poitiers, Centre d'Etudes Aérodynamiques et Thermiques, 86000 Poitiers
Tu, S.W.	USA	Institute of Hydraulic Research, University of Iowa, Iowa City, Iowa

Weinstein, L.M.	USA	Nasa Langley Research Center, Hampton, Virginia 23665
Zaman, K.B.M.Q.	USA	Department of Mechanical Engineering, University of Houston, Texas 77004

Contents

★) Extended lecture

Introduction

R. Legendre , Haut Conseiller Scientifique à l'ONERA, Membre de l'Académie des Sciences

Onera, 29 avenue de la Division Leclerc, 92320 Chatillon

Some six years ago, I was a member of the IUTAM Bureau and participated in General Assemblies where the possible subjects for future symposia were discussed. As boundary layer problems are dealt with in many international meetings, there was not much enthousiasm for their selection, until Professor Michel made his proposal for this symposium devoted to unsteady turbulent boundary layers, which is already a wide domain of interest for aerodynamicists.

This was approved by the General Assembly in Herrenhalb where Professor Maurice Roy and myself supported Professor Michel's proposal, underlining that viscosity has a greater effect on unsteady flows than on steady ones. That is the reason why I have the pleasure to welcome in Toulouse this bright assembly of aerodynamicists.

Before I started thinking on it and discussing with Professor Michel, I imagined that unsteadiness had no serious effect except for a periodical flow inside the Tolmien-Schlichting instability domain. It implied, for instance, that the flow would have been quasi-steady around a fluttering airplane. I was wrong and you know it : you will discuss many curious phenomena appearing during not so quick flow reversals.

But I am personally more interested in the influence of viscosity on separation, because there are important manifestations of hysteresis which are significant for applications in aeronautics. The most evident is the loop in the curve of lift versus incidence for a helicopter blade. It seems that you have considered that such a behaviour was accessory regarding your main topic of boundary layer unsteadiness since only a few papers are dealing with separation. It would not be easy to obtain soon a new IUTAM symposium to discuss at length unsteady separation which is closely connected with boundary layer evolution and is one of its more complex consequence. I mention long time lags, multivalued solutions, present unpredictability outside empiricism.

For the time being, as I am retired and free to choose the matter of my thoughts, I am discussing with my colleague Sirieix the mechanisms of vortex breakdowns which, as you probably consider, are no longer consequences of boundary layer evolutions.

Vortex breakdown appears inside the fluid, far from the solid surfaces, but generally inside wakes. Even the vortex starting from a focus near the apex of a delta wing is to be considered as a particular wake, prolongating inside the fluid, the boundary layer on the suction side. It implies that you will be obliged, sooner or later, to extend your investigations to include vortex breakdown in a broad family of viscosity consequences either in laminar or in turbulent flows.

So far, vortex breakdown was mainly studied in steady conditions, but I am sure that many interesting phenomena are to be discovered by a careful and methodical study of its behaviour in unsteady conditions. Already, aerodynamicists who observed vortex breakdowns inside glass tubes showed that hysteresis has complex and curious effects.

I am not so much interested in what happens inside tubes. I should prefer to observe evolution of breakdown over a delta wing in a wind tunnel with the diversity of pitch, roll and yaw.

When I searched for and succeeded in observing vortex sheets rolling up from leading edges, I did not pay enough attention to the vortex breakdown which was already conspicuous on some photographs by my colleague Werlé. Works in Great Britain and in Germany demonstrated, theoretically and experimentally, the importance of the phenomenon.

However, you were wise in choosing not too wide a field for discussion and leaving the topics needing deeper investigations for later meetings.

I wish your meeting a great success.

A Review of Unsteady Turbulent Boundary-Layer Experiments

LAWRENCE W. CARR

U.S. Army Aeromechanics Laboratory
Research and Technology Laboratories (AVRADCOM)
Ames Research Center, Moffett Field, Calif., U.S.A.

Summary

The essential results of a comprehensive review of existing unsteady turbulent boundary-layer experiments are presented. Different types of unsteady flow facilities are described, and the related unsteady turbulent boundary-layer experiments are cataloged and discussed. The measurements that have been obtained in the various experiments are described, and a complete list of experimental results is presented. All the experiments that measured instantaneous values of velocity, turbulence intensity, or turbulent shear stress are identified, and the availability of digital data is indicated. The results of the experiments are analyzed, and several significant trends are identified. An assessment of the available data is presented, delineating gaps in the existing data, and indicating where new or extended information is needed. Guidelines for future experiments are presented.

Introduction

During the past few years, there has been a significant increase in the level of effort directed toward the analysis of unsteady turbulent boundary layers. A wide range of theoretical methods have proliferated during this period, while the existing experimental data base has been meager, scattered, and disparate. Several experimental programs are presently under way to produce further experimental data for use in comparison to theory, but the data base is still widely dispersed.

Since such a wide range of experimental data exists without a strong common pattern, there is an increasing need for central documentation of the various results. In this way, the various research efforts would be more readily available, and comparison of the results can be facilitated. Several workshops on unsteady turbulent boundary-layer experimental research have been organized by the present author. During these workshops, it has become increasingly clear that a careful review of the existing data, as well as a documentation of the current experimental programs in a single source, would be of great value to future endeavors in this area.

To satisfy this need, an AGARDograph has been prepared which catalogs all the pertinent sources, much of the relevant data, and indications of future studies. A comprehensive international literature search has been performed, identifying those groups who have actually published work in the subject area, as well as disclosing sources that have valuable but unpublished data appropriate to the present subject. Selected research personnel in the United States and several European countries have been visited to discuss and obtain pertinent data sets and descriptions of experiments. The data from these various sources are now cataloged and prepared in a form appropriate for general distribution and analysis; more than 40 pertinent experiments are reviewed.

In the present paper, highlights from the AGARDograph are presented, including description of both past and present experimental programs. The types of experimental data that are available are discussed, and experimentally observed characteristics of unsteady turbulent boundary layers are assessed. Guidelines for future experiments are presented.

Types of Experimental Facilities

The procedure for experimentally modeling an unsteady viscous flow problem in a laboratory is always a difficult task. In fact, the ingenuity that has been demonstrated by the various experimentalists is quite impressive. A brief review of some examples of tunnel design will indicate the range of techniques that have been employed. The first type of facility, shown in Fig. 1, was used by Karlsson (1958) for his pioneering experiment studying the response of an unsteady turbulent boundary layer on a flat plate. The basic facility is an open-return wind tunnel; the flow oscillation is produced by a set of rotating vanes installed near the exit of the tunnel. As these vanes rotate, they produce a variable blockage that causes the tunnel flow to pulsate. Variations of this technique have included controlled-speed vanes installed upstream of the test section (Simpson et al., 1978), a slotted cylinder at the tunnel exit (Acharya and Reynolds, 1975), a rotating butterfly valve at the exit (Cousteix et al., 1977), and several others. The technique of variable blockage has also been used in unsteady pipe flow by Schultz-Grunow (1940), Ramaprian and Tu (1980), Mizushina et al. (1973), Lu et al. (1973), and others; it remains one of the most common of the experimental techniques for creating pulsation in the free-stream flow.

Some other, more esoteric techniques are also of interest. One successful approach incorporates the tunnel wall as a part of the oscillation mechanism. Brembati (1975) installed a flexible section in the ceiling of an open-return wind tunnel (Fig. 2), and sinusoidally oscillated this movable ceiling, thus producing a combination of variable free-stream velocity and adverse pressure gradient. The technique used by Patel (1977), Kenison (1977), and Pericleous (1977) in their studies incorporates the tunnel structure in still another way. In this case, as shown in Fig. 3, the flow from the contraction section of the tunnel enters a partially open test section. The ceiling and floor of the test section are removed; the upper and lower surfaces of the entrance section of the tunnel are continued into the test section, and are carefully constructed to permit smooth deflection of these surfaces as flaps. These flaps are sinusoidally oscillated in pitch; they induce a series of traveling vortices which move down the test section, creating an oscillatory perturbation velocity on the test section.

Still another technique for producing an unsteady flow has been devised by Parikh et al. (1981). In this case (Fig. 4), the entrance flow is maintained at a constant value by holding the total mass-flow rate constant and an oscillating flow with varying magnitude of adverse pressure gradient is produced in the test section by removing fluid from the wall opposite the test surface in a programmed manner. The tunnel surface opposite the test surface is partitioned into two porous sections, one directly below the test surface, the other downstream. A slotted plate controls the amount of fluid drained from each section. As the plate moves back and forth, varying amounts of fluid exit from the tunnel through the forward or aft sections of the porous surface, while the total fluid flow remains constant through the cycle.

These are only a few examples of the techniques used to produce oscillatory flow in the laboratory. The interested reader is referred to the AGARDograph (Carr, 1981) for descriptions of the many other facilities that have been devised. These techniques demonstrate the novelty of the various designs; they also show that the generation of unsteady flows in the laboratory is a very difficult and complex task. Each of the facilities discussed has both benefits and limitations; no one design is clearly better than the others. It is important to realize that results obtained in facilities having such diverse design and performance characteristics as these should be compared with special care.

Types of Flows Reviewed

Each engineering application has had its own set of requirements. For example, the information needed for the analysis of an unsteady heat-transfer problem is significantly different from the information needed for accurate prediction of dynamic stall. The design of a fluidic device depends on parameters much different from those required for design of a compressor blade. Thus, each of these engineering applications has placed a strict limitation on the type of flow result that was sought. The basic fluid mechanics common to all of these problems has always been of interest. However, parametric variation of flow conditions has not been possible in most of the facilities. Instead, many of these experiments have been exploratory in nature, attempting to identify potential areas of interest rather than studying the behavior of the unsteady turbulent boundary-layer itself. No single experiment has been able to study all the parameters that are necessary to define the behavior of unsteady turbulent boundary layers.

Thus, there are gaps in the existing data, even though many types of flows have been studied. The many laminar, transitional, and turbulent unsteady flow experiments that have been performed are fully referenced in the AGARDograph. Only the unsteady turbulent boundary-layer experiments will be discussed here. These include flat plate flows, with and without pressure gradient, two-dimensional channel, pipe, diffuser, airfoil, and compressor blade flows. Jet and wake flows have not been included since the survey was limited to viscous flows in contact with a solid boundary.

The existing turbulent boundary-layer experiments have been summarized in Fig. 5. Note that certain authors' names are presented in bold type — these experiments are documented in Carr (1981), and contain instantaneous measurements of the unsteady turbulent boundary-layer characteristics. The light-faced type denotes experiments of general interest, but without instantaneous data. In Fig. 6, pipe, airfoil, and cascade experiments are presented, as well as a list of new experiments from which results have not yet been acquired by the present author.

The data for the experiments that have been included in the AGARDograph are presented in the form supplied by the original author whenever possible; no smoothing or modification of the data has been performed. Although every effort has been made to ensure a complete list of available experiments, particularly those with instantaneous ensemble-averaged data, there certainly are experiments that have not been discussed or were overlooked completely. These omissions were definitely not intentional. Please send documentation of these experiments to the present author for inclusion in the data bank and catalog. Figure 7 shows the format used to document the various experiments presented in the AGARDograph. The information indicated in this figure is recommended as a minimum level of documentation that should be recorded for any future unsteady turbulent boundary-layer experiment.

Data Acquisition and Analysis

The acquisition of data for an unsteady turbulent boundary layer can be a formidable task. For example, the velocity in an unsteady turbulent boundary layer can be measured in a variety of ways: electrochemically (Mizushina et al., 1973); by use of a micropropeller (Jonnson and Carlsen, 1976); hot wire anemometers (Cousteix et al., 1977); single-beam lasers (Reynolds et al., 1981); dual-beam lasers (Simpson et al., 1980); as well as other techniques. The unsteady velocity signal is a combination of mean, periodic, and random fluctuations of varying magnitude, and extraction of the pertinent components requires varying levels of sophistication. Since the various experiments have differing goals, data analysis techniques vary as well. As shown in Fig. 8, there are several levels of sophistication which can be employed for analysis of the resultant signal. The least difficult — the time-averaged mean velocity — can be obtained by performing a digital or analog long-time average of the turbulent velocity signal. This approach is also used to obtain the RMS value for the turbulence intensity. The next level of sophistication is the measurement of the periodic component of velocity. There are several ways this information can be obtained, including cross-correlation of the turbulent velocity signal with a sine wave having a frequency equal to the driving frequency. Another approach is to Fourier transform or harmonically analyse the unsteady turbulent signal and extract the Fourier coefficients associated with the fundamental and higher harmonics of the oscillation. If even more information is desired about the flow, a phase-averaging device can be used which samples the turbulent signal at fixed phases in each cycle, storing the value of the signal at each specified phase and retaining the summed signal either by analog or digital means. Each of these methods can produce an output containing the amplitude and phase of the first harmonic response of the boundary layer to the imposed oscillation. In addition, the phase-averaged signal contains detailed information about the time history of the velocity signal during a cycle. This information can be of great value when complex flow phenomena are being studied, because all the harmonic content of the original signal potentially can be retained.

The existing turbulent boundary-layer experiments have been classified in Figs. 9 and 10, based on these different levels of analysis: time-averaged mean (level I), periodic amplitude and phase (level II), single-component ensemble-average (level III), and dual-component ensemble- or phase-average (level IV). In these figures, bold type indicates data recorded by the originating author; light type denotes information that can be reconstructed from data presented in the supporting documents (e.g., Tomsho (1978) supplied ensemble-averaged data for velocity; time-averaged mean data can then be reconstructed from this information).

Evaluation of Experimental Results

As noted earlier, over 40 unsteady turbulent boundary-layer experiments have been identified. This quantity precludes individual analysis in the present paper. However, the large number of experiments offers a unique opportunity for comparison of results. In particular, several significant observations can be made.

Time-Mean Averages: For all the flows examined, the experiments demonstrate that the time-averaged mean velocity, $\bar{U}(y)$, is the same as the value expected for a steady flow having a velocity corresponding to the mean of the oscillatory outer flow, $U_m(y)$. This has been observed on a flat plate by Karlsson (1958), where $\bar{U}(y)$ was demonstrated to be the same as $U_m(y)$ over a wide range of frequencies and amplitudes. At the other end of the range of experimental complexity, $\bar{U}(y)$ on a stator blade in a jet engine compressor was demonstrated by Evans (1978) to be the same as the steady $U_m(y)$ (Fig. 11).

There are certainly conditions and situations where the fact that the $\bar{U}(y)$ of the unsteady flow is the same as the $U_m(y)$ from steady flow is of significant value — unsteady heat transfer, mean diffuser behavior — situations where only the mean performance characteristics are needed for analysis of the problem. However, this equivalence, as significant as it is, can be very misleading if the purpose of the research is to identify the fluid mechanics of the unsteady flow field in question. A good example is Karlsson's experiment itself, where he observed regions of reversed flow on the flat plate, even though $\bar{U}(y)$ was the same as $U_m(y)$. Evans (1978) demonstrates that even though $\bar{U}(y)$ is the same as $U_m(y)$, no assumption can be made about the unsteadiness of the flow itself. In his experiments, the flow changed from laminar to turbulent through the cycle (Fig. 12). This change was completely masked by the time-averaging process (see Fig. 11). Another example, the diffuser study by Schachenmann (1974), showed the time averages to be the same for conditions in the boundary layer which varied dramatically with frequency. (The periodic velocity fluctuation in the boundary layer varied from 1 to 100% of the oscillation magnitude at the center of the diffuser, while the mean velocity in the boundary layer remained the same.) Thus, the observation that $\bar{U}(y)$ is the same as $U_m(y)$ has merit, but should not be used as a basis for describing the dynamics of the flow field itself.

Turbulence Structure: A variety of experiments have been performed to study the turbulence intensity in unsteady turbulent flow. Several of these show the turbulence structure unaffected by oscillation of the flow field. A study by Cousteix et al. (1977) demonstrates this conclusion. Figure 13 presents the measured turbulence intensity and Reynolds shear stress at various parts of a cycle of oscillation. Note that even though significant variations appear in these quantities, the ratio of the shear stress to its component turbulence intensities remains constant at a value equivalent to that of steady flow (Fig. 14). Thus, under certain conditions, steady flow turbulence models can be used to predict unsteady turbulent boundary-layer behavior. Indeed, several experiments have been accurately represented by conventional numerical techniques. These include Lu et al. (1973) for flow in a pipe, Johnson and Carlsen (1976) for purely oscillatory flow, Cousteix and his colleagues (1977, 1979) for both zero- and adverse-pressure gradient flows on a flat plate, and Parikh et al. (1981) for a time-varying adverse pressure gradient (predicted by Lyrio et al., 1981).

However, there are cases where substantial changes in the turbulence intensity can occur. As the frequency of oscillation is increased, a critical frequency can be reached at which there can be a significant interaction between the oscillatory motion and the turbulent structure. An example of this can be seen in work done by Mizushina et al. (1973) for fully developed flow in a pipe. For frequencies below this critical frequency, the ensemble-averaged turbulence intensity is very similar to the turbulence intensity that would appear at that particular point in the cycle for the corresponding steady velocity (Fig. 15). However, if the frequency of oscillation is increased beyond a critical frequency, the situation is significantly altered. The turbulence intensity no longer has a pattern similar to that which would be associated with the steady flow (Fig. 15) and significant variations appear in the velocity distribution obtained from ensemble-averaging (Fig. 16). Mizushina et al. determined that the behavior was associated with a critical frequency related to turbulent bursts in the flow; this kind of behavior was also observed by Ramaprian and Tu (1980). This result is very important for those who wish to model turbulent unsteady flows. When these interactions occur, they can significantly change the structure of the turbulence, seriously compromising the validity of the model that is being used to predict the flow behavior.

Strong Interaction Effects: In many of the experiments that have been reviewed, unsteady viscous effects were present but did not cause any significant variation in the overall behavior of the flow field. However, when turbulent boundary layers near separation are exposed to oscillation, the situation can be dramatically altered. Under these conditions, significant global changes can occur in the boundary layer, resulting in major alteration of the shape factor and displacement thickness. A good example of this phenomenon is shown in Fig. 17, from Houdeville et al. (1976). Here the adverse pressure gradient has combined with oscillation to produce clearly defined changes in the evolution of displacement thickness. This variation in displacement thickness will be quite important if prediction of the coupled viscid-inviscid interaction is attempted.

Unsteady Flow Near the Wall: When an oscillating external velocity is imposed on a viscous flow, the flow near the wall responds quite readily to this unsteadiness. In many of the experiments that have been performed, the unsteady viscous reaction to the imposed flow variations is completely confined to the Stokes layer near the wall; the outer region of the boundary layer behaves as if it were "frozen." This is both a benefit and a problem. If the goal of the research is to predict the global flow behavior of an unsteady flow with well-defined initial and boundary conditions, the Stokes layer region can often be virtually ignored without serious detriment to the accuracy of the prediction (Lyrio et al., 1981).

However, there is a class of problems that depends strongly on the character of the Stokes layer. In many situations, no data other than wall shear stress and pressure distributions can be measured. In these cases, prediction of the boundary-layer behavior will directly depend on the ability to relate the wall shear stress to the flow in the central region of the boundary layer. The experimental studies that have emphasized study of the wall region show major phase changes near the wall (e.g., Simpson et al., 1980; Binder and Kueny, 1981). These measurements are extremely difficult to perform, and the results are limited. However, they clearly demonstrate that the flow behavior near the wall can vary dramatically during oscillation. Thus, future research should emphasize the near-wall region of unsteady turbulent boundary layers, matching the unsteady wall shear and Stokes layer behavior with the boundary-layer behavior that occurs away from the wall.

Amplitude and Frequency Effect: Low amplitude or low frequency does not necessarily mean quasisteady behavior. The values of amplitude and frequency used in selected experiments are shown in Fig. 18. There is obviously a wide range of values that can result in unsteady effects. It is quite probable that no single dimensionless factor can be chosen to represent all the effects of unsteadiness: there are different time scales for the wall region compared to the outer flow; the eddy structure changes rapidly in adverse pressure gradient; the flow responds to temporal variation in velocity differently than it does to spatial variations. In addition, many experiments contribute only a single point to Fig. 18. Various dimensionless parameters have been suggested (e.g., Strouhal number based on x, δ, δ^*, L, etc.). The results for one of these, $S_\delta = f\delta/U$, are shown in Fig. 19 for the same set of experiments as presented in Fig. 18 (S_δ is based on local velocity and boundary-layer thickness). The shaded region shows that there is a small range of amplitude and frequency for which no unsteady effects have been reported. As the frequency or the amplitude increases, unsteady effects appear in the outer region of the boundary layer, especially for adverse pressure gradient flows. Note that the data from the Parikh et al. (1981) experiment show outer flow effects for the low range of S_δ, but only inner-layer variation at high S_δ.

Another parameter that has been considered significant for determining the possibility of unsteady effects is the burst frequency. This burst frequency (F_b) has been developed from steady flow (Offen and Kline, 1973; Rao et al., 1971), and acts as an indicator of the frequency at which the turbulent eddy structure will respond to external forcing function. This value is defined as $F_b = U/5\delta$ for a flat plate; it has been modified in the present report to reflect changes in structure due to adverse pressure gradient (local values are used for U and δ, as measured at the downstream end of the test surface of the related experiments). Figure 20 presents the tested frequencies for some existing experiments compared to the corresponding burst frequencies.

Note that the zero pressure gradient flows show unsteady effects only near the wall (with the exception of Mizushina et al., 1973). Acharya and Reynolds (1975) found sublayer effects when oscillating at the burst frequency, but not at 60% F_b. On the other hand, Karlsson (1958) found the largest phase change to occur in the sublayer for frequencies less than 40% of F_b; Ramaprian and Tu (1980) found significant effects at only 27% of F_b; Mizushina et al. (1973) found a major change occurred across the full pipe flow for F_{crit} less than 20% of F_b.

The adverse pressure gradient flows, even when related to a corrected burst frequency, all show unsteady effects for frequencies well below the burst frequency: Cousteix et al. (1979) at 28% F_b, Parikh et al. (1981) at 12% F_b, Simpson et al. (1980) at 6% F_b. Thus, for most of the experiments that have been reported, the unsteady effects have occurred at frequencies significantly lower than the burst frequency of the boundary-layer structure. This result is true whether in air or water, channel or boundary layer, zero or adverse pressure gradient. Again, the shaded region shows that there is only a relatively small range of oscillation amplitude and frequency for which unsteady effects are not detected.

Importance of Initial Conditions: Most of the unsteady turbulent boundary-layer experiments that have been performed suffer from a lack of sufficient data to accurately determine the flow development along the surface being studied. Experiments in unsteady transition show major effects of oscillation on the development of the resultant turbulent boundary layer. However, in many of the recorded unsteady turbulent boundary-layer experiments, measurements were made at only one x location; in others no trip was used at the start of the test surface. Without data measured at other x stations, the task of isolating local unsteady viscous effects from upstream history is very difficult, if not impossible. Therefore, future experiments should document the character of the flow at several stations. This requirement should also be applied to supposedly "fully developed" flows such as those in pipes; without such documentation, the true contribution of the unsteady viscous effects cannot be isolated.

Concluding Remarks and Recommendations

1. Existing experiments on unsteady turbulent boundary layers have been reviewed and documented. These include flat plate, diffuser, pipe, airfoil, and cascade flows; 27 experiments containing instantaneous data and 12 more containing time-averaged data have been identified.

2. The experiments that provide instantaneous boundary-layer measurements are described in detail in an AGARDograph (Carr, 1981). This AGARDograph contains all the digital data presently available for these experiments. However, many of the experimental results no longer exist in digital form.

3. There are certain trends which can be determined based on the existing experiments.

(a) The time-averaged mean velocity profile is almost always the same as the velocity profile that would occur in a steady flow having an equivalent mean external flow velocity. However, even though these mean profiles are the same, there may be strong local unsteady viscous flow effects present.

(b) In many cases, the turbulent structure in the oscillating flow is not changed from the equivalent steady-state counterpart.

(c) The unsteady effects are often confined to a thin layer near the wall, while the outer region of the boundary layer is not strongly affected.

(d) Several experiments have been accurately predicted using conventional turbulence models.

(e) When existing data are plotted using the dimensionless frequency, S_δ, quasisteady results occur for only a small range of amplitude or frequency.

(f) Unsteady effects occur even when the imposed oscillation frequency is significantly lower than the local turbulence burst frequency, especially in adverse pressure gradient flow.

4. The following recommendations are offered:

(a) Any future experiments studying unsteady turbulent boundary-layer behavior should document the results in digital form, using the format outlined in the present paper.

(b) Documentation of the initial condition of the boundary layer at upstream stations is required. This information may be as important as the results obtained at the nominal test position, even for "fully developed" flows. Unless information about the character of the flow at these earlier stations is recorded, the effects of unsteadiness are very difficult to separate from the effects of upstream history.

(c) Experimental studies of the flow near the wall in unsteady turbulent boundary layers must be emphasized since, in many applications, no information will be available except for the wall values. The ability of a technique to correlate these wall values with the rest of the boundary layer will be a major test of proposed computational schemes.

References

1. Acharya, M.; Reynolds, W. C.: Measurements and prediction of a fully developed turbulent channel flow with imposed controlled oscillations. Rept. TF-8, Mechanical Engineering Dept., Stanford Univ., Stanford, Calif., May 1975.

2. Banner, M. L.; Melville, W. K.: On the separation of air flow over waves. J. Fluid Mech. 77 (1976) 825-842.

3. Binder, G.; Kueny, J. L.: Measurement of the periodic velocity oscillations near the wall in unsteady turbulent boundary layers. Presented at the IUTAM Symposium on Unsteady Shear Flows, Toulouse, France, May 1981.

4. Brembati, F.: An investigation of an unsteady turbulent boundary layer. Project Report 1975-17, Von Karman Inst. of Fluid Dynamics, Rhode Saint Genese, Belgium, June 1975.

5. Brown, F. T.; Margolis, D. L.; Shah, R. P.: Small-amplitude frequency behavior of fluid lines with turbulent flow. J. Basic E., 91D (Dec. 1969) 678-693.

6. Carr, L. W.: A compilation of unsteady turbulent boundary layer data, AGARDograph AG265 (in print) 1981.

7. Carr, L. W.; McAlister, K. W.; McCroskey, W. J.: Analysis of the development of dynamic stall based on oscillating airfoil experiments. NASA TN D-8382 (1977).

8. Charnay, G.; Melinand, J. P.: Investigation of the intermittent region of steady or unsteady turbulent boundary layer. Presented at Eurovisc 77 — Unsteady Turbulent Boundary Layers and Shear Flows, Toulouse, France, Jan. 1977.

9. Cousteix, J.; Desopper, A.; Houdeville, R.: Recherches sur les couches limits turbulentes instationaires (in French) (Research on Unsteady Turbulent Boundary Layers). ONERA/DERAT TP 1976-147, Toulouse, France, 1976.

10. Cousteix, J.; Desopper, A.; Houdeville, R.: Investigation of the structure and of the development of a turbulent boundary layer in an oscillating external flow. Presented at Symposium on Turbulent Shear Flows, Pennsylvania State Univ., University Park, Pa., April 1977 (Also ONERA TP N1977-14).

11. Cousteix, J.; Houdeville, R.; Raynaud, M.: Oscillating turbulent boundary layer with strong mean pressure gradient. Presented at 2nd Symposium on Turbulent Shear Flows, Imperial College, London, 1979 (Also ONERA TP 1979-89).

12. Cousteix, J.; Houdeville, R.; Javelle, J.: Experiments on an oscillating turbulent boundary layer with and without pressure gradient. Presented at IUTAM Symposium on Unsteady Turbulent Shear Flows, Toulouse, France, May 1981.

13. DeRuyk, J.; Hirsch, C.: Turbulence structure in unsteady boundary layers and wakes on an oscillating airfoil. Dept. of Fluid Mech., Vrije Universiteit Brussel, Brussels, Belgium, 1981.

14. Ehrensberger, M.: Experiments in unsteady separating boundary layers. Presented at Euromech 135 — Unsteady Separation and Reversed Flow in External Fluid Dynamics, Marseilles, France, Oct. 1980.

15. Evans, R. L.: Boundary layer development on an axial-flow compressor stator blade. Trans. ASME, J. of Eng. for Power, 100, 2 (1978) 287-293.

16. Foresman, J. L.: Turbulent boundary layer separation characteristics with blowing in an oscillating flow. M.S. Thesis, Naval Postgraduate School, Monterey, Calif., Sept. 1974.

17. Gerrard, J. H.: An experimental investigation of pulsating turbulent water flow in a tube. J. Fluid Mech. 46, 1 (1971) 43-64.

18. Gostelow, J. P.: A new approach to the experimental study of turbomachinery flow phenomena. ASME Paper 76-GT-47. Presented at the ASME Gas Turbines Conference and Products Show, New Orleans, La., March 1976.

19. Ho, C. M.; Chen, S. H.: Unsteady wake of a plunging airfoil. AIAA Paper 80-1446, 1980.

20. Houdeville, R.; Cousteix, J.: Turbulent boundary layers in oscillating flow with a mean adverse pressure gradient. Presented at 15th AAAF Applied Aerodynamics Colloquium, Marseilles, France, Nov. 1978.

21. Houdeville, R.; Desopper, A.; Cousteix, J.: Experimental analysis of average and turbulent characteristics of an oscillatory boundary layer. ONERA TP 1976-30, 1976 (Also Rsch. Aerosp. 1976-4).

22. Hussain, A. K. M. F.; Reynolds, W. C.: The mechanics of an organized wave in turbulent shear flow — part 2, experimental results. J. Fluid Mech. 54 (1972) 241-261.

23. Jacobs, G. K.: Intensity distribution in the oscillating turbulent boundary layer on a flat plate. M.S. Thesis, Naval Postgraduate School, Monterey, Calif., March 1968.

24. Jonnson, I. G.; Carlsen, N. A.: Experimental and theoretical investigations in an oscillatory boundary layer. J. Hydraulic Res. 14, 1 (1976) 45-60.

25. Karlsson, S. K. F.: An unsteady turbulent boundary layer. Ph.D. Thesis, Johns Hopkins Univ., Baltimore, Md., 1958.

26. Kendall, J. M.: The turbulent boundary layer over a wall with progressive surface waves. J. Fluid Mech. 41, 2 (1970) 259-281.

27. Kenison, R. C.: Measurements of a separating turbulent boundary layer with an oscillating free stream. Ph.D. Thesis, Queen Mary College, Univ. of London, England, 1977.

28. Kita, Y.; Hirose, K.; Adachi, Y.: Periodically oscillating turbulent flow in a pipe. Trans. Japanese Soc. of Mech. Eng., Bull. 23 (May 1980) 656-664.

29. Kobashi, Y.; Hayakawa, M.: Structure of a turbulent boundary layer on an oscillating flat plate. IUTAM Sympsoium on Unsteady Turbulent Shear Flows, Toulouse, France, May 1981.

30. Lu, S. Z.; Nunge, R. J.; Erian, F. F.; Mohajery, M.: Measurements of pulsating turbulent water flow in a pipe. Proceedings of 3rd Symposium on Turbulence in Liquids, Univ. of Missouri, Rolla, Mo. (1973) 375-392.

31. Lyrio, A. A.; Ferziger, J. H.; Kline, S. J.: An integral method for the computation of steady and unsteady turbulent boundary layer flows, including the transitory stall regime in diffusers. Dept. of Mech. Eng., Stanford Univ. Rept. ME-PD23, Stanford, California, May 1981.

32. Mainardi, H.; Panday, P. K.: A study of turbulent pulsating flow in a circular pipe. Eurovisc 77 — Unsteady Turbulent Boundary Layers and Shear Flows, Toulouse, France, Jan. 1977.

33. Marvin, J. G.; Levy, L. L.; Seegmiller, H. L.: On turbulence modeling for transonic flows. AIAA Paper 79-0071, New Orleans, La., Jan. 1979.

34. Miller, J.: Heat transfer in the oscillating boundary layer. Trans. of ASME, J. of Eng. for Power, 91 (Oct. 1969) 239-244.

35. Mizushina, T.; Maruyama, T.; Hirasawa, H.: Structure of the turbulence in pulsating pipe flows. J. Chem. Eng. Japan, 8, 3 (1975).

36. Mizushina, T.; Maruyama, T.; Shiozaki, Y.: Pulsating turbulent flow in a tube. J. Chem. Eng. Japan, 6, 6 (1973).

37. Morrissey, J. E.: The effect of large amplitude flow oscillations on turbulent forced-convection heat transfer from a flat plate. M.S. Thesis, Naval Postgraduate School, Monterey, California, 1967.

38. Norris, H. L.; Reynolds, W. C.: Turbulent channel flow with a moving wavy boundary. Report TF-7, Dept. of Mech. Eng., Stanford Univ., Stanford, California, 1975.

39. Offen, G. R.; Kline, S. J.: Experiments on the velocity characteristics of "bursts" and on the interactions between the inner and outer regions of a turbulent boundary layer. Rept. MD-31, Mech. Eng. Dept., Stanford Univ., Stanford, California, 1973.

40. Ohmi, M.; Usui, T.; Tanaka, O.; Toyama, M.: Pressure and velocity distributions in pulsating turbulent pipe flow, Part 2: Experimental investigations. Trans. JSME 41, 349 (Dec. 1974) 2632-2641 (in Japanese). Also Bulletin JSME 19, 134 (Aug. 1976) 951-957 (in English).

41. Ostrowski, J.; Wojciechowski, J.: The generation of a turbulent boundary layer in unsteady flow. Presented at Symposium on Turbulence, Berlin, W. Germany, 1977. Published as Structure and Mechanism of Turbulence, Springer-Verlag.

42. Parikh, P. G.; Reynolds, W. C.; Jayaraman, R.: On the behavior of an unsteady turbulent boundary layer. In, Numerical and Physical Aspects of Aerodynamic Flows. Cebeci, T. (ed.), Univ. of Calif. at Long Beach, Long Beach, Calif., Jan 1981.

43. Patel, M. H.: On turbulent boundary layers in oscillatory flow. Proc. Roy. Soc. (London) A353 (1977) 121-144.

44. Pericleous, K. A.: An oscillatory turbulent boundary layer in an adverse pressure gradient. Ph.D. Thesis, Queen Mary College, Univ. of London, England, 1977.

45. Pittaluga, F.: Oscillatory flow phenomena in diffusers in the turbulent regime. Mech. Res. Comm. 2, 5-6 (1975) 283-288.

46. Rakowsky, E. L.: The effect of freestream oscillations on the incompressible turbulent boundary layer on a flat plate with pressure gradient. Ph.D. Thesis, Stevens Inst. of Tech., Hoboken, N.J., 1966.

47. Ramaprian, B. R.; Tu, S. W.: Periodic turbulent pipe flow at high frequencies of oscillation. Presented at IUTAM Symposium on Unsteady Turbulent Shear Flows, Toulouse, France, May 1981.

48. Ramaprian, B. R.; Tu, S. E.: An experimental study of oscillatory pipe flow at transitional Reynolds numbers. J. Fluid Mech. 100, 3 (1980) 513-544.

49. Rao, K. N.; Narasimha, R.; Badri Narayanan, M. A.: The bursting phenomenon in a turbulent boundary layer. J. Fluid Mech. 48, 2 (1971) 339-352.

50. Reynolds, W. C.; Parikh, P. G.; Jayaraman, R.; Carr, L. W.: Dynamic behavior of an unsteady turbulent boundary layer. Presented at IUTAM Symposium on Unsteady Turbulent Shear Flows, Toulouse, France, May 1981.

51. Richter, K.; Ronneberger, D.: Turbulence modulated by a coherent shear wave in a wall boundary layer. Presented at IUTAM Symposium on Unsteady Turbulent Shear Flows, Toulouse, France, May 1981.

52. Ronneberger, D.; Ahrens, C. D.: Wall shear stress caused by small amplitude perturbations of turbulent boundary layer flow: An experimental investigation. J. Fluid Mech. 83, 3 (Dec. 1977) 433-464.

53. Saxena, L. S.: An experimental investigation of oscillating flows over an airfoil. Ph.D. Thesis, Illinois Inst. of Tech., Chicago, Ill., May 1977.

54. Schachenmann, A. A.: Oscillating turbulent flow in a conical diffuser. Ph.D. Thesis, Lehigh Univ., Bethlehem, Pa., 1974.

55. Schultz-Grunow, F.: Pulsierender durchfluss durch rohre. Forschung 11, 4 (1940) 170-187. Translated as "Pulsation flow through pipes." NASA Technical Translation, NASA-TT-F-14881, 1973.

56. Simpson, R. L.; Chew, Y. T.; Shivaprasad, B. G.: Measurements of unsteady turbulent boundary layers with pressure gradients. Final Report. U.S. Army Research Office Grant DAAG29-76-G-0187. Also SMU Rept. WT-6, Southern Methodist Univ., Dallas, Tex., Aug. 1980.

57. Simpson, R. L.; Shivaprasad, B. G.; Chew, Y. T.: Some important features of unsteady separating turbulent boundary layers. Presented at IUTAM Symposium on Unsteady Turbulent Shear Flows, Toulouse, France, May 5-8, 1981.

58. Simpson, R. L.; Sallas, J. J.; Nasburg, R. E.: Tailoring the waveform of a periodic flow with a programmable damper, J. Fluids Eng. 100 (1978) 287-290.

59. Stenning, A. H.; Schachenmann, A. A.: Oscillatory flow phenomena in diffusers at low Reynolds numbers. Trans. ASME, J. Fluids Eng. (Sept. 1973) 401-407.

60. Thomas, L. C.; Shukla, R. K.: Theoretical and experimental study of wall region periodicity for turbulent pulsatile flow. Trans. ASME, J. Fluids Eng. 98 (March 1976) 27-32.

61. Tomsho, M. E.: The oscillating turbulent boundary layer in a conical diffuser. Ph.D. Thesis, Lehigh Univ., Bethlehem, Pa., 1978.

62. Weinstein, L. M.: Effect of driven-wall motion on a turbulent boundary layer. Presented at IUTAM Symposium on Unsteady Turbulent Shear Flows, Toulouse, France, May 5-8, 1981.

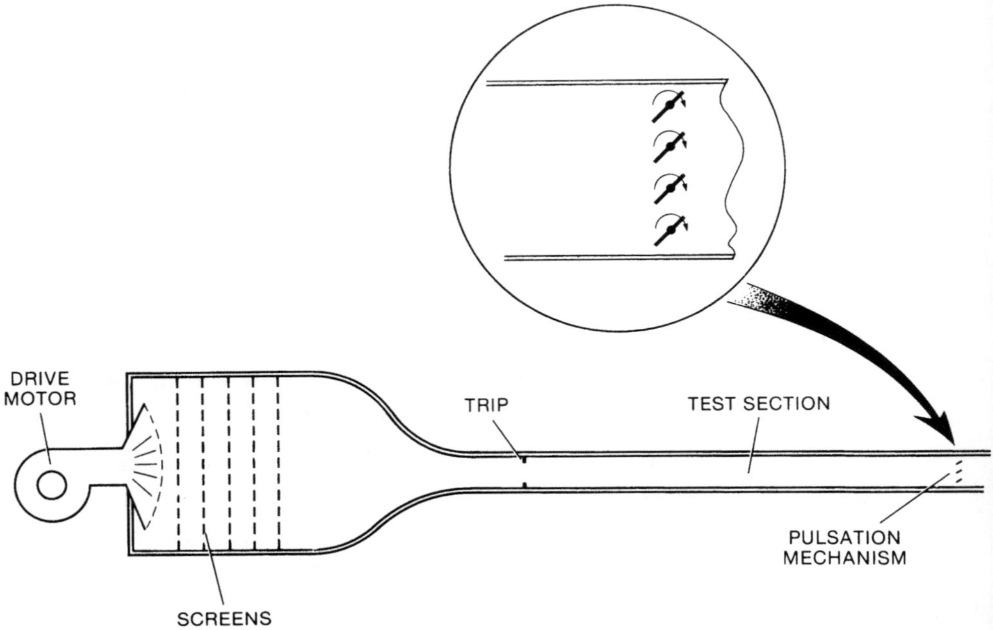

Fig. 1. Oscillating flow facility — variable blockage, from Karlsson

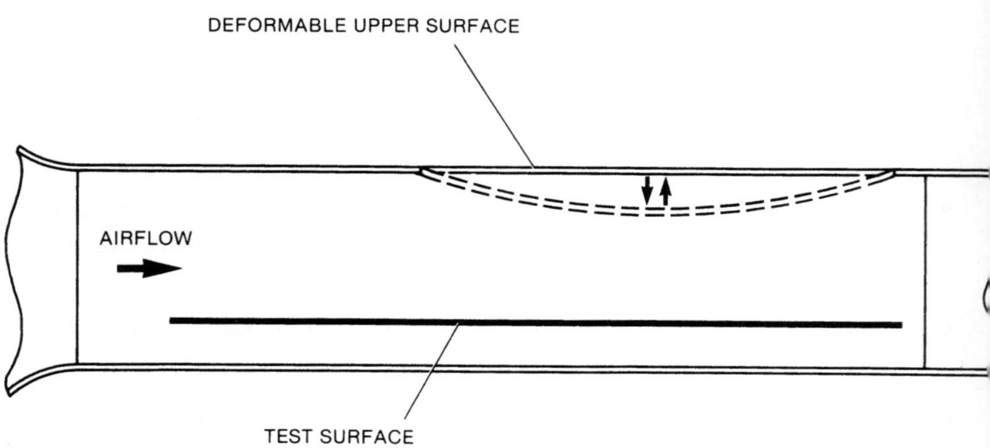

Fig. 2. Oscillating flow facility — deformable wall, from Brembati

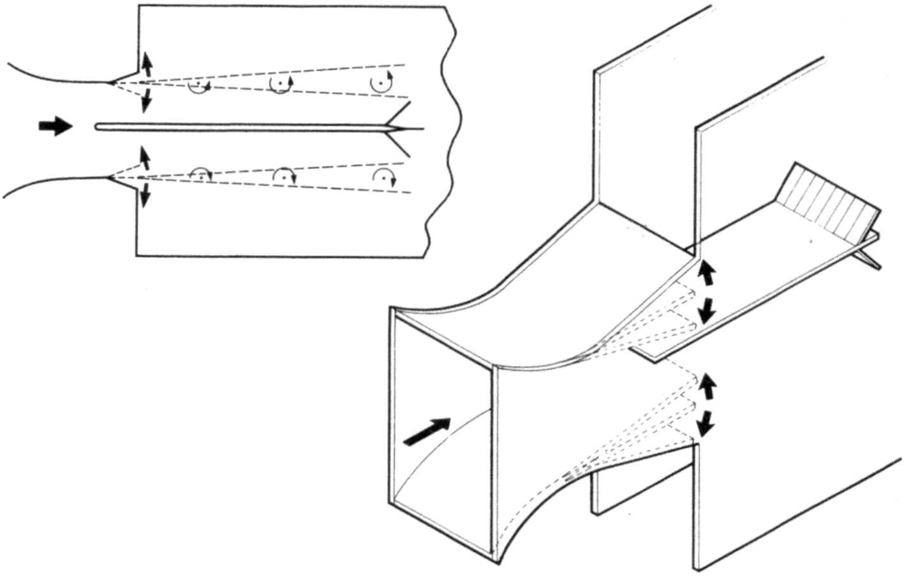

Fig. 3. Oscillating flow facility — vortex-induced motion, from Kenison

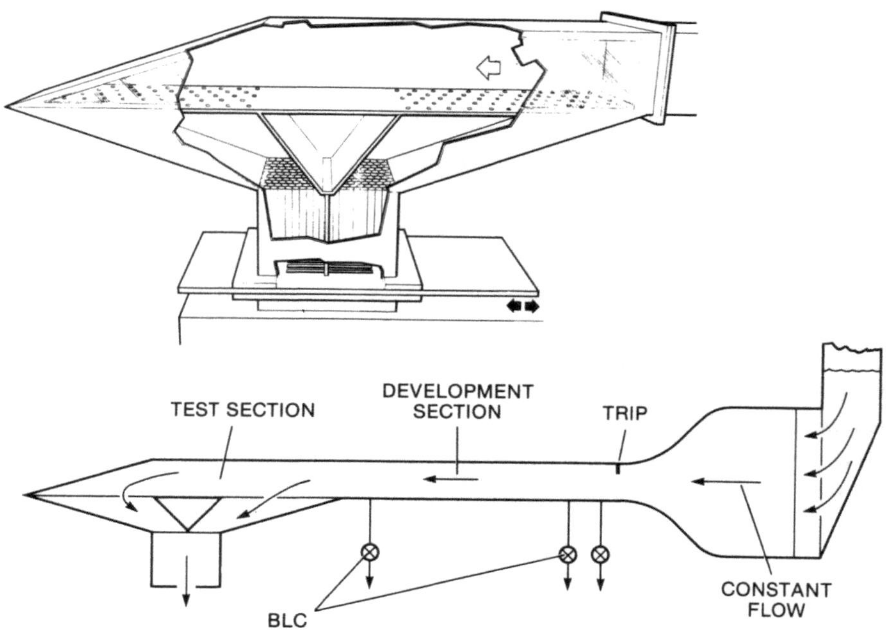

Fig. 4. Oscillating flow facility — steady incoming flow, from Parikh et al.

dP/dx = 0	dP/dx > 0	SEPARATION
ACHARYA & REYNOLDS*	**BREMBATI***	BANNER & MELVILLE
BINDER & KUENY	**COUSTEIX et al. (1979)***	**COUSTEIX et al. (1979)***
COUSTEIX et al. (1976)*	FORESMAN	**KENISON**
COUSTEIX et al. (1977)*	**KENISON**	**SIMPSON et al. (1980)***
CHARNAY & MELINAND	**OSTROWSKI & WOJCIECHOWSKI**	
HUSSAIN & REYNOLDS	**PARIKH et al.***	
JACOBS	**PERICLEOUS***	
JONNSON & CARLSEN*	PITTALUGA	
KARLSSON*	RAKOWSKY	
KENDALL	**SCHACHENMANN**	
MILLER	**SIMPSON et al. (1980)***	
MORRISSEY	STENNING & SCHACHENMANN	
NORRIS & REYNOLDS	THOMAS & SHUKLA	
PATEL*	**TOMSHO***	
RONNEBERGER & AHRENS		

Fig. 5. Existing unsteady turbulent boundary-layer experiments. Bold type indicates experiments that are reviewed in detail in AGARDograph. Asterisk ( )* indicates that digital data are available on magnetic tape

PIPE	AIRFOILS & CASCADES	NEW EXPERIMENTS
BROWN et al.	CARR, McALISTER & McCROSKEY	COUSTEIX et al. (1981)
GERRARD	**EVANS**	DE RUYCK & HIRSCH
KITA et al.	GOSTELOW	EHERENSBERGER
LU et al.	HO & CHEN	KOBASHI & HAYAKAWA
MAINARDI & PANDAY	**MARVIN et al.***	LORBER & COVERT
MIZUSHINA et al. (1973)	**SAXENA**	RAMAPRIAN & TU (1981)
MIZUSHINA et al. (1975)		REYNOLDS et al.
OHMI et al.		RICHTER & RONNEBERGER
RAMAPRIAN & TU (1980)*		SIMPSON et al. (1981)
SCHULTZ-GRUNOW		WEINSTEIN

Fig. 6. Additional unsteady viscous flow experiments. Bold type indicates experiments that are reviewed in detail in AGARDograph. Asterisk ( )* indicates that digital data are available on magnetic tape

EXPERIMENT

FLOW TYPE
PRIMARY REFERENCE
PRINCIPAL INVESTIGATOR
FACILITY
- LOCATION
- APPARATUS
- TEST SECTION
- OSCILLATION MECHANISM
- PRESSURE GRADIENT MECHANISM
- MEASUREMENT SURFACE
- TRIP
- WALL BOUNDARY LAYER CONTROL
- NATURAL FREQUENCY
- MAXIMUM WALL DEFLECTION
- POSITIONAL ACCURACY
- FREE STREAM TURBULENCE
- VERIFICATION OF TWO-DIMENSIONALITY
- MEASUREMENT ACCURACY

DATA

MEASUREMENT TECHNIQUE
DATA REDUCTION TECHNIQUE
NOMINAL TEST CONDITIONS
AVAILABILITY OF DATA
- GRAPHIC DATA PRESENTED IN REPORT(S)
- DATA ON MAGNETIC TAPE

COMMENTS BY ORIGINAL AUTHORS
COMMENTS BY LWC
NUMBER OF CARD IMAGES ON TAPE
PERTINENT REFERENCES

Fig. 7. Standard format for review of experiments

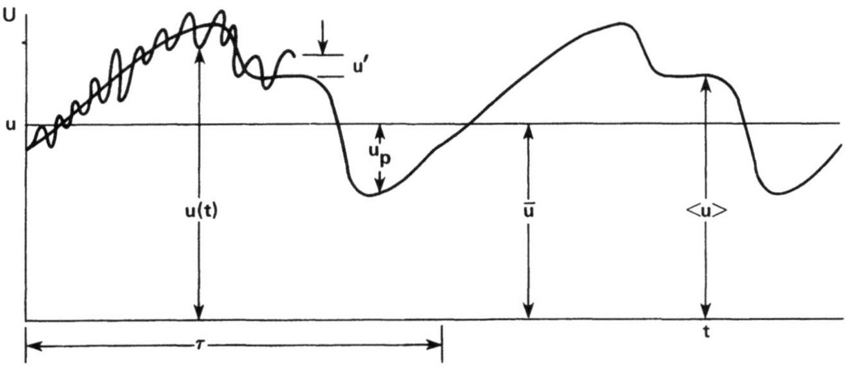

u(t) – INSTANTANEOUS MEASURED VELOCITY $u(t) = \bar{u} + u_p + u'$

ū – TIME AVERAGED MEAN VELOCITY $\bar{u} = \frac{1}{T}\int_0^T u(t)dt$

<u> – ENSEMBLE AVERAGED VELOCITY $\langle u(t) \rangle = \frac{1}{N}\sum_{n=0}^{N} u(t+n\tau)$

u_p – PERIODIC COMPONENT OF VELOCITY $u_p(t) = \langle u(t) \rangle - \bar{u}$

u′ – RANDOM FLUCTUATIONS OF VELOCITY $u'(t) = u(t) - \langle u(t) \rangle$

Fig. 8. Analysis of unsteady turbulent velocity signal

I		II
$\bar{u}$	$\overline{u'^2}$	$u_p(A;\phi)$
ACHARYA & REYNOLDS	**ACHARYA & REYNOLDS**	**ACHARYA & REYNOLDS**
BREMBATI	COUSTEIX et al. (1976)	**BINDER & KUENY**
COUSTEIX et al. (1976)	COUSTEIX et al. (1977)	**COUSTEIX et al. (1976)**
COUSTEIX et al. (1977)	COUSTEIX et al. (1979)	**COUSTEIX et al. (1977)**
COUSTEIX et al. (1979)	**JACOBS**	**COUSTEIX et al. (1979)**
EVANS	**KARLSSON**	EVANS
JACOBS	KENDALL	**HUSSAIN & REYNOLDS**
JONNSON & CARLSEN	MIZUSHINA et al. (1973)	**KARLSSON**
KARLSSON	MIZUSHINA et al. (1975)	**KENISON**
KENDALL	**PARIKH et al.**	JONNSON & CARLSEN
KENISON	**PATEL**	KENDALL
LU, et al.	**PERICLEOUS**	**LU et al.**
MILLER	RAMAPRIAN & TU (1980)	MIZUSHINA et al. (1973)
MIZUSHINA et al. (1973)	**SCHACHENMANN**	MIZUSHINA et al. (1975)
MIZUSHINA et al. (1975)	**STENNING & SCHACHENMANN**	**OHMI et al.**
OHMI et al.		**PATEL**
PARIKH et al.		**PERICLEOUS**
PATEL		RAMAPRIAN & TU (1980)
RAMAPRIAN & TU (1980)		**SIMPSON et al. (1980)**
SCHACHENMANN		**TOMSHO**
SIMPSON et al. (1980)		
STENNING & SCHACHENMANN		
TOMSHO		

Fig. 9. Summary of available unsteady turbulent boundary layer data — levels I and II (Bold type indicates data that are available directly from reports; light type indicates data that can be reconstructed from the published data)

III		IV
$\langle u \rangle$	$\langle u'^2 \rangle$	$\langle u'v' \rangle$
BREMBATI	BREMBATI	COUSTEIX et al. (1977)
COUSTEIX et al. (1976)	COUSTEIX et al. (1977)	COUSTEIX et al. (1979)
COUSTEIX et al. (1977)	COUSTEIX et al. (1979)	KENDALL
COUSTEIX et al. (1979)	KENDALL	MIZUSHINA et al. (1975)
EVANS	MIZUSHINA et al. (1973)	RAMAPRIAN & TU (1980)
GOSTELOW	MIZUSHINA et al. (1975)	SIMPSON et al. (1980)
JONNSON & CARLSEN	RAMAPRIAN & TU (1980)	
KENDALL	SIMPSON et al. (1980)	
MIZUSHINA et al. (1973)		
MIZUSHINA et al. (1975)		
OHMI et al.		
OSTROWSKI & WOJCIECHOWSKI		
RAMAPRIAN & TU (1980)		
SAXENA		
SIMPSON et al. (1980)		
TOMSHO		

Fig. 10. Summary of available unsteady turbulent boundary layer data — levels III and IV (Bold type indicates data that are available directly from reports; light type indicates data that can be reconstructed from the published data)

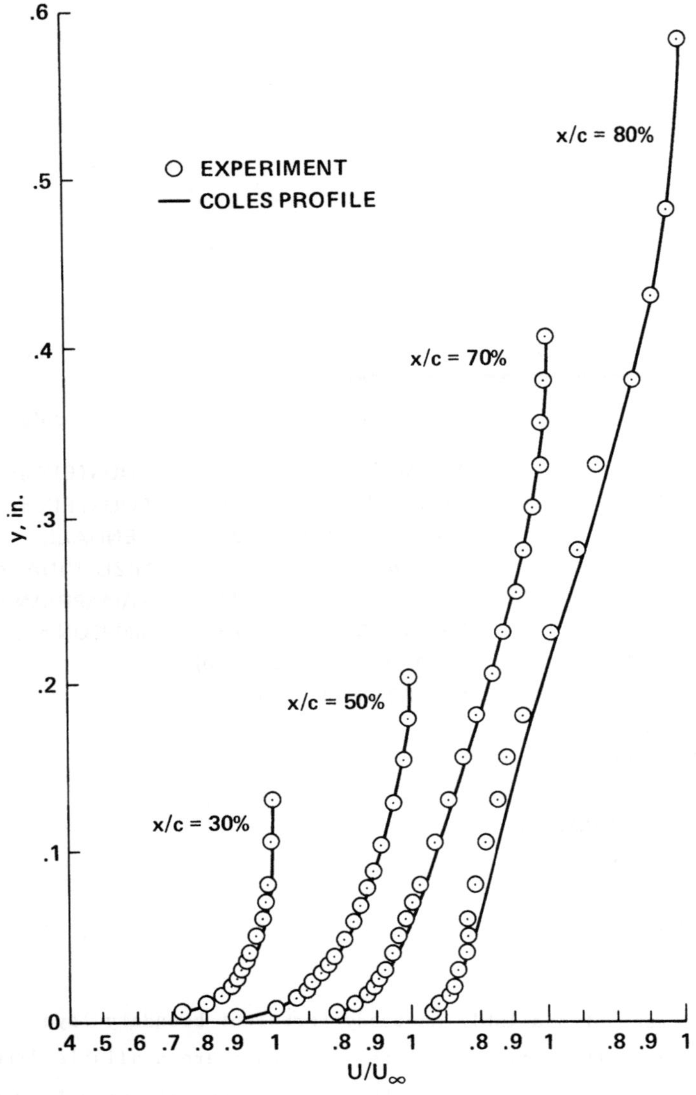

Fig. 11. Time-averaged mean profiles — from Evans

$$\langle u(t)\rangle = \bar{u} + |u| \sin \omega t + \ldots$$

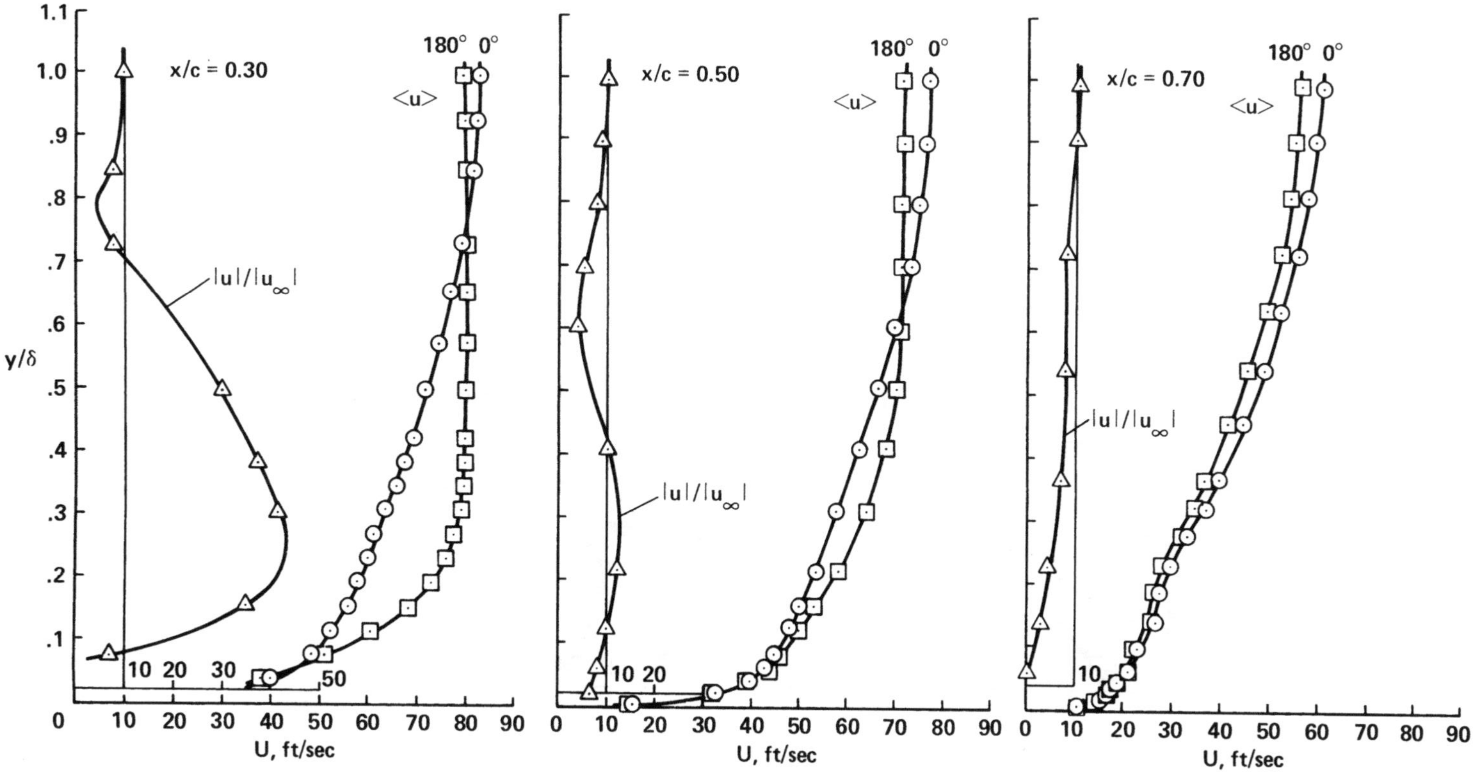

Fig. 12. Ensemble-averaged velocity profiles – from Evans

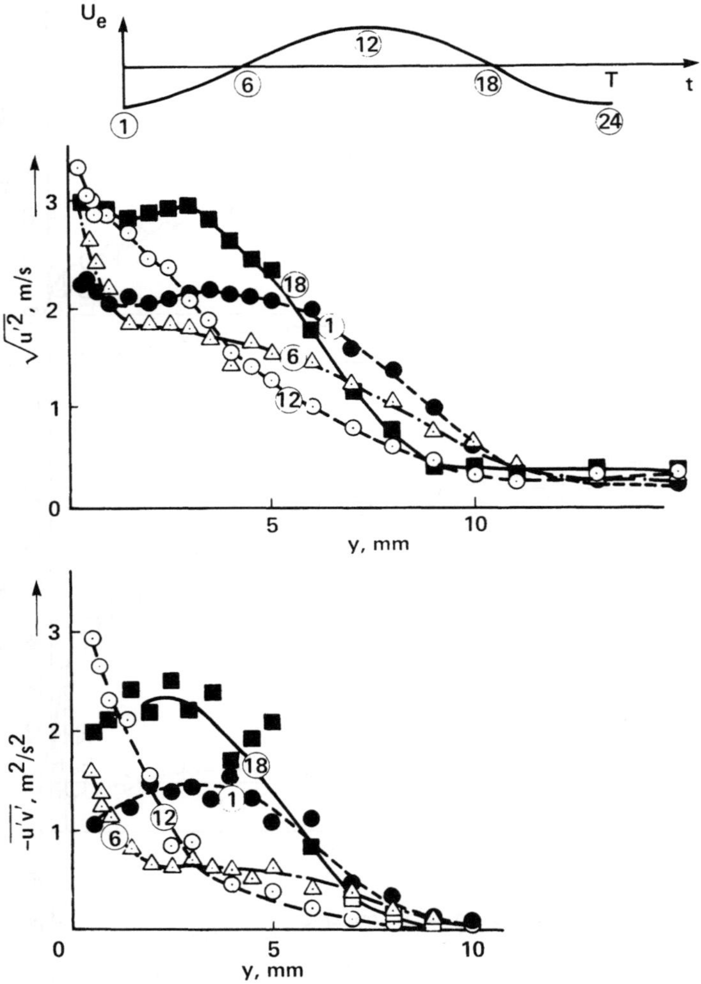

Fig. 13. Profiles of turbulence intensity and Reynolds shear stress — from Cousteix et al. (1977)

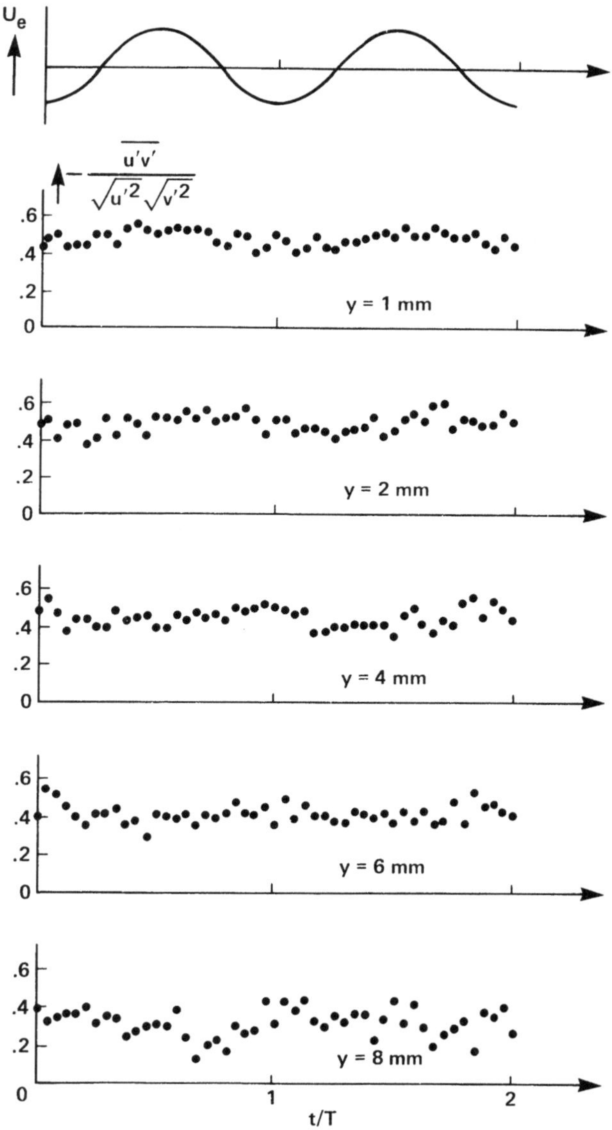

Fig. 14. Statistical distribution of velocity fluctuation — from Cousteix et al. (1977)

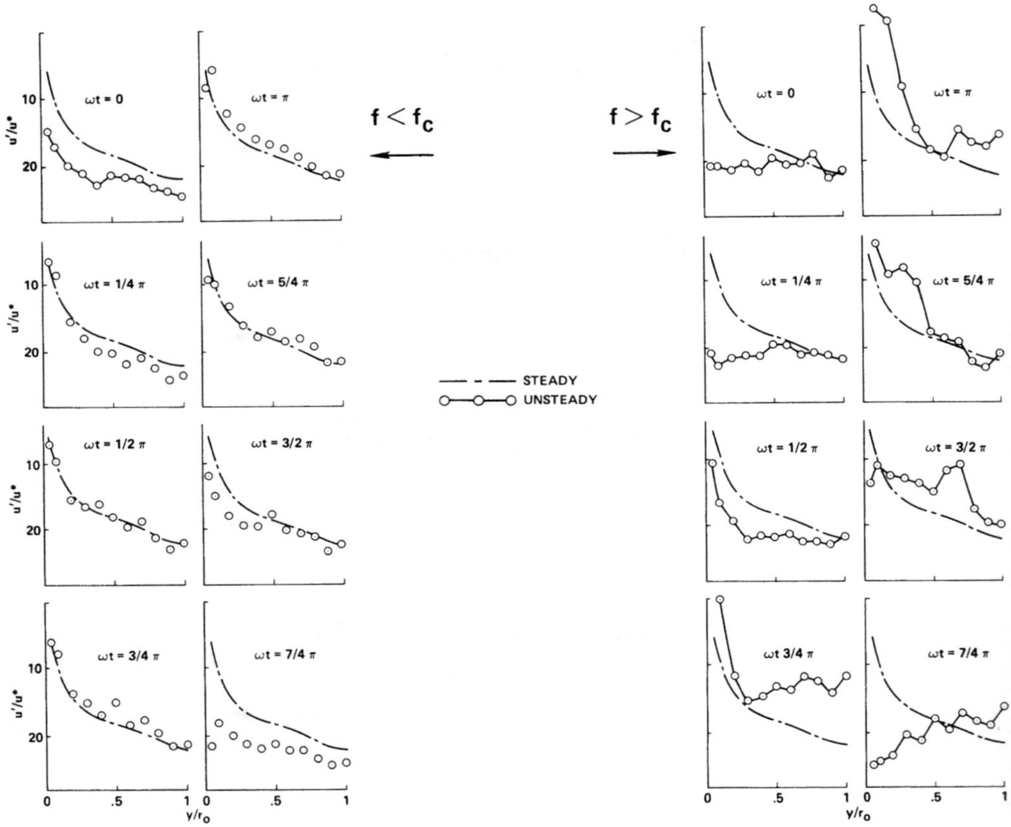

Fig. 15. Effect of frequency on turbulence intensity — from Mizushina et al. (1973)

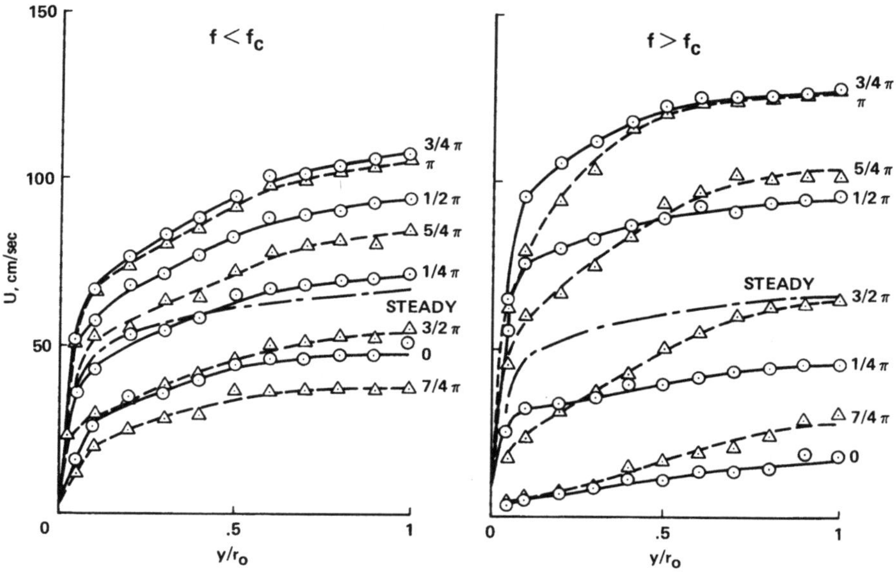

Fig. 16. Effect of frequency on instantaneous velocity — from Mizushina et al. (1973)

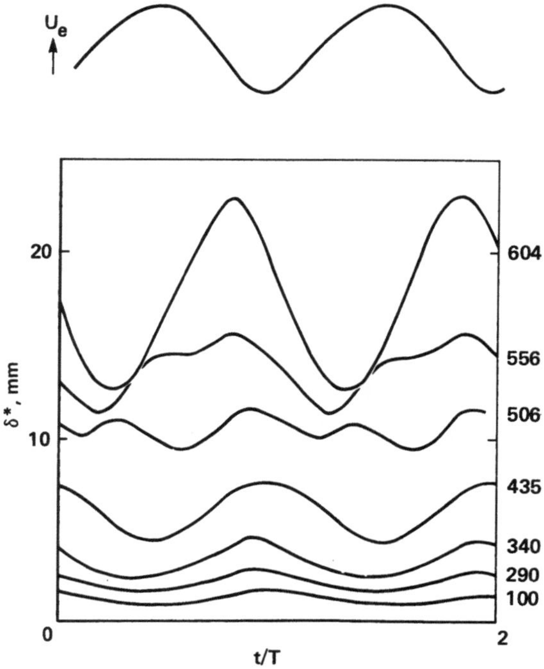

Fig. 17. Variation of δ* in adverse pressure gradient — from Houdeville and Cousteix (1978)

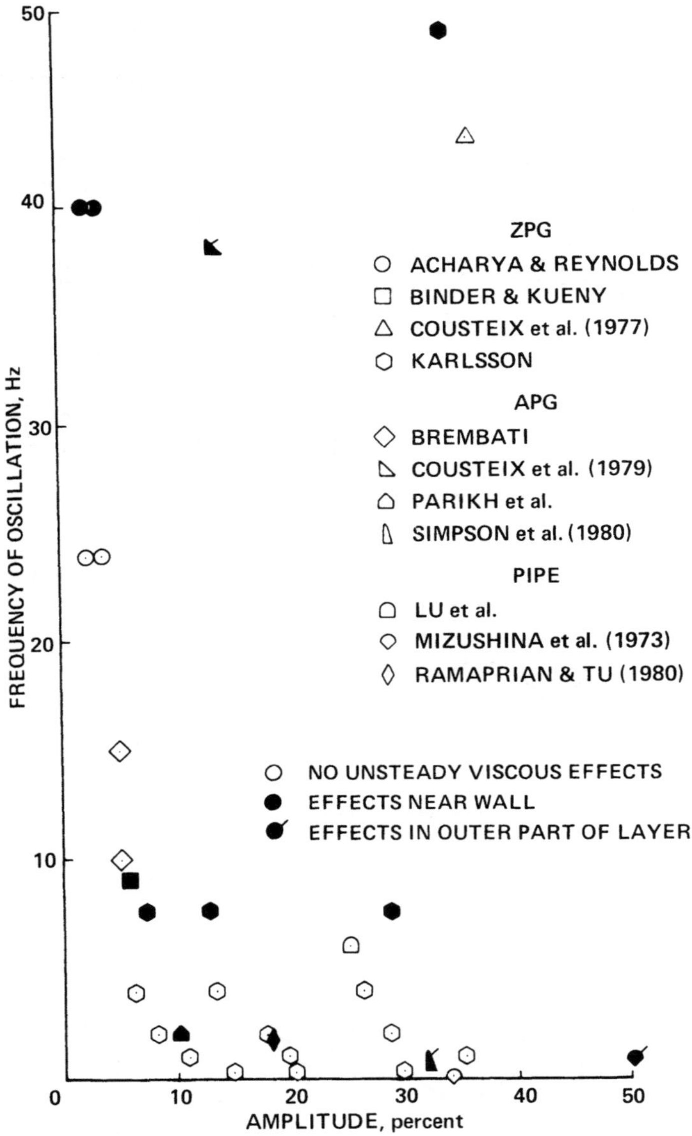

Fig. 18. Values of amplitude and frequency used in selected experiments

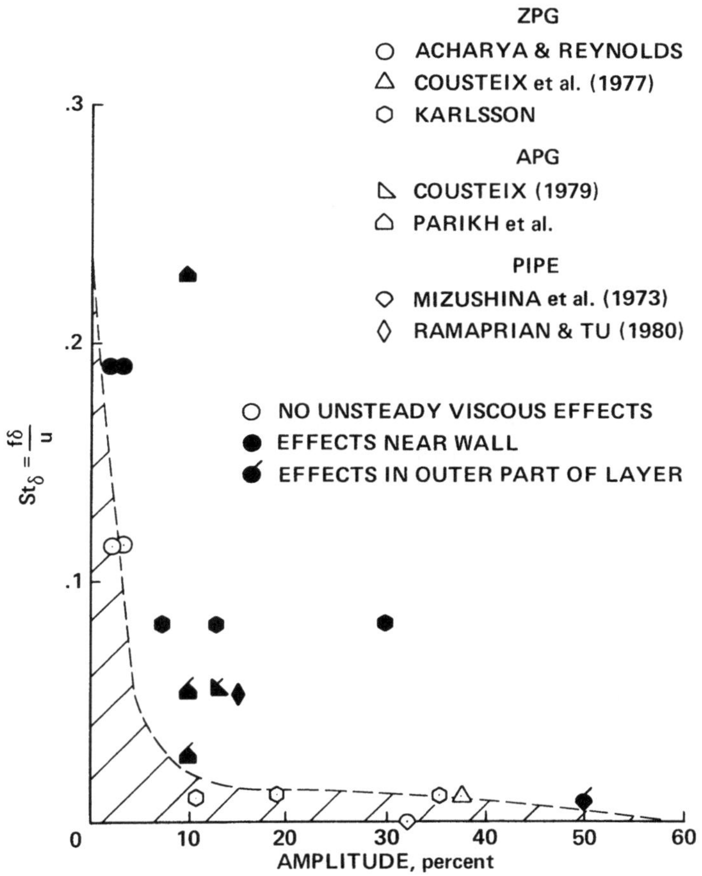

Fig. 19. Strouhal number and amplitude for selected experiments

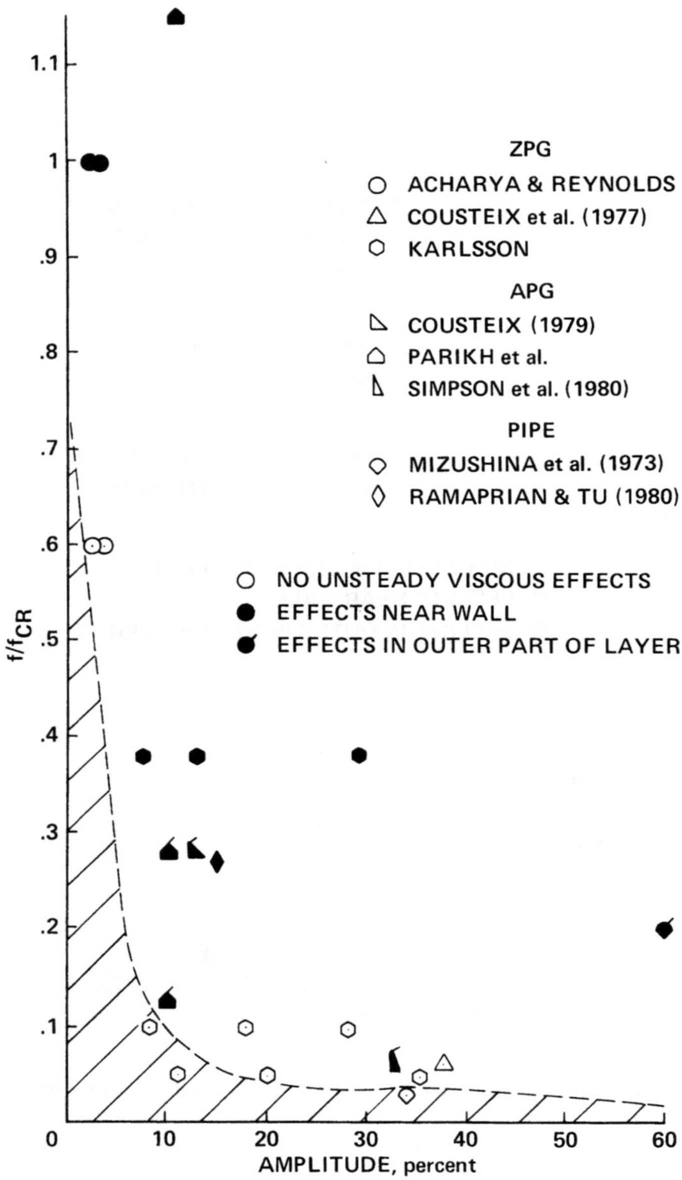

Fig. 20. Ratio of oscillation frequency to turbulent burst frequency for selected experiments

Dynamic Behavior of an Unsteady Turbulent Boundary Layer

P. G. Parikh, W. C. Reynolds, R. Jayaraman, and L. W. Carr*
Department of Mechanical Engineering, Stanford University
Stanford, California 94305

Summary

This paper reports experiments on an unsteady turbulent boundary layer. The upstream portion of the flow is steady (in the mean). In the downstream region, the boundary layer sees a linearly decreasing free-stream velocity. This velocity gradient oscillates in time, at frequencies ranging from zero to approximately the bursting frequency. Considerable detail is reported for a low-amplitude case, and preliminary results are given for a higher amplitude sufficient to produce some reverse flow. For the small amplitude, the mean velocity and mean turbulence intensity profiles are unaffected by the oscillations. The amplitude of the periodic velocity component, although as much as 70% greater than that in the free stream for very low frequencies, becomes equal to that in the free stream at higher frequencies. At high frequencies, both the boundary layer thickness and the Reynolds stress distribution across the boundary layer become frozen. The behavior at larger amplitude is quite similar. Most importantly, at sufficiently high frequencies the boundary layer thickness remains frozen at its mean value over the oscillation cycle, even though flow reverses near the wall during a part of the cycle.

Introduction

The objectives of the Stanford Unsteady Turbulent Boundary Layer Program are: to develop a fundamental understanding of such flows, to provide a definitive data base which can be used to guide turbulence model development, and to provide test cases which can be used by computors for comparison with predictions.

Due to space limitations, work of other investigators will not be summarized here, except to note that all the previous experiments are characterized by unsteady flow at the inlet to the unsteady region. For a comparison of the present experimental parameter range with those of other investigations, see Reference 1. The distinctive feature of the present experiments is that the boundary layer at the inlet to the unsteady region is a standard, steady, flat-plate turbulent boundary layer. It is then subjected to controlled oscillations of the free stream. This feature is especially important from the point of view of a computor, who needs precise specification of boundary conditions for computation of the flow.

---

*U. S. Army Aeromechanics Laboratory, Moffett Field, CA 94035

Free-Stream Boundary Condition of the Present Experiment

The desired free-stream velocity $u_\infty(x,t)$ in the water tunnel built for this work is shown in Fig. 1. u_∞ remains steady and uniform for the first two meters of boundary layer development. It then decreases linearly in the test section; the magnitude of the velocity gradient varies sinusoidally from zero to a maximum value during the oscillation cycle. The mean free-stream velocity distribution in the test section is thus linearly decreasing and corresponds to the distribution at the cycle phase angle of 90°, while the amplitude of imposed free-stream oscillations grows linearly in the streamwise direction, starting at zero at the entrance to a maximum value of a_o at the exit. Hence,

$$u_\infty(x,t) = u_{\infty,o} \qquad x < x_o$$

$$= u_{\infty,o} - \frac{a_o(x-x_o)}{L}\left[1 - \cos\omega t\right], \qquad x_o < x < x_o + L$$

The important parameters of this problem are the amplitude parameter $\alpha = a_o/u_{\infty,o}$ and the frequency parameter: $\beta_\delta = f\delta_o/u_{\infty,o}$. Here $f = \omega/(2\pi)$ and δ_o is the thickness of the boundary layer at the inlet to the unsteady region. In the present experiments:

$u_{\infty,o} = 0.73$ m/s, $\delta_o = 0.05$ m, $0 < f < 2$ hz, $0 < \alpha < 0.25$, $0 < \beta_\delta < 0.14$

It should be mentioned that the value of the frequency parameter β_δ at the so-called "bursting frequency" in turbulent boundary layers is about 0.2 [2]. Thus the imposed oscillation frequencies used in the present experiments cover the range from quasi-steady ($f \simeq 0$) to values approaching the bursting frequency. The results reported here are for two non-dimensional amplitudes, $\alpha = 0.05$ and 0.25 (nominally). The latter is sufficient to cause reverse flow in a turbulent boundary layer at the end of the test section during a part of the oscillation cycle.

Experimental Facility

Figure 2 is a schematic of the facility. The 16:1 nozzle contraction is followed by a 2 m long development section, where the test boundary layer is grown on the top wall. A constant head and a constant flow resistance provide a constant flow . The free-stream velocity in the development section is maintained uniform along x by bleed from the bottom wall.

The linear decrease in free-stream velocity in the test section is accomplished by uniformly bleeding off some flow through the bottom wall in the test section. The remainder of the flow exits downstream. Each of these two flows exits the tunnel through slots in an oscillating plate. The design ensures that, regardless of the position of the oscillating plate, the total flow area of the slots remains the same. The slots are the controlling resistance of the entire fluid circuit, hence the constant flow. By sinusoidally oscillating the plate, a linearly decreasing periodic free-stream distribution is established in the test section, while the upstream flow in the development section remains steady.

Measurement and Data-Processing Techniques

Pitot tubes are used for mean velocity measurements in steady flow regions. Unsteady velocity measurements use a single-channel, forward-scatter, Bragg-shifted DISA laser anemometer in the tracking mode.

Following Hussain and Reynolds [3], the instantaneous velocity signal from an unsteady turbulent flow may be decomposed into three parts:

$$u = \bar{u} + \tilde{u} + u' \qquad (1)$$

where $\bar{u}$ is the mean, $\tilde{u}$ is the time-dependent, organized (deterministic) component, and u' is the random fluctuation. $\bar{u}$ is determined by long-time averaging of u. Here $\tilde{u}$ is of a periodic nature and may be determined by first phase-averaging the instantaneous velocity signal and then subtracting out the mean. Thus,

$$\tilde{u} = \langle u \rangle - \bar{u} \qquad (2)$$

Here $\langle u \rangle$, the phase average velocity, is determined by averaging over an ensemble of samples taken at a fixed phase in the imposed oscillation. In the present experiments, with harmonic oscillation of the free stream, the response at points within the boundary layer is almost sinusoidal, with higher harmonics contributing less than 5%. Hence, $\tilde{u}$ may also be extracted from the instantaneous signal u by cross-correlation with a sine wave in phase with the oscillation. A digital correlator (HP 3721A) was used to determine cross-correlations leading to the $\tilde{u}$ data reported here. Currently a DEC MINC-11 laboratory minicomputer system is used for automatic data acquisition and processing, allowing the determination of phase averages of u and u'^2.

The measurements reported here were taken at a fixed streamwise location near the end of the test section at $x - x_o = 0.568$ m.

Behavior at Small Amplitude of Imposed Oscillations

The mean velocity profiles measured with the oscillating plate in fixed positions $\theta = 0, 90°, 180°$ are fit by dashed curves in Fig. 3. These represent phase-averaged profiles at zero frequency, i.e., quasi-steady profiles. At this amplitude ($\alpha = 0.05$), the response of the boundary layer is almost linear, so that the profile corresponding to $\theta = 90°$ lies nearly midway between the $\theta = 0$ and $180°$ profiles. The $90°$ profile represents the mean profile for quasi-steady oscillations. The difference between the 0 and $90°$ profiles at a fixed y-location represents the amplitude of quasi-steady oscillations at that location in the boundary layer. Note that the quasi-steady amplitudes in the boundary layer are larger than the free-stream amplitude.

The mean velocity profiles measured under oscillatory conditions at 0.5 hz and 2.0 hz are shown as data points in Fig. 3. Note that the mean velocity profiles at various frequencies are identical with the profile measured under stationary condition with pulser angle set at $\theta = 90°$. It may be concluded that the mean velocity profile (at a fixed amplitude $\alpha = 0.05$) is independent of the imposed oscillation frequency in the entire range $0 \leq f \leq 2$ hz. The same behavior persists all the way up to the wall.

This behavior of the mean velocity profile may be explained by an examination of the governing equations. Use of (1) in the momentum equation and time-averaging yields

$$\bar{u}\frac{\partial \bar{u}}{\partial x} + \bar{v}\frac{\partial \bar{u}}{\partial y} = -\frac{1}{\rho}\frac{\partial \bar{p}}{\partial x} + \nu\frac{\partial^2 \bar{u}}{\partial y^2} - \frac{1}{\rho}\frac{\partial}{\partial y}\left[\overline{u'v'} + \overline{\tilde{u}\tilde{v}}\right] \quad (3)$$

Equation (3) may be recognized as the equation governing an ordinary turbulent boundary layer, except for the addition of the term $\overline{\tilde{u}\tilde{v}}$, which represents Reynolds stresses arising from the organized oscillations.

The time-mean pressure gradient $\partial\bar{p}/\partial x$ may be shown to be independent of the imposed oscillation frequency and the same as that obtained for $f = 0$ at $\theta = 90°$. Therefore, the mean velocity field will be frequency-dependent if and only if one or both of the following happen:

- The distribution of Reynolds stress $\overline{u'v'}$ is altered under oscillatory conditions and is dependent on the frequency of imposed oscillations.
- The Reynolds stress $\overline{\tilde{u}\tilde{v}}$ arising from organized fluctuations becomes significant compared with $\overline{u'v'}$.

We shall now argue that neither of the above requirements is met. Figure 4 shows the measured distribution of u'_{rms} under stationary condition with the pulser at $\theta = 90°$ (the mean position) as well as those measured under oscillatory conditions at frequencies up to 2 hz. Note that u'_{rms} is independent of the imposed oscillation frequency and, further, that it is the same as that measured at $f = 0$ and $\theta = 90°$. We believe that the same would be true for $\overline{u'v'}$, which at present we cannot measure. Figure 5 gives a comparison between measured values of $\overline{\tilde{u}\tilde{v}}$ at 2 hz with data on $\overline{u'v'}$ obtained by Anderson [4] in a steady adverse pressure gradient boundary layer at comparable conditions. The present data on $\overline{\tilde{u}\tilde{v}}$ were obtained by separate LDA measurements of $\tilde{u}$ and $\tilde{v}$ and their respective phases. It may be seen that the contribution of $\overline{\tilde{u}\tilde{v}}$ to total Reynolds stress is insignificant over almost the entire boundary layer. Hence, $\overline{u'v'}$ is independent of frequency and $\overline{\tilde{u}\tilde{v}}$ is negligible, and so the mean velocity profile is also independent of frequency and is the same as that found at $f = 0$ with $\theta = 90°$.

The behavior of the periodic component $\tilde{u}$ will next be examined. We denote

$$\tilde{u} = a_1(y) \cos[\omega t + \phi(y)] \tag{4}$$

The profiles of amplitudes a_1 measured in the boundary layer and normalized by the free-stream amplitude $a_{1,\infty}$ are shown in Fig. 6. The profile for quasi-steady $(f \simeq 0)$ oscillations was determined, as explained earlier, from the mean velocity profiles measured at $f = 0$ with $\theta = 0$, $90°$, and $180°$ (see Figs. 3(a), (b)). Note that, during quasi-steady oscillations, the amplitude in the boundary layer exceeds the free-stream amplitude by as much as 70%. It may be mentioned that data for $f = 0.1$ hz, not shown on Fig. 6, do indeed come very close to the quasi-steady behavior.

As the frequency is increased, the amplitude within the boundary layer is attenuated. The amplitude appears to drop as f is increased and then rise again. At high frequencies, the amplitude in most of the boundary layer is the same as in the free-stream; near the wall the amplitude of the periodic component rapidly drops to zero.

The phase differences between the boundary layer oscillations and free-stream oscillations are shown in Fig. 7. For $f = 0$ there is no phase difference. The largest phase lags in the outer region of the boundary layer were observed at $f = 0.25$ hz. The effect of increasing the frequency is to reduce the phase lag in the outer region, but to introduce large phase leads in the region very close to the wall. Clearly, the asymptotic behavior of the outer region for high frequencies is once again a zero phase lag with respect to free-stream oscillations, as in the quasi-steady case.

At high frequencies, the combination of the asymptotic behaviors of $a_1/a_{1,\infty}$ and ϕ in the outer region together with the fact that the mean velocity profile is unaffected by imposed oscillations, has the effect of freezing the boundary layer thickness. This is shown in Fig. 8, where the phase-averaged boundary layer thickness $\langle \delta_{.99} \rangle$ is plotted as a function of the cycle phase angle for several frequencies. The quasi-steady behavior of $\langle \delta_{.99} \rangle$ is quite obvious: at $\theta = 0$, the boundary layer in the test section continues to develop under a zero pressure gradient and is the thinnest at this point in the entire cycle. As the phase angle is increased, pressure gradients of increasing adversity are imposed on the boundary layer, causing it to thicken. The maximum thickness is attained at $\theta = 180°$ under the maximum adverse pressure gradient. Hence, at $f = 0$, δ oscillates 180° out of phase with u_∞.

Under oscillatory conditions at $f = 0.25$, 0.5, and 2.0 hz, two things happen: a significant phase lag develops from quasi-steady behavior and the amplitude attenuates with increasing frequency. For the $f = 2.0$ hz case, the variation over the complete cycle is less than 1% and the boundary layer thickness is practically frozen during the oscillation cycle.

It may be shown by a simple argument based on a mixing length model of boundary layer turbulence that the freezing of the boundary layer thickness at high frequencies is also accompanied by freezing of the Reynolds stress over the oscillation cycle. To prove this, we hypothesize that the phase-averaged Reynolds stress distribution may be related to the phase-averaged velocity profile in the same manner as for a steady boundary layer, i.e.,

$$- \langle u'v' \rangle = \varepsilon_m \frac{\partial \langle u \rangle}{\partial y} \quad , \quad \varepsilon_m = \ell^2 \left| \frac{\partial \langle u \rangle}{\partial y} \right| \tag{5}$$

Now, in the outer region of the boundary layer, the mixing length ℓ may be modeled as

$$\ell = \lambda \langle \delta_{.99} \rangle \tag{6}$$

where λ is nearly a constant. Now,

$$\langle u \rangle = \bar{u} + \tilde{u} = \bar{u} + a_1(y) \cos[\omega t + \phi(y)] \tag{7}$$

However, in the high-frequency limit,

$$a_1(y) = a_{1,\infty} = \text{const} \; ; \quad \phi(y) = 0 \quad \text{and} \quad \langle \delta_{.99} \rangle = \overline{\delta}_{.99} = \text{const.} \tag{8}$$

Therefore

$$\frac{\partial \langle u \rangle}{\partial y} = \frac{\partial \overline{u}}{\partial y} \tag{9}$$

Combining the above, one finds

$$-\langle u'v' \rangle = \lambda^2 \left(\overline{\delta}_{.99}\right)^2 \left[\frac{\partial \overline{u}}{\partial y}\right]^2 = -\overline{u'v'} \tag{10}$$

i.e., the phase-averaged Reynolds stress in the outer region also becomes frozen at $-\overline{u'v'}$.

Experimental evidence of this stress-freezing behavior was obtained by measurements of phase-averaged normal turbulent stress $\langle u'^2 \rangle$. The quasi-steady ($f = 0$) profiles of $\langle u'^2 \rangle$ are shown in Fig. 9 for three phase angles $\theta = 0°$, $90°$, and $180°$. Note that the distribution for 90° lies nearly midway between those for 0° and 180°. The distribution of $\langle u'^2 \rangle$ for 90° is the same as the distribution of $\overline{u'^2}$, as seen earlier. Therefore, the difference between the 0° and 90° curves in Fig. 9 represents the amplitude of quasi-steady oscillations of $\langle u'^2 \rangle$ at any point in the boundary layer. This amplitude was determined graphically from Fig. 9 and is plotted in Fig. 10 for the case of $f = 0$. Under oscillatory conditions, the amplitude of the normal stress oscillations in the boundary layer attenuates as the frequency of imposed oscillations is increased from $f = 0$. At $f = 2.0$ hz, the amplitude of stress oscillations across the boundary layer is almost zero over the outer region, as seen in Fig. 10, i.e., the stress is almost frozen over the oscillation cycle.

Behavior Under Large Amplitudes of Imposed Oscillations

We now discuss the case of $\alpha = 0.25$. All data reported for this case are preliminary and subject to revision. They are included here because of their special interest to this meeting. Also, because of apparatus peculiarities, α varies somewhat with f in this case, hence 0.25 is only a nominal value.

The behavior is qualitatively similar to the $\alpha = 0.05$ case. The mean velocity profiles for $f = 0$, 0.25, 0.5, and 2.0 Hz are shown in Fig. 11. Note that the profiles are identical for the cases of $f = 0$, 0.5, and 2.0 hz. For the case of $f = 0.25$, however, there is a significant deviation in the outer part of the boundary layer. This deviation results from excessive thickening of the boundary layer during a part of the oscillation cycle around the phase angle of 180°. The blockage effect of an excessively thick boundary layer causes an increase in the local free-stream velocity in the test section. Therefore, the desired linearly decreasing free-stream velocity distribution is not achieved over a part of the cycle. At higher frequencies, though, the boundary layer thickness over the entire oscillation cycle deviates very little from its mean value, corresponding to the $\theta = 90°$, $f = 0$ condition.

The behavior of the amplitude ratio and phase difference with respect to free stream, as shown in Figs. 12 and 13, is quite similar to that for the lower-amplitude case. At high frequency, the overshoot in the amplitude ratio disappears and phase angles over most of the boundary layer approach zero. Very close to the wall, there is a tendency to develop phase leads.

The phase-averaged velocity profiles for $f = 2.0$ hz are shown in Fig. 14. Note that at $\theta = 180°$ there is a small region of reversed flow close to the wall. Despite this flow reversal, the boundary layer thickness remains close to its mean value, as seen in Fig. 15. This behavior is in contrast to that of a steady boundary layer, where excessive thickening of the boundary layer occurs as flow reversal is approached. At low frequency ($f = 0.25$ hz), the thickness oscillates as much as ± 40% about the mean value; however, at $f = 2.0$ hz this variation is only about ± 5%.

Conclusions

The conclusions from our experiments to date may be summarized as follows:

1. The mean velocity profile in the boundary layer is unaffected by imposed free-stream oscillations in the range of frequencies employed, and it is the same as the one measured with a free-stream velocity distribution held steady at its mean value.
2. This behavior of the mean velocity field is a consequence of two observations: (a) the time-averaged Reynolds stress distribution across the boundary layer is unaffected by the imposed oscillations and is indeed the same as the one measured with the free-stream velocity distribution held steady at the mean value; and (b) the Reynolds stresses arising from the organized velocity fluctuations under imposed oscillatory conditions are negligible compared to the Reynolds stresses due to the random fluctuations.

3. The amplitude of the periodic component in the boundary layer under quasi-steady oscillations ($f \rightarrow 0$) is as much as 70% larger than the imposed free-stream amplitude. However, at higher frequencies the peak amplitude in the boundary layer is rapidly attenuated toward an asymptotic behavior where amplitudes in the outer region of the boundary layer become the same as the free-stream amplitude, dropping off to zero in the near-wall region.

4. Quasi-steady boundary layer velocity response is in phase with the imposed free-stream oscillations. As the frequency is increased, phase lags begin to develop in the outer region of the boundary layer. The magnitude of this phase lag reaches a maximum and then decreases with increasing frequency until an asymptotic limit is reached where the outer region once again responds in phase with the free stream. Near the wall, however, large lead angles are present at higher oscillation frequencies.

5. A consequence of (3) and (4) above is that the boundary layer thickness becomes nearly frozen over the oscillation cycle at higher frequencies. This remains true even if flow reversal takes place in the near-wall region over a part of the oscillation cycle, as in the large-amplitude case.

6. A consequence of (3), (4), and (5) above is that the Reynolds stress distribution in the outer region of the boundary layer also becomes frozen over the oscillation cycle at higher frequencies.

Acknowledgments

This research is carried out at Stanford in cooperation with and under the sponsorship of the Army Aeromechanics Laboratory, the NASA-Ames Research Center, and the Army Research Office. The authors wish to express their gratitude to James McCroskey (AML), Mr. Leroy Presley (NASA-Ames), and Dr. Robert Singleton (ARO) for their continued assistance.

References

1. Parikh, P. G., et al.: "On the Behavior of an Unsteady Turbulent Boundary Layer." To appear in Numerical and Physical Aspects of Aerodynamic Flows, Tuncer Cebeci, ed., Springer-Verlag Publishing Co. (1981).

2. Rao, K. N., et al.: "The 'Bursting Phenomenon' in a Turbulent Boundary Layer." J. FLuid Mech., 48, 339-352 (1971).

3. Hussain, A.K.M.F., and Reynolds, W. C.: "The Mechanics of an Organized Wave in Turbulent Shear Flow." J. Fluid Mech., 41, 241-258 (1970), and 54, 241-288 (1972).

4. Andersen, P. S., Kays, W. M., and Moffat, R. J.: "The Turbulent Boundary Layer on a Porous Plate: An Experimental Study of the Fluid Mechanics for Adverse Free-Stream Pressure Gradients." Report No. HMT-15, Dept. of Mech. Engrg., Stanford University (1972).

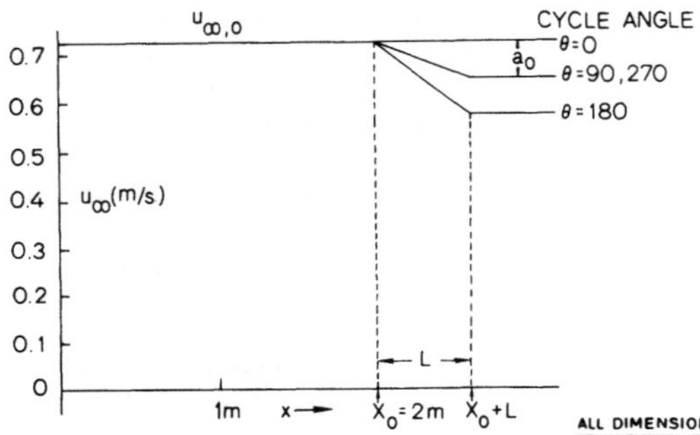

Fig. 1. Free-stream velocity variation.

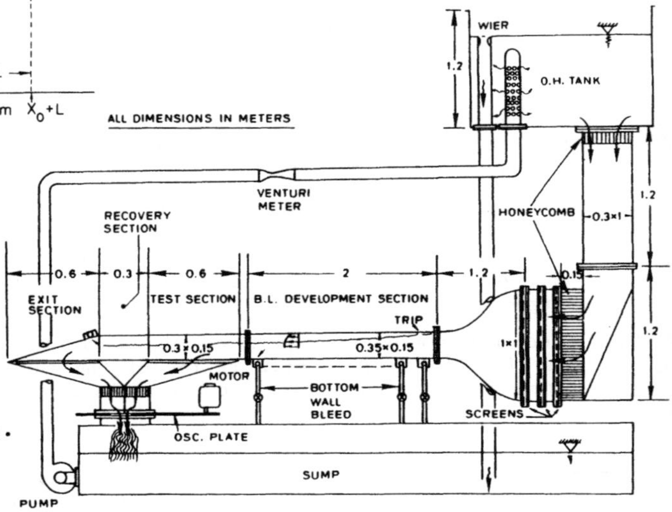

Fig. 2. The apparatus.

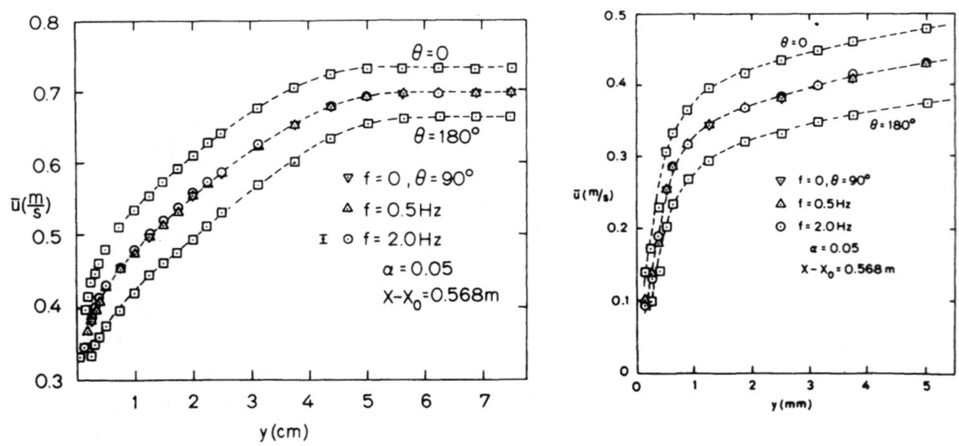

Fig. 3. Mean velocity profiles at three frequencies and profiles at the extremes of the oscillation at zero frequency for α = 0.05.

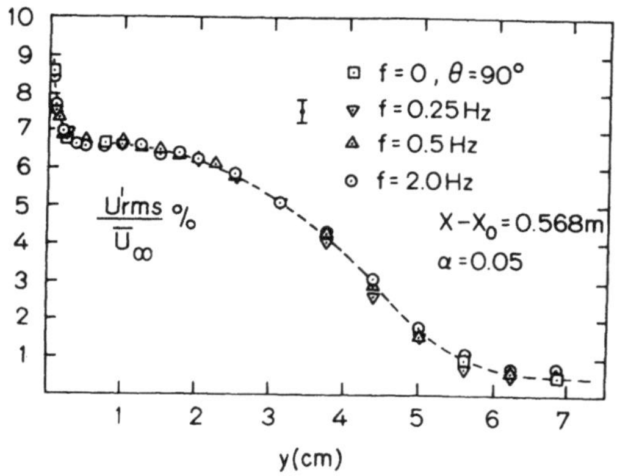

Fig. 4. Mean turbulence profiles for $\alpha = 0.05$.

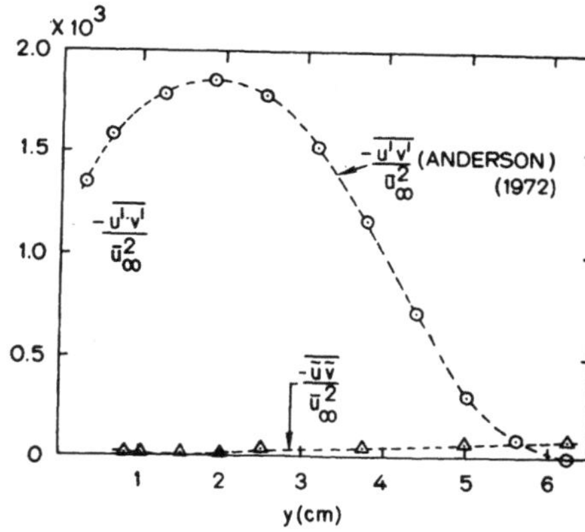

Fig. 5. Turbulent and organized Reynolds stresses for $\alpha = 0.05$.

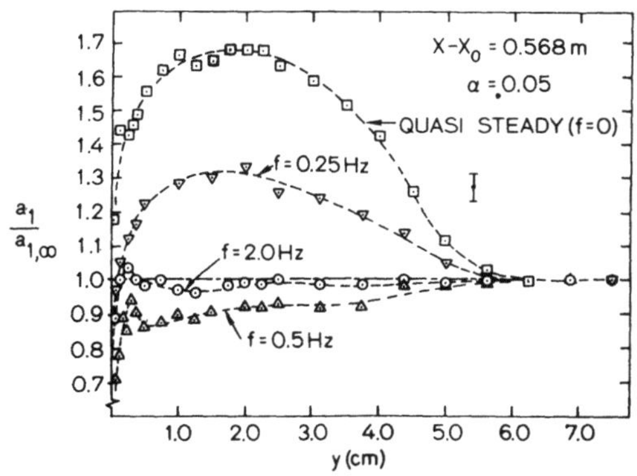

Fig. 6. Amplitude of organized disturbance for $\alpha = 0.05$.

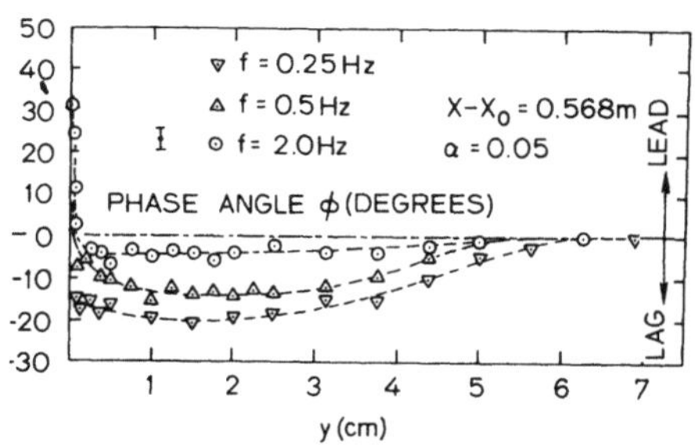

Fig. 7. Phase of organized disturbance for $\alpha = 0.05$.

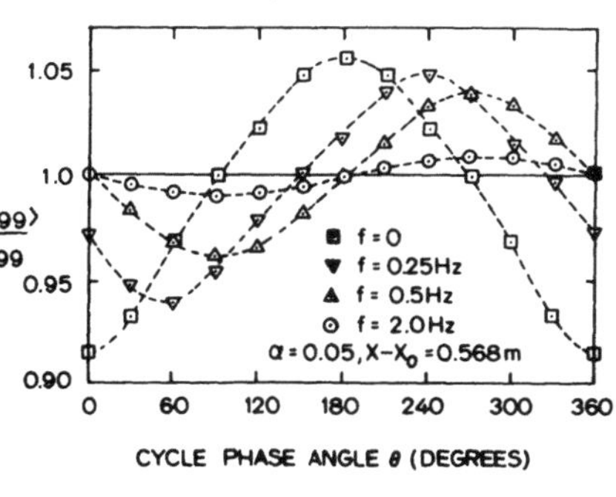

Fig. 8. Boundary layer thickness variation for $\alpha = 0.05$.

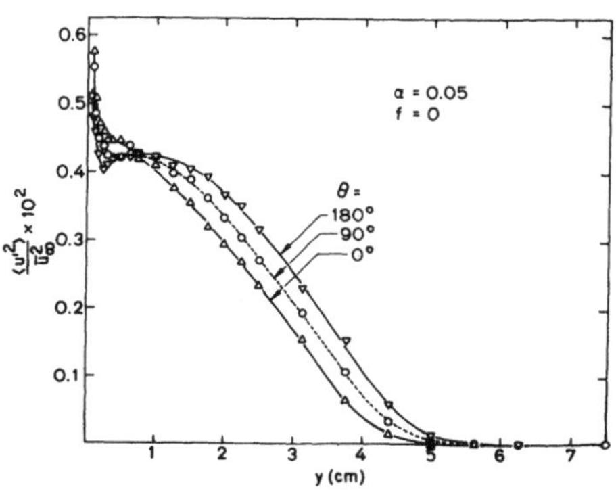

Fig. 9. Phase-average longitudinal fluctuation for $\alpha = 0.05$, $f = 0$.

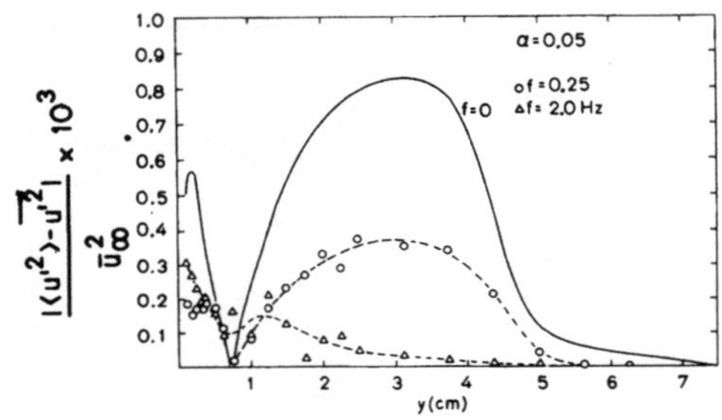

Fig. 10. Reynolds stress oscillations at α = 0.05.

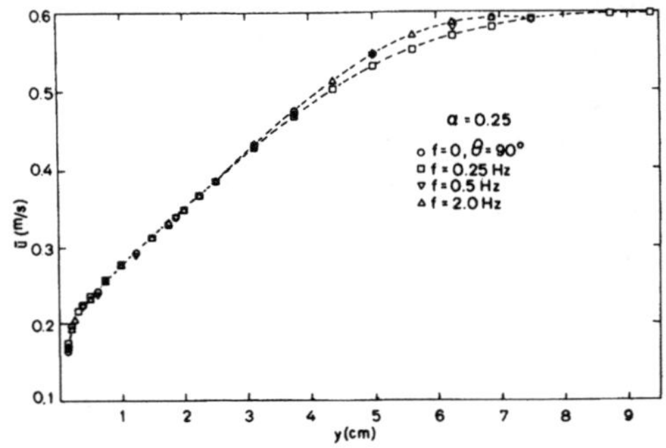

Fig. 11. Mean velocity at α ≃ 0.25 (preliminary).

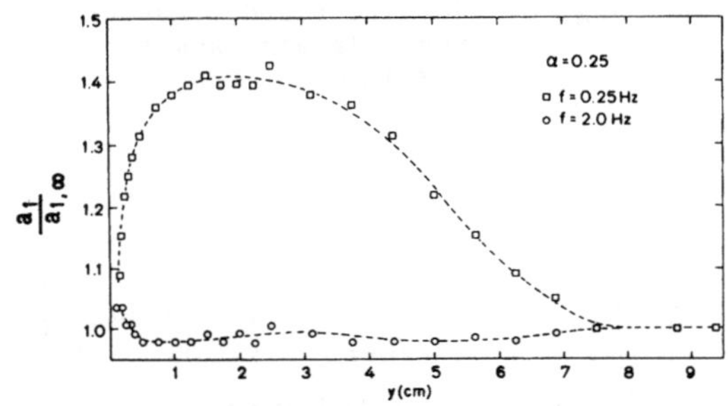

Fig. 12. Amplitude of organized disturbance for α ≃ 0.25 (preliminary)

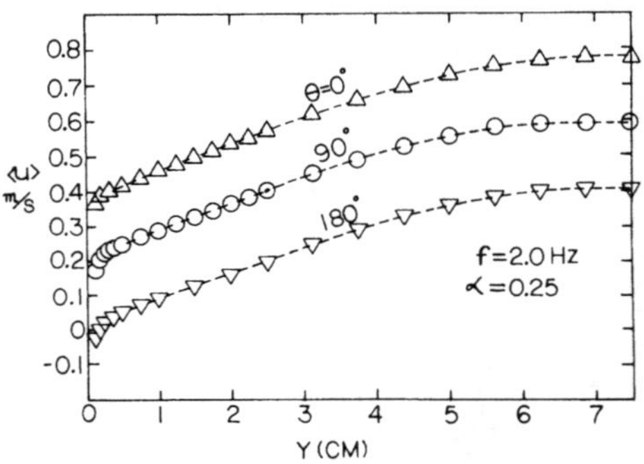

Fig. 14. Phase average velocity profiles for α ≃ 0.25 (preliminary).

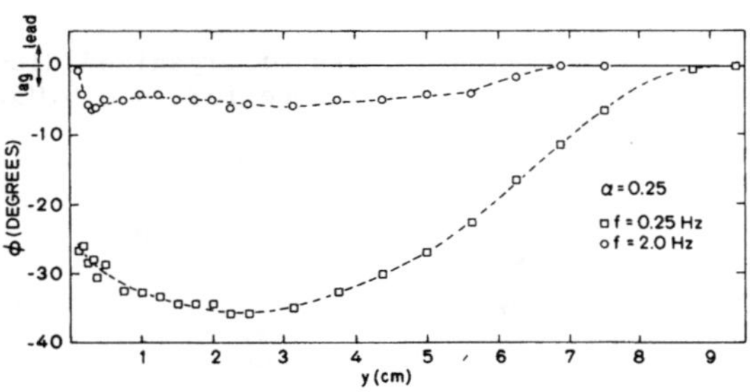

Fig. 13. Phase of organized disturbance for α ≃ 0.25 (preliminary).

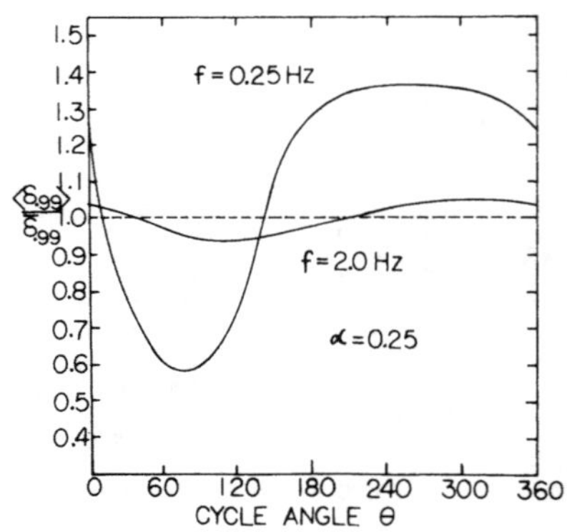

Fig. 15. Boundary layer thickness variation for α ≃ 0.25 (preliminary).

Periodic Turbulent Pipe Flow at 'High' Frequencies of Oscillation

B.R. RAMAPRIAN AND S.W. TU

Institute of Hydraulic Research
The University of Iowa, Iowa City, Iowa U.S.A.

Summary

The paper reports some of the results of an ongoing study of the effect of imposed unsteadiness on the time-mean behavior as well as the detailed structure of unsteady turbulent flows. First, experimental data obtained in a very simple unsteady flow namely periodic turbulent flow in a long circular pipe are presented. Secondly, a calculation procedure based on an existing one-equation steady flow turbulence model has been developed and used to predict periodic pipe flow with a view to studying the implications of such models in the calculation of rapidly varying unsteady flows. The important difference between this study and those reported in the earlier literature is that attention is now focussed on flows in which the imposed oscillation frequency is of the same order as the characteristic frequency of turbulence.

Introduction

Experiments on periodic turbulent flow in long circular pipes have been reported by a few researchers in the past; e.g., [1,2,3,4,5,]. Only a few of these have measured the detailed turbulence structure in such flows. Further, in a majority of these experiments, the frequency of the periodic oscillation of the flow was very small compared to the characteristic frequency of turbulence in the flow. The experiments reported by Mizushima et. al. [4,5] are the only ones that the authors are aware of, in which oscillation frequencies of the order of turbulent frequency have been studied. At such frequencies, turbulent motions and impressed oscillations interact with each other. It is very difficult to produce such interactive unsteady flows in wind tunnels. However, it is possible to produce relatively easily such flows in water or oil flow facilities.

Description of the Present Experiments

Two series of experiments are reported in this paper. The first of these two was performed in fully developed turbulent flow of oil in a long circular pipe of 50 mm diameter (D) in which the flow was sinusoidally modulated at a frequency (f) of 1.75 Hz. The mean Reynolds number, $\overline{R}_e$ was about 2100 and the modulation amplitude was about 30%. The periodic flow was found to remain turbulent throughout the oscillation cycle. The estimated characteristic turbulent frequency (f_t) in this flow was about 2 Hz. Laser Doppler anemometry was used to measure the instantaneous velocity in the flow. Wall shear stress was obtained from the gradient of the instantaneous velocity profile at the wall. In the second series of experiments water was used as the working fluid in place of the oil. In these experiments measurements were obtained at a mean Reynolds number of 50,000 (with $f_t \simeq 4$ Hz) and an oscillation frequency of 3.6 Hz, with a modulation amplitude of about 15 per cent. Information on instantaneous velocity distribution and wall shear stress (τ_w) were obtained. The latter was measured using a flush mounted skin friction gauge.

In both the experiments, digital phase averaging technique was used to process the instantaneous data. For this purpose, the instantaneous value, say U at any radial location r can be written as

$$U(r,\theta,t) = \overline{U}(r) + U_p(r,\theta) + u(r,\theta,t) \quad \ldots\ldots\ldots\ldots\ldots (1)$$

where $\overline{U}$ is the time-mean velocity, U_p the periodic (phase averaged) velocity, u the random velocity and θ the phase angle. The phase-averaged value U_p as well as the phase-averaged rms value u'_p of the turbulent velocity u were obtained by sampling a large number of cycles. In the experiments with oil, 300 cycles were used for averaging and information was obtained at 48 phase positions during a cycle. The details of data acquisition and processing are provided in Ramaprian and Tu [6,7]. In the experiments with water, 1000 cycles and 100 phase intervals were used in the processing of the data. In both series of experiments, the phase averaged Reynolds shear stress $\langle uv \rangle$ was not directly measured but obtained indirectly from the velocity profile and phase-averaged wall shear stress (τ_{wp}) via the momentum equation. The details of this procedure are described in Ramaprian and Tu [6].

Experimental Results

The results of the low Reynolds number experiments will be discussed first. Figures 1, 2, and 3 show some of the most important results from experiment. In figure 1, the time-mean velocity profile $\overline{U}$ in unsteady flow is compared with the quasi-steady profile (corresponding to infinitesimally low frequency of oscillation) at the same mean Reynolds number. The unsteady flow exhibits a small but significant kink in the velocity profile indicating that at interactive frequency of oscillation, there is a net effect on the time-mean flow (which, in this case, is an increase in the time-average wall shear stress $\overline{\tau}_w$). Figures 2 and 3 show the effect on the turbulent structure. Figure 2 shows that the phase-averaged turbulence intensity u'_p is virtually frozen during the oscillation cycle. The phase-averaged Reynolds shear stress <uv> normalized with respect to $\overline{u}_*$ ($\overline{u}_*$ being defined as $\sqrt{\overline{\tau}_w/\rho}$) is seen from figure 3 to follow the oscillation cycle at very small distances from the wall (specifically, within the Stokes layer) beyond which it also tends to remain frozen during the cycle. The phase-averaged shear velocity u_{*p} (defined by $\sqrt{\tau_{wp}/\rho}$), however, was found to follow the oscillation cycle (with some phase lead) with the result that u'_p/u_{*p} and $<uv>/u^2_{*p}$ in figures 2 and 3 are seen to vary continuously through the cycle indicating a complete destruction of structural equilibrium.

Some of the important results from the high Reynolds number experiments are shown in figures 4-8. Figure 4 shows that at this Reynolds number and frequency there is very little observable effect on the mean velocity profile. Likewise, it is seen from figure 5 that there is no significant effect on the time-averaged intensity of turbulence, defined, for unsteady flow, as

$$u' = \left[\frac{1}{2\pi}\int_0^{2\pi} u'^2_p \, d\theta\right]^{1/2} \quad \ldots\ldots\ldots\ldots\ldots\ldots\ldots (2)$$

Incidentally, figure 5 shows that the present measurements of u' are in reasonable agreement with the data of Laufer [8]. Figure 6 shows that the phase-averaged turbulent intensity $u'_p(\theta)$ remains nearly frozen through the oscillation cycle, just as in the first experimental series. Figure 7 shows a few typical distributions of the phase-averaged Reynolds shear stress <uv> (normalized with respect to the time-mean cross sectional average velocity $\overline{U}_m$). It is seen that near the wall ($\eta \lesssim 0.2$) <uv> oscillates with a significant amplitude. In the central part of the flow, the amplitude is seen to be considerably attenuated, thereby indicating a trend similar to the low Reynolds number experiments. This trend is also seen from the typical phasewise distributions of <uv> shown in figure 8.

This figure also shows the distribution of the phase-averaged wall shear stress τ_{wp}. It is seen that very near the wall (η = 0.0157), the Reynolds shear stress <uv> follows the wall shear stress τ_{wp} with similar amplitude while, far-away from the wall (η = 0.7638), the amplitude of oscillation of <uv> is significantly attenuated. The kinks in the distributions of τ_{wp} and <uv> near the wall appear to be associated with higher harmonics of the oscillation frequency for which the authors have no explanation at this time. In fact, the results for the second series of experiments are still being analyzed at the time of writing.

Prediction of Periodic Pipe Flow

It would be interesting to study how a turbulence model based on steady flow experience predicts an interactive unsteady flow of this type. Such a calculation method using a Prandtl-Kolmogorov type turbulence model in the form $\langle uv \rangle = k_p^{1/2} L_k \, \partial U_p/\partial y$ (where k_p is the phase-averaged turbulent kinetic energy per unit mass, L_k the length scale and y the distance from the wall) was developed. A one-equation model approach with specified distribution of L_k and with model assumptions similar to those of Acharya and Reynolds [9] was used. The time-dependent one-dimensional momentum equation was solved by an implicit finite difference procedure. One of the features of this method is that no wall functions were used and the finite difference representation was carried upto the wall.

The above calculation procedure was used to predict the high Reynolds number water flow experiments. Some of the typical results of the calculation are shown along with the experimental results in figures 4 and 7. The details of the calculation procedure and more extensive comparisons with experiment will be presented in a forthcoming report. It is seen from figure 4 that the calculation predicts the mean velocity profile accurately at 3.6 Hz (in fact, the same profile is predicted even at a very low frequency 0.018 Hz). There are some quantitative differences between the predicted and experimental values of <uv> as seen from figure 7. This is presumably due to the inadequacy of quasi-steady turbulence modeling at the interactive frequency. Even so, the time-averaged shear stress is reasonably well predicted.

Lastly, figure 9 shows a comparison between the predicted and measured values of the amplitude $(U_p)_{max}$ of the periodic velocity, normalized with respect to the amplitude $(U_{mp})_{max}$ of the cross-sectional average periodic velocity. It can be seen that, while both prediction and experiment indicate an overshoot in the amplitude, there is a significant discrepancy between prediction and experiment in respect of both the magnitude and location of this overshoot.

Conclusion

In conclusion, it can be said that experimental data obtained so far seem to indicate that when turbulent pipe flow is disturbed at high frequencies (of the order of the characteristic frequency of turbulence), the structural equilibrium of the flow breaks down. Hence, it may not be possible to predict all aspects of the flow using a quasi-steady turbulence model, even though such a model seems to predict the time-mean behavior of the flow reasonably well. Further work related to periodic turbulent shear flows is currently being pursued by the authors at the Iowa Institute of Hydraulic Research

References

1. Lu, S.Z., Nunge, R.J., Eriam, F.F., and Mohajery, S.M., "Measurements of pulsating turbulent water flow in a pipe." Proceedings of the Third Symposium on Turbulence in Liquids, The University of Wisconsin, Rolla, (1973) 375-392.

2. Schulz-Grunnow, "Pulsation flow through pipes", NASA-TT-F-14881, (1973).

3. Hino, M., Sawamoto, M. and Takasu, S., "Experiments on transition to turbulence in oscillatory pipe flow," Journal of Fluid Mechanics, 75, Pt. 2, (1975) pp. 193-207.

4. Mizushina, T., Maruyama, T., Shiozaki, Y., "Pulsating turbulent flow in a tube," Journal of Chemical Engineering of Japan, 6, No. 6 (1973) pp. 487-494.

5. Mizushina, T., Maruyama, T., Hirasawa, H., "Structure of the turbulence in pulsating pipe flows," Journal of Chemical Engineering of Japan, 8, No. 3 (1975) pp. 210-216.

6. Ramaprian, B.R., and Tu, S.-W., "Experiments on transitional oscillating pipe flow". IIHR Report No. 221, University of Iowa, Institute of Hydraulic Research (1980).

7. Ramaprian, B.R., and Tu, S.W., "An experimental Study of oscillatory pipe flow at transitional Reynolds numbers," Journal of Fluid Mechanics, 100 (1980) 513-544.

8. Laufer, J., "The structure of turbulence in fully developed pipe flow," NACA Rep. 1174 (1954).

9. Acharya, M. and Reynolds, W.C., "Measurements and predictions of a fully developed channel flow with imposed controlled oscillations." Stanford University, Mechanical Engineering Department, Thermosciences Division, Report No. TF-8 (1975).

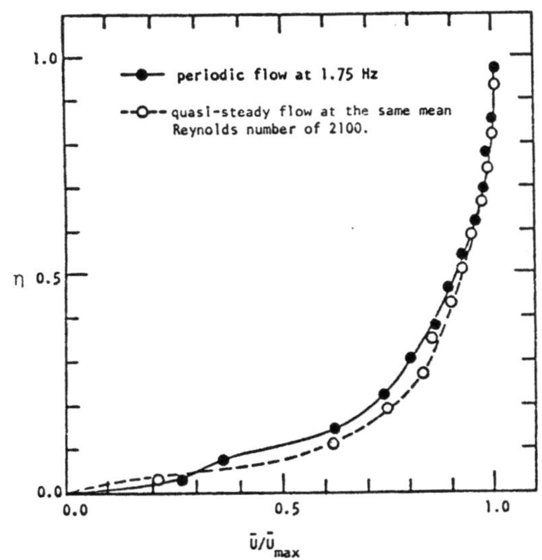

Figure 1. Distribution of the time-mean velocity profile across the pipe. $\eta = 1-2r/D$.

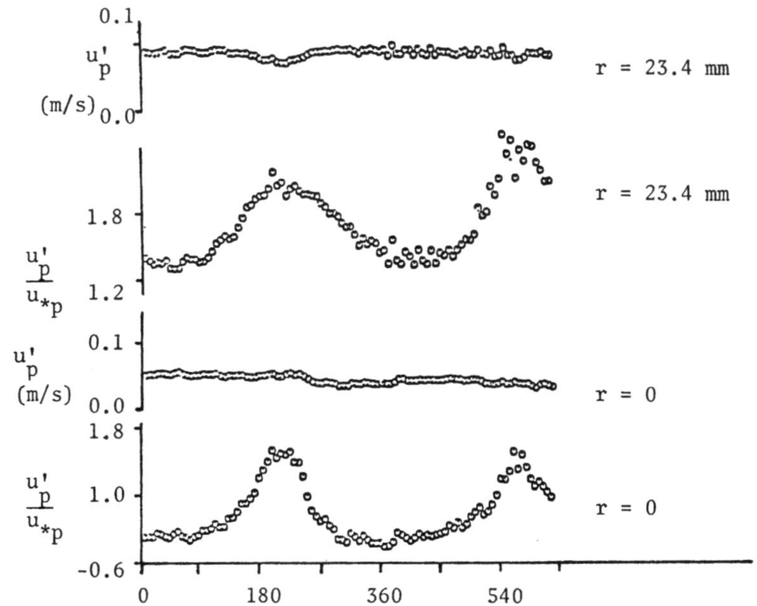

Figure 2. Behavior of u'_p during the oscillation cycle. r = 0 refers to the pipe centerline and r = 23.4 refers to a point 2 mm away from the wall. The Stokes layer thickness, $4\sqrt{\nu/\omega}$ was about 6 mm in this experiment. f = 1.75 Hz, $\overline{R}_e$ = 2100.

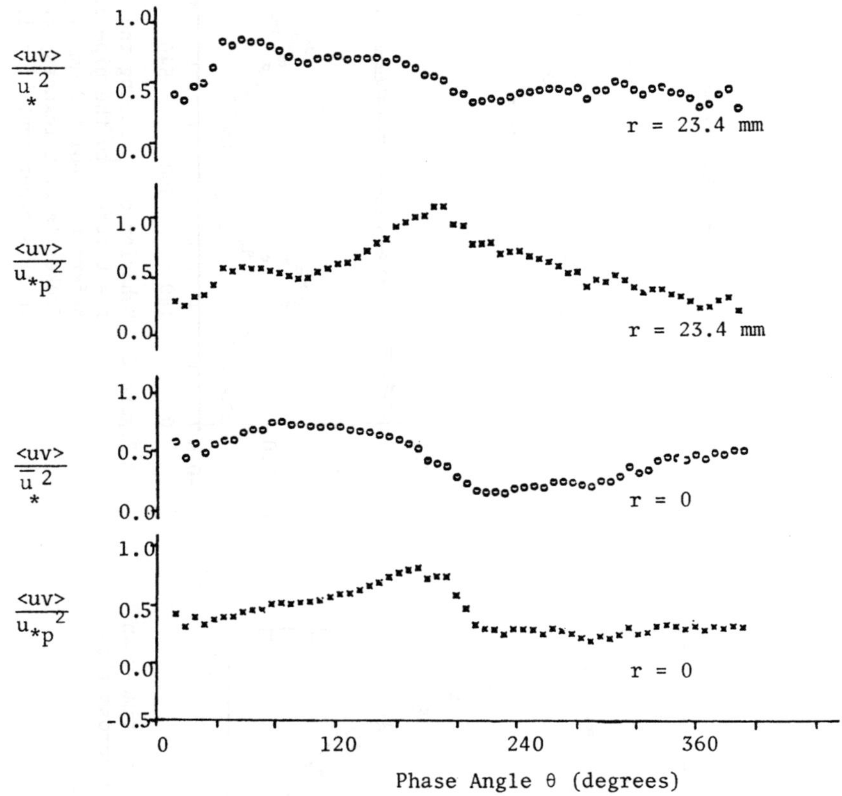

Figure 3. Behavior of <uv> during the oscillation cycle. f = 1.75 Hz, $\overline{R}_e$ = 2100. For other comments see caption for Figure 2.

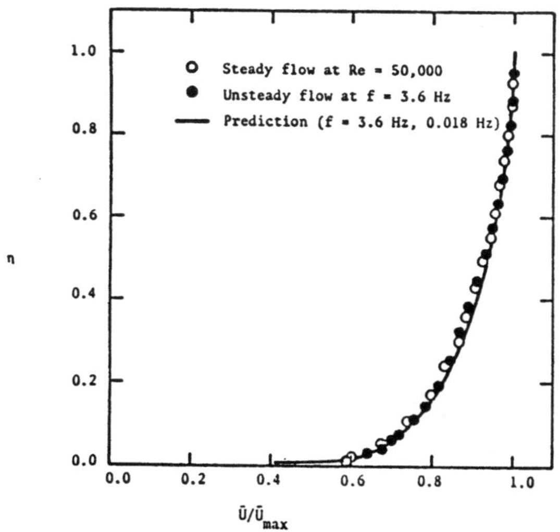

Figure 4. Time-mean velocity profile in the pipe ($\overline{R}_e$ = 50,000)

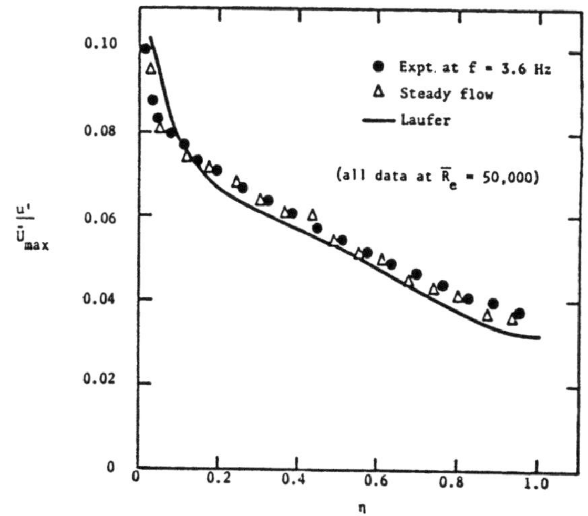

Figure 5. Distribution of u' across the pipe.

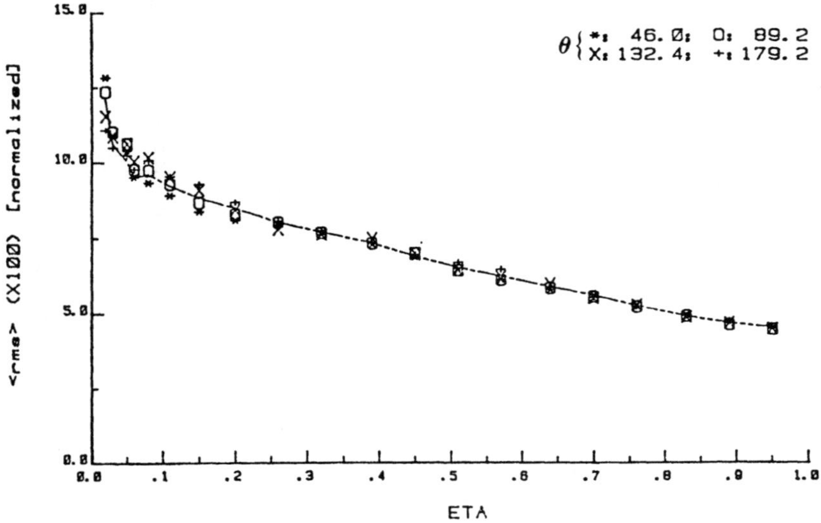

Figure 6. Turbulent Intensity Distributions across the pipe $f = 3.6$ Hz, $\overline{R}_e = 50{,}000$

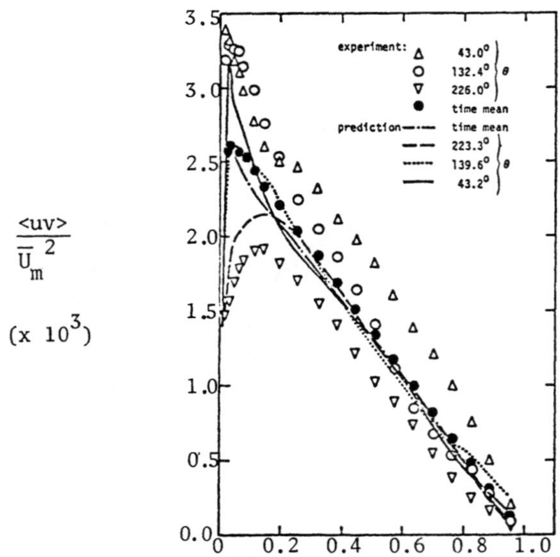

Figure 7. Reynolds shear stress distribution across the pipe ($f = 3.6$ Hz, $\overline{R}_e = 50{,}000$)

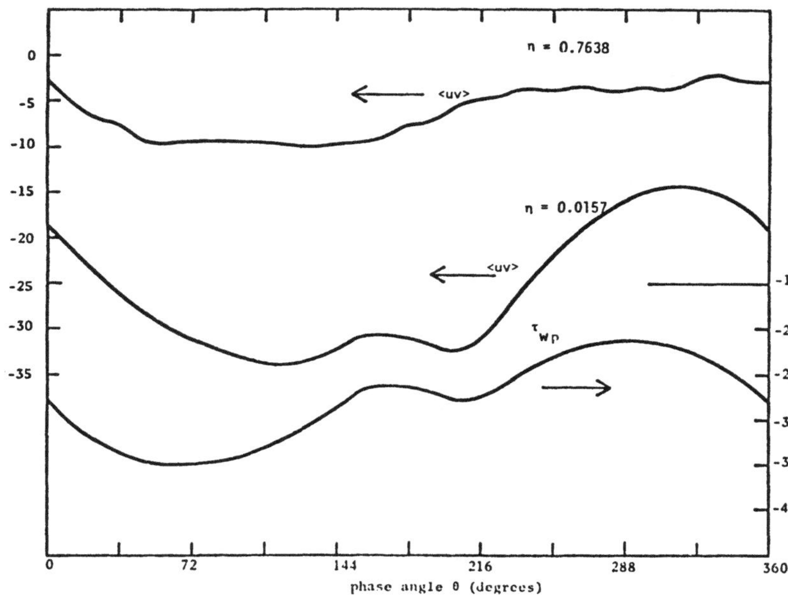

Figure 8. The variation of shear stress during the oscillation cycle (f = 3.6 Hz, $\overline{R}_e$ = 50,000)

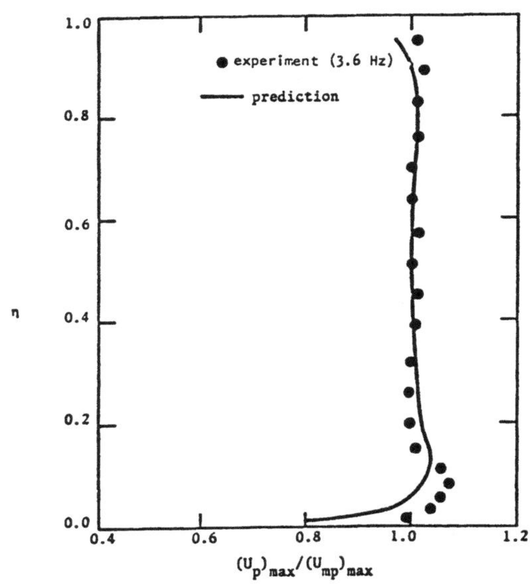

Figure 9. Distribution of the oscillation amplitude across the pipe ($\overline{R}_e$ = 50,000)

Effect of Driven-Wall Motion on a Turbulent Boundary Layer

LEONARD M. WEINSTEIN

NASA Langley Research Center
Hampton, VA 23665

Summary

An experimental boundary layer study was conducted in low-speed flow over a driven wall model. Acoustic horn drivers were used to vibrate a thin membrane surface to obtain two-dimensional standing wave motion. The purpose of the study was to examine the influence of the unsteady wall boundary motion on the structure of a zero pressure gradient turbulent boundary layer. Measured quantities included velocity profiles, turbulence quantities (including Reynolds stress), and smoke wire flow visualization.

The turbulent boundary-layer flow was examined at several velocities. This gave a combination of flow and wall induced disturbances. Longitudinal and vertical surveys were made to determine the variations relative to the surface.

At small distances from the driven wall there were large variations in fluctuation quantities along each model wave. These longitudinal variations decreased rapidly with increasing distance from the driven wall. However, the average level of these fluctuations only decreased slightly even outside the boundary layer. The effects of the acoustically induced quantities (due to the driven wall) seemed to be essentially independent of the state of the boundary layer, but with the level related to the mean velocity. The acoustic field did not alter significantly the turbulent boundary layer structure for the model surface motions examined in the present study. When the acoustic fluctuations were subtracted out, the remaining fluctuations were essentially the same as for the undriven model.

Nomenclature

f	driven frequency
U_ℓ	local longitudinal velocity
U_∞	free stream longitudinal velocity
u'	longitudinal fluctuation velocity
v'	vertical fluctuation velocity
x	distance from start of test section
y	vertical distance above model surface
$\delta_{.995}$	location of $U_\ell/U_\infty = 0.995$

λ	wave length of model
$(\tilde{\ })$	root mean square value
$(\bar{\ })$	time average value

Introduction

In the last 20 years there have been a large number of theoretical and experimental studies on the use of flexible or compliant walls for turbulent boundary layer drag reduction. The results of these studies were inconclusive and indicated the need for controlled experiments, including detailed wall motion information. Studies conducted at NASA Langley Research Center (Ref. 1) indicate that for low-speed air boundary layers, passive walls appear to be incapable of producing the very short wave length and moderate amplitude believed necessary for skin friction reduction. In Reference 2 it was suggested that driven wall surface motion could be used to assess possible turbulent boundary layer modifications due to wall motion.

Previous studies using active walls to examine turbulent boundary layers have generally used mechanical drivers (e.g., Refs. 3-7). Mechanical drivers are, in practice, limited to low frequency motion (typically < 50 Hz). In Reference 8 an electrostatically driven wall was described, which was capable of operating over a frequency range of 0-10 KHz, and had wave lengths as small as 0.18 cm. However, the amplitude of the wave motion for these models was not adequate ($< 10^{-4}$ cm for the case described), due to limitations of electrostatic models (i.e., maximum allowable voltage, material properties, etc.). An acoustic driven model was used by Schiltz (9), but also had limited amplitude motion.

The present study presents results of an acoustically driven model, with amplitudes greater than 0.01 cm. The model, which had a 1 cm wavelength, was tested in the frequency range of 500 Hz to 2.2 KHz. The goal of this study was to examine the influence of unsteady wall boundary conditions on the structure of a zero pressure gradient turbulent boundary layer.

Experimental Apparatus and Test Procedures

The model, shown in Figure 1(a), was mounted to the floor of a closed circuit wind tunnel with a 17.8 x 27.4 cm cross section. The tunnel was operated from 6 to 36 m/s for the current study. A trip wire was placed 46 cm ahead of the model. For speeds of 12 m/s or higher, the boundary layer was fully turbulent, with the virtual origin very close to the trip wire.

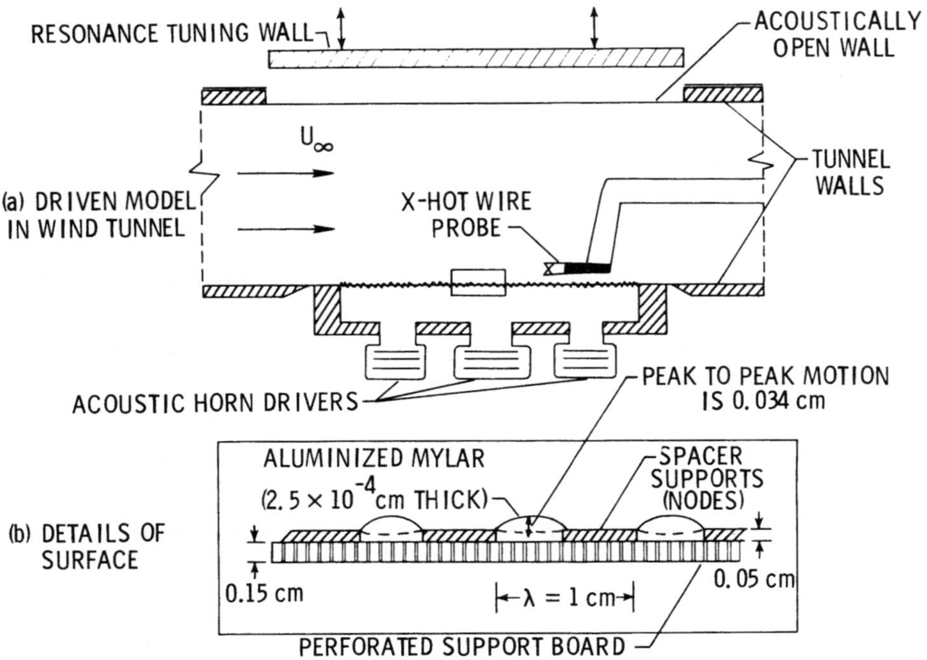

Figure 1. - Acoustic driven model setup

The model was 40 cm long by 20 cm wide, and had six acoustic horn drivers mounted on the bottom of a 2.5 cm high chamber below the model surface. The horn drivers were all driven in phase. The upper wall of the wind tunnel was replaced by an acoustically permeable surface (Ref. 10) to allow the sound field to pass out of the wind tunnel. Two test cases were examined: (1) a sound absorbing surface above the acoustically permeable plate resulted in a "free field," and (2) a movable, solid plate above the tunnel gave a "resonant field" which could be tuned for maximum intensity. A schematic of the model surface is shown in Figure 1(b). The membrane was bonded to lateral spacers 0.5 cm long, with 0.5 cm gaps over a perforated substrate (to pass the driving pressure field). The mylar over the gaps had pre-slack to allow motion with amplitude greater than 0.01 cm.

Data were obtained with a pitot probe and a hot-wire "x" probe, which were laterally separated 2 cm and attached to a traverse mechanism. Both longitudinal and vertical surveys were made throughout the boundary layer and in the region just above the boundary layer. The pitot data gave the mean velocity profiles, while the hot wire was used to examine the fluctuation quantities. A smoke wire was used for flow visualization studies.

Results and Discussion

Figure 2 shows the details of the flow structure obtained using the smoke wire. The flow velocity was 12 m/s. The undriven wall case shows the turbulent boundary layer at the bottom of the photograph, with a region of the free stream extending to about half the test section height. At a driven frequency of 750 Hz, the flow outside the boundary layer becomes irregular, with slight but visible disturbances through the entire flow field. At 2,200 Hz, the shorter wavelength makes these disturbances show up more clearly. These show up in the free stream streak lines as small bumps. In both driven cases shown, the turbulent boundary layer was not noticeably altered.

Pitot surveys made at several locations along the driven model were also not noticeably changed from the undriven case. The absence of change in mean velocity profiles indicated no change in momentum thickness, and thus no change in total drag for the driven model.

Longitudinal surveys were made with the hot-wire probe positioned 0.16 cm above the model. The survey region extended over five waves, and showed the near wall variations of fluctuation quantities over the driven waves. Figure 3 shows results for U_∞ = 12 and 24 m/s for the undriven wall and for the driven wall at 750 Hz. The average level of $\tilde{v}/U_\infty$ is larger for the driven wall, with large variations along the waves. The values of $\tilde{u}/U_\infty$ for the driven and undriven cases are not noticeably different. This is due to the fact that the wall induced fluctuations in $\tilde{u}/U_\infty$ are generally smaller than the turbulent fluctuations near the wall, while the wall induced fluctuations in $\tilde{v}/U_\infty$ are larger than the turbulent fluctuations in $\tilde{v}/U_\infty$ for these cases. The Reynolds shear stress variations over the model also shows significant effects of wall motion. As the tunnel velocity is increased from 12 to 24 m/s while keeping the wall motion the same, the fluctuation quantities show less change for the driven wall compared to the undriven case. Since the turbulent fluctuations increase with increased test velocity, and since the wall induced fluctuations close to the wall appear to have a large component which is directly porportional to the wall motion, this result would be expected. At 6 m/s the boundary layer is laminar;

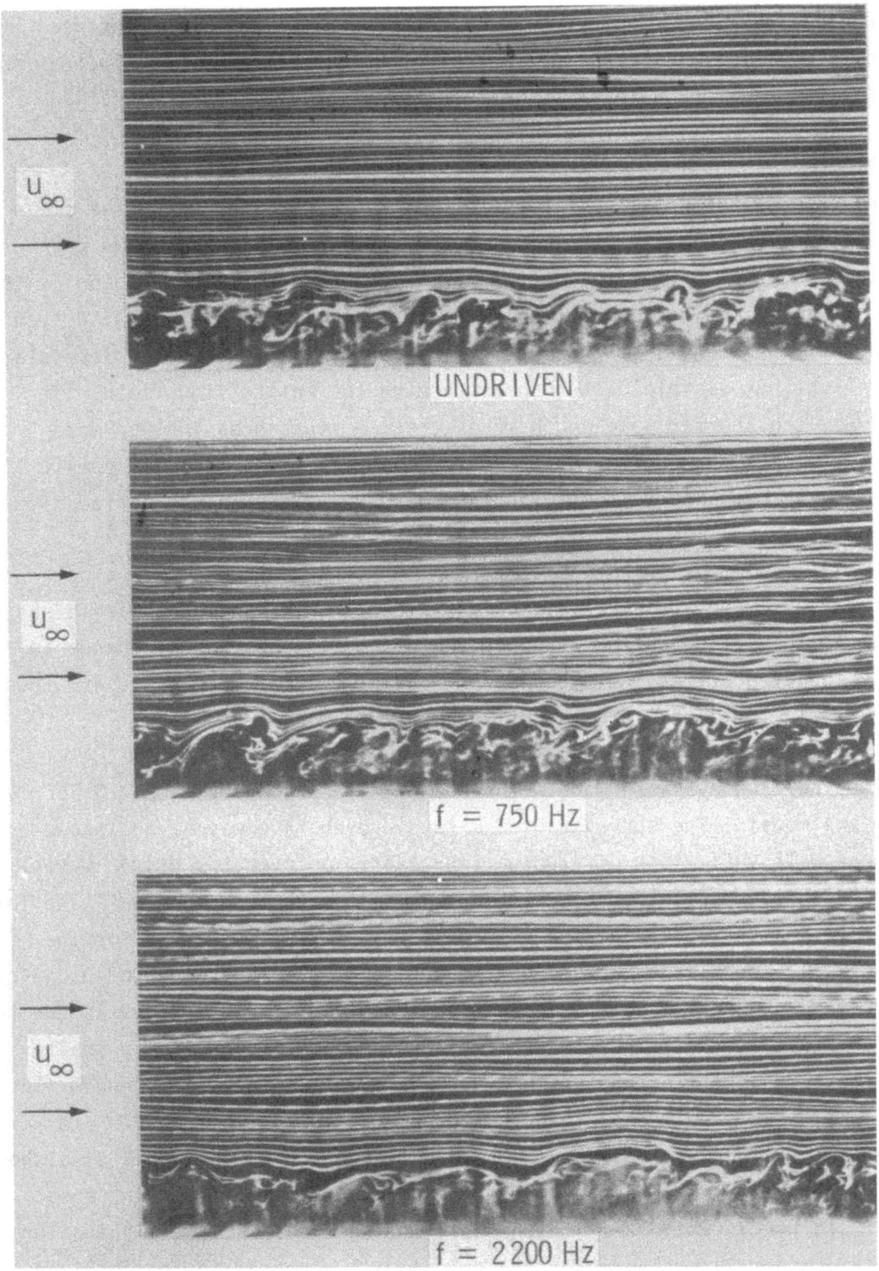

Figure 2. - Smoke wire visualization of flow x = 40 to 55 cm, U_∞ = 12 m/s

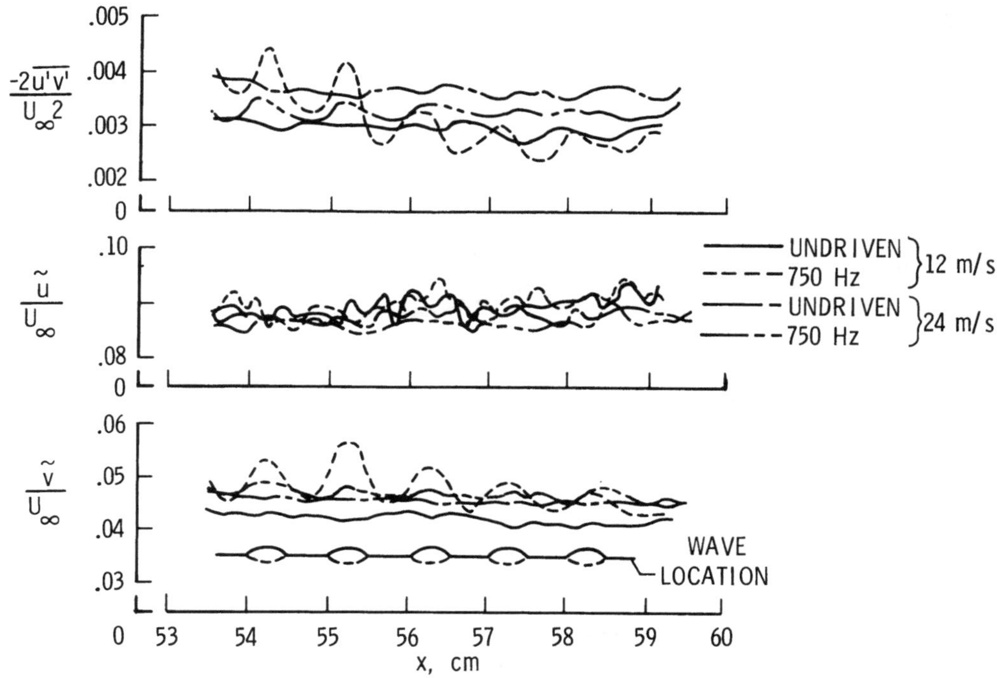

Figure 3. - Effect of driven wall on near wall fluctuation quantities over several waves. y=0.16 cm

consequently, the fluctuations induced by wall motion alone can be measured. If the wall induced fluctuations and turbulent fluctuations are assumed to be uncorrelated, and the fluctuations due to wall motion (for 6 m/s, from Ref. 10) are subtracted from the data of Figure 3, the large longitudinal variations are removed. A shift in average fluctuations levels remains; this will be examined later.

Vertical surveys were made at the two locations shown in Figure 4, which were the locations of the largest variations in Reynolds stress (locations (2) and (4)). At U_∞ = 12 m/s, Reynolds stress shows variations near the wall, but these variations die off by $y/\delta_{.995} \simeq 0.3$ (which about $\lambda/2$). This indicates the range of the dipole contribution of surface motion which is the cause of the longitudinal variations in fluctuation quantities. There is a significant amount of Reynolds stress at $y/\delta_{.995} = 1.4$, and this is believed to be due to the piston component of wall motion interacting with the mean flow. The values of $\tilde{u}/U_\infty$ and $\tilde{v}/U_\infty$ are the same through the boundary layer for the two locations examined. At 24 m/s the trends are the same but less extensive for the Reynolds stress.

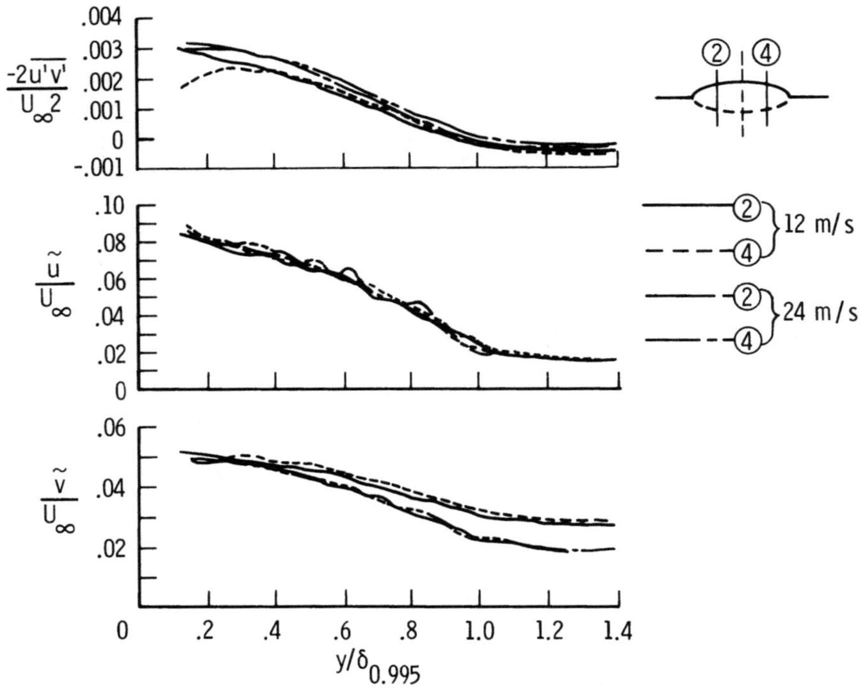

Figure 4. - Variation of fluctuation quantities through boundary layer over driven wall. $x \approx 54$ cm., f=750Hz.

The piston component of induced fluctuations should be nearly constant in the flow field examined, since its length scale is the order of the length of the entire model. In order to determine the magnitude and velocity dependence of this effect, longitudinal surveys were made at velocities of 6 to 24 m/s over the full length of the model, completely outside the boundary layers. The fluctuation quantities were averaged over the model length for both tuned and free field cases. These average values are shown in Figure 5 (in dimensional form) for 750 Hz. The variations of these averages with tunnel velocity are approximately linear in the form shown. The large difference in level for tuned as compared with free field data reflects the greater wall motion and thus larger fluctuation quantities available with a resonant setup.

If the Reynolds stress value outside the boundary layer is adjusted to zero, the same shift will move the longitudinally averaged level of the data to the undriven wall values. This indicates that the piston component of the induced fluctuations is uncorrelated with the turbulence, and does not interact significantly with the turbulent boundary layer. This agrees with the lack of change in mean velocity profiles previously mentioned.

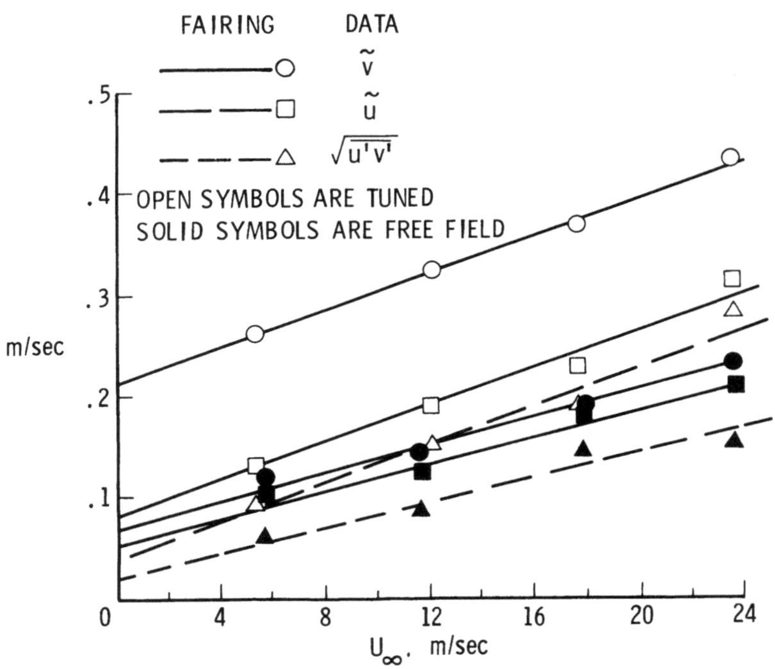

Figure 5. - Fluctuation quantities 2.5 cm above wall. Average over model. f=750 Hz

Conclusions

This study showed that for the driven model, hot-wire measurements show large variations in fluctuation quantities (including Reynolds stress) compared to the undriven model. However, these variations are due to the superposition of the acoustic and turbulent fluctuations which indicates that there is no significant interaction between the acoustic and turbulent fluctuations.

References

1. Ash, R. L., Bushnell, D. M.; Weinstein, L. M.; and Balasubramanian, R.: Compliant Wall Surface Motion and Its Effect on the Structure of a Turbulent

Boundary Layer. Fourth Biennial Symposium on Turbulence in Liquids, University of Missouri-Rolla, September 1975.

2. Bushnell, D. M.; Hefner, J. N.; and Ash, R. L.: Effect of Compliant Wall Motion on Turbulent Boundary Layers. The Physics of Fluids, Vol. 20, No. 10, Pt. 11, October 1977, pp. 531-538.

3. Kendall, J. M.: The Turbulent Boundary Layer Over a Wall With Proresive Surface Waves. Journal Fluid Mechanics, Vol. 41, Pt. 2, 1970. pp. 259-281.

4. Saeger, J. C.; and Reynolds, W. C.: Perturbation Pressures Over Traveling Sinusoidal Waves With Fully Developed Turbulent Stress Flow. Technical Report FM-9, Dept. of Mech., Eng., Stanford University (1971).

5. Merkulov, V. I.; and Savchenko, Yu. I.: Experimental Investigation of Fluid Flow Along a Traveling Wave. Hydrodynamic Problems of Bionics, Bionica, Keiv, 4, pp. 3-120, (1970).

6. Mattout, R.: Reduction de trainee par parois souples (Reduction of Drag by Flexible Walls). Association Technique Maritime et Aeronautique, Bulletin No. 72, pp. 207-227, 1962.

7. Mattout, R.; and Cottenceau, B.: Etude experimentale d'une paroi souple activee en tunnel hydrodynamique measures globale. Societe Bertin and Cie. Note Technique, No. 71-C1-09, 1972.

8. Weinstein, L. M.; and Balasubramanian, R.: An Electrostatically Driven Surface for Flexible Wall Drag Reduction Studies. Second International Conference on Drag Reduction, August 31-September 1977. Held at St. Johns College, Cambridge, England.

9. Schiltz, Von W.: Untersuchunger Ulsen Den Einfluss Beige fon migan Wnadschivingungln Aug Die Entivicblung Derstnom ungrunschicht. Acoustics, V. 15, 1965, pp. 6-10.

10. Weinstein, L. M.: Effect of Driven Wall Motion on Boundary Layer Flow. D.Sc. Dissertation, George Washington University, 1981.

Structure of Turbulent Boundary Layer on an Oscillating Flat Plate

Y. KOBASHI and M. HAYAKAWA

Faculty of Engineering
Hokkaido University
Sapporo, 060, Japan

Summary

In order to study the effects of the unsteadiness of the flow on the boundary layer flow and its turbulence structure, an oscillating turbulent boundary layer has been investigated experimentally in a wind tunnel in which a flat plate is driven sinusoidally parallel to the constant freestream. Turbulent patches, which appear in the transition region, were also examined to compare its structure with that of the fully turbulent layer.

Introduction

In spite of the importance in the practical problems, very little experimental data are available regarding to the effects of the flow unsteadiness on turbulent boundary layers. The pioneering study of an oscillating turbulent boundary layer was carried out in 1959 by Karlsson [1], who measured the mean properties and the harmonic component under the wide range of freestream amplitude and oscillation frequency with rather disappointing results that the mean characteristics of the boundary layer are not so much different from those of the steady state. Since that time, many calculations based on various turbulence models have been presented. But none of these models has been checked sufficiently because of the lack of experimental data [2]. Quite recently some experimental efforts have been reported by the ONERA group [3,4] and Patel [5], the latter being concerned with a traveling wave type oscillation. To our knowledge, Houdeville et al.[3] were the first who studied the turbulence characteristics by using a statistical treatment. It is felt, however, that more experimental informations are needed in this context.

The present study was conducted with the aim of clarification of the effects of flow unsteadiness on the behaviour and structure of boundary layer both in the transition and the fully turbulent regions. The experiments were made on the moving plate in the low turbulence wind tunnel, which can realize the precisely sinusoidal motions at any frequency without producing undesirable disturbances.

Apparatus and Measurements

The experiments were carried out in a blow-down type wind tunnel whose test section has the cross-section of 0.2m x 0.4m and 2.1m long. A flat plate of 8mm thick, 0.38m wide and 2.3m long is installed vertically in the tunnel center section and is driven back and forth sinusoidally by a crank mechanism [6]. A trip-wire of the suitable size was attached across the plate near its leading edge in order to induce earlier transition to turbulence.

Hot-wire anemometers with linearizer units were used. A single wire and X-wire probes fixed in the flow stream were employed for the velocity-components measurement. The wall shear stress was measured by using a heated surface sensor, which is 0.95mm wide and 0.34mm long and is mounted flush with the wall surface. The signals of both the hot-wire and the surface sensor were stored on analogue tapes together with the timing signals which were synchronized with the plate motion. All the data recorded were digitized and processed by either a mini-computer or a wave form eductor to perform the periodic averaging and the conditional sampling analyses.

Both the transition regime and the fully turbulent regime were measured under the following conditions:

U_0	(external flow velocity)	5.4 – 6.6 m/s
f_w	(oscillation frequency of plate)	1.5 – 6.5 Hz
r	(length of crank arm)	10 and 30 mm
Δ	(velocity amplitude ratio = $2\pi f_w r/U_0$)	0.03 – 0.12.

Since the hot-wire probes were inserted from a side wall of the test section, the measured streamwise velocity was transformed into the relative value to the flat plate in the course of the data deduction process. The reference phase position($T = 0$) is defined as the time at which the relative flow speed attained its maximum value. Therefore, the instantaneous freestream velocity can be given by

$$U_e(T) = U_0(1 + \Delta \cos T); \quad T = \omega t = 2\pi f_w t$$

where Δ is the velocity amplitude ratio.

Results and Discussions

(1) Structure of Turbulent Patch

The most pronounced feature in the transition regime of an oscillating boundary layer is that the disturbances are highly amplified within some limited phase range due to the localization effect of the disturbances. This causes the turbulent patch of two dimensional shape to be appeared periodically in space and time instead of the turbulent spot of the steady boundary layer [6,7]. Fig.1 shows some of the velocity signals detected simultaneously by two hot-wire probes, one of which is fixed at the center section of the plate(z = 0), while the other is moved at the various spanwise positions. The velocity in the outer part of the patch is smaller than the surrounding non-turbulent flow but appreciably larger near the wall. Turbulence intensity has its peak near the front part of the patch except adjacent to the wall. These behaviours are essentially the same as the case of the turbulent spot.

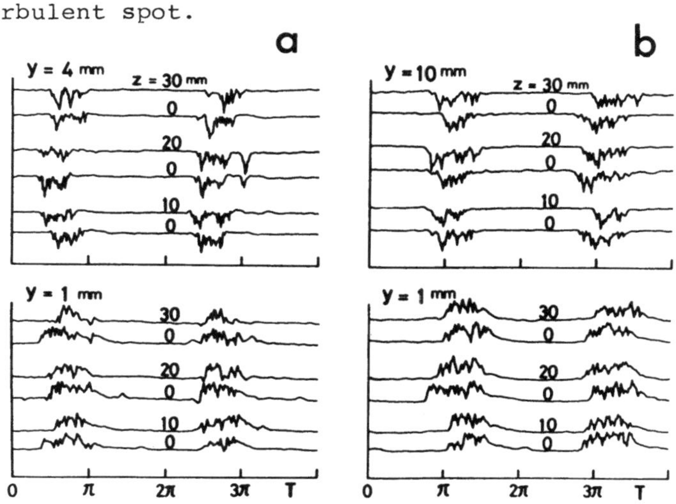

Fig.1. Velocity signals in a transition region
U_0= 5.50m/s, f_w= 2.24Hz, Δ= 0.076
(a) x= 500mm, (b) x= 900mm

Although the appearence of the patch is somewhat random with respect to the phase of the flow oscillation, it is possible to determine the average boundary between the turbulent and the non-turbulent region if we define its boundary as the phase position at which the local intermittency factor γ_L falls on the prescribed value, say $\gamma_L = 0.5$. Here γ_L is provided by the phase averaged value of the intermittency functions which were calculated from the out-put signals of the turbulence discriminator. The patch contour thus obtained is shown in Fig.2, together with the velocity profiles and the distributions of the normal velocity component V. It is apparent that the V-component becomes negative in some limited region near the leading interface and is positive elsewhere. In the figure, the broken line denotes the demarcation line between the positive and the negative V-component, and the shaded areas represent the active parts for the turbulence production infered from the distributions of the Reynolds stress and the turbulence intensity(not shown here).

The overall picture of the patch is quite similar to that of the plane of symmetry of the turbulent spot, and is characterized by the existance of a large-scale circulatory flow, accompanied with the relatively large entrainment at the leading interface and the upward flow along the trailing interface. It is also noted that the flow pattern associated with the patch resembles that

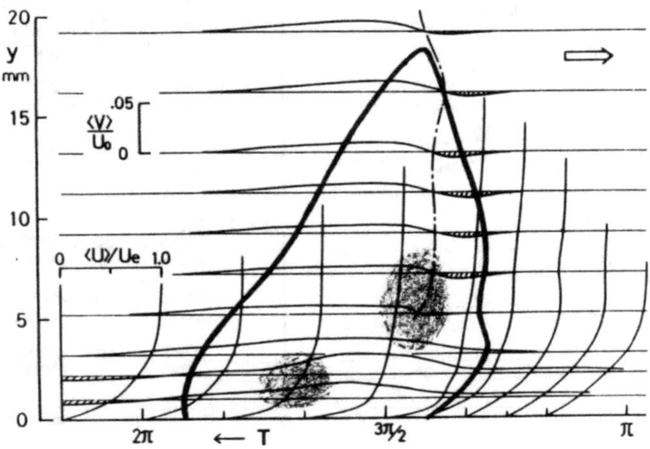

Fig.2. Structure of a turbulent patch
$U_0 = 5.55\text{m/s}$, $f_w = 3.43\text{Hz}$, $\Delta = 0.116$, $x = 700\text{mm}$

of the turbulent bulge in the intermittent region of the turbulent boundary layer. The space-time correlation analysis, not shown here, also suggests the presence of some coherent structure inside the patch which inclines to the downstream direction.

(2) Fully Turbulent Region

Fig.3 shows the variations of U and v' (RMS value) of the turbulent boundary layer refered to the phase of the flow oscillation. Both the velocity and the turbulence intensity vary almost sinusoidally about their mean values according to the flow oscillation. The mean profiles do not deviate very much from the corresponding steady case, as was already reported by Karlsson.

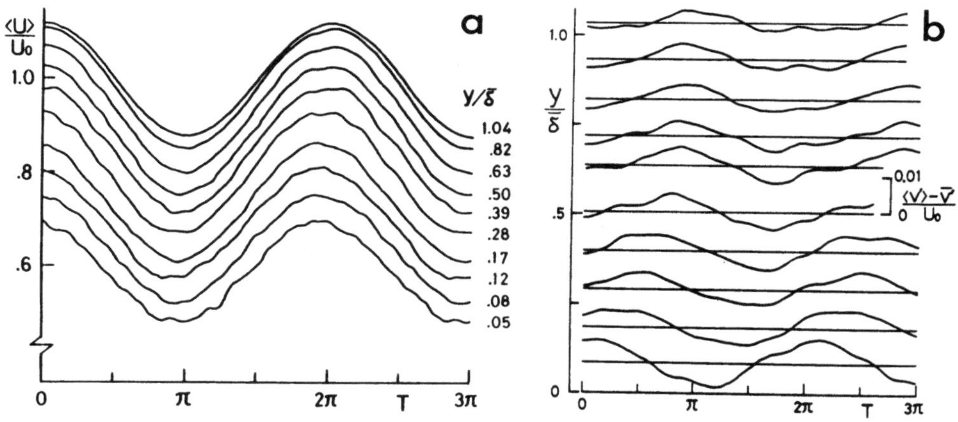

Fig.3. Phase averages of U and v' in a fully turbulent region
U_0= 5.45m/s, f_w= 3.43Hz, Δ= 0.119

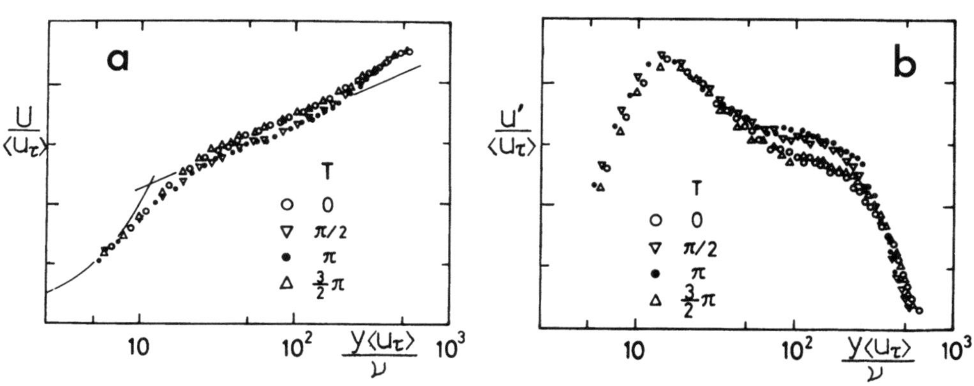

Fig.4. Distributions of U and u'
U_0= 6.34m/s, f_w= 3.45Hz, Δ= 0.103

The phase-averaged profiles, however, depend weakly on the instantaneous flow conditions whether they are in the accelerating phase or in the decelerating phase; the velocity tends to become higher in the accelerating phase and lower in the decelerating phase, while the turbulence intensity behaves reversely to the velocity change, as is shown in Fig.4.

Fig.5 shows the distributions of the reduced amplitude and the phase angle of the fundamental frequency component, in which the phase shift relative to the external flow is denoted by ϕ . For the case of the flow oscillation at lower frequency, the phase angle is positive(phase advance) in the most part of the layer, while it becomes negative in the outer region as the frequency increases. Within the frequency range of our tests, the phase angle in the close proximity to the wall or $yu_\tau/\nu < 10$ (u_τ: wall shear velocity) are all negative, whose value, however, tends to decrease as the frequency increases(Fig.5(b)). These results give a qualitative agreement with those of Karlsson and the ONERA group. The phase of the turbulence intensity variation coincides with that of the velocity change near the wall, but delays gradually outwards to even more than 180 degrees at the outer edge of the boundary layer.

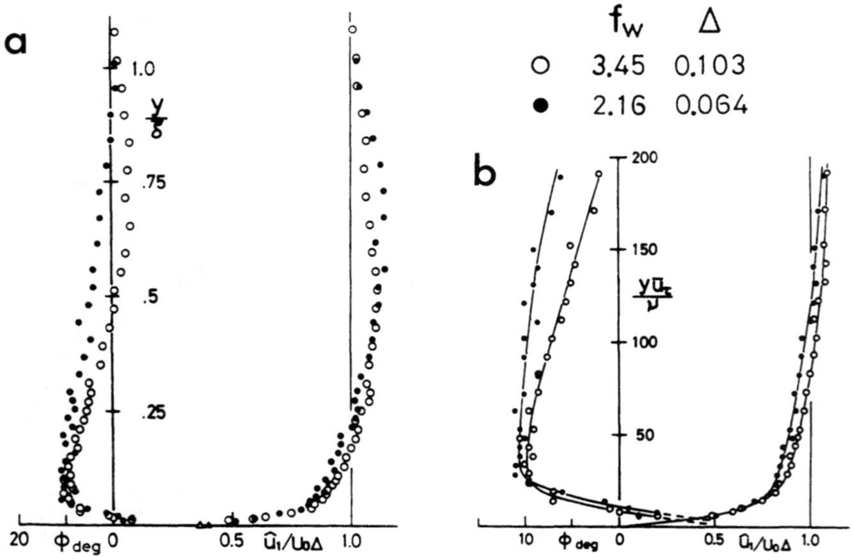

Fig.5. Amplitude and phase angle of fundamental frequency component of U, $U_0 \simeq 6.4$m/s

The intermittency factor distributions are shown in Figs.6(a) and 6(b), in which the distance from the wall is normalized by the mean and the phase-averaged boundary layer thickness($\bar{\delta}$ and $\langle\delta\rangle$), respectively. The result of the steady case which was obtained under the same flow condition is also shown for comparison. It is seen from Fig.(a) that the intermittent region tends to be displaced outwards as the flow unsteadiness increases. There is no marked change in the γ-distribution at each phase position, but it seems to become wider in the accelerating phase, the tendency of which may be related to the fact that the boundary layer subjected to a favourable pressure gradient is accompanied with a spread of the intermittent region [8].

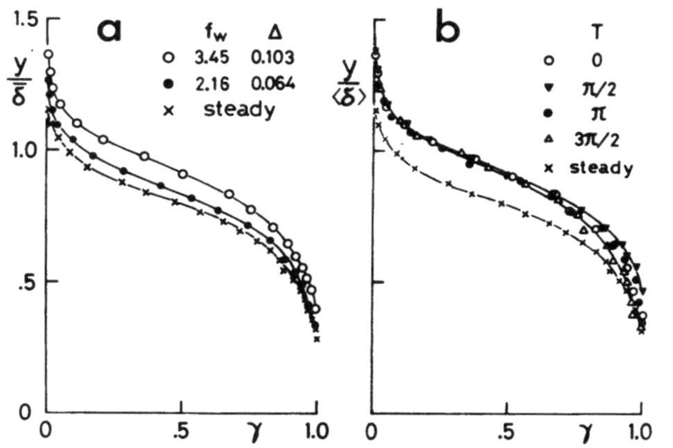

Fig.6. Distributions of intermittency factor
(a) $U_0 \simeq 6.4$m/s, (b) $f_w = 3.45$Hz, $\Delta = 0.103$

It must be another interesting problem if the so-called bursting phenomenon could be affected by the unsteadiness of the flow. In this connection, the histogram of the burst occurrences against the phase of the flow oscillation was calculated from the streamwise velocity fluctuations by applying the variable interval time averaging method [9]. This technique is based on the fact that the bursting process is always accompanied with the characteristic change of the streamwise velocity with a relatively weak deceleration(ejection phase) followed by a strong acceleration(sweep phase). The result is shown in Fig.7, together with the ensemble averaged velocity pattern of u and v,

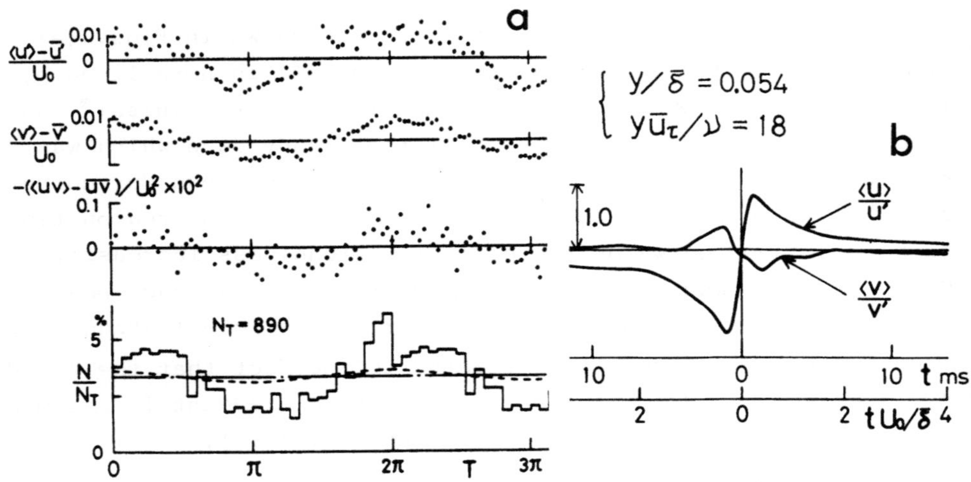

Fig.7. (a) Histgram of burst occurrences and phase averages of u', v' and -uv, (b) Ensemble averaged velocity pattern of bursts. U_0 = 5.45m/s, f_w = 3.43Hz, Δ = 0.119

and the phase averages of u', v' and the cross product –uv. The figure clearly shows that the occurrences of the bursting phenomena depend on the phase of the flow oscillation. The phase position where the bursts are observed most frequently lies at around T = 0, at which the turbulence intensity and the Reynolds stress have their maximum values. This selective occurrence of the bursts is also observed at the other test runs. However, the mean period of the burst appearences T_B does not change from the value of the corresponding steady case($T_B U_0/\delta^* = 36$, δ^*: displacement thickness) within the experimental accuracy.

The measurements of the wall shear stress τ_w and its fluctuation were made by the heated surface sensor. In Fig.8, the amplitude $\hat{\tau}_{w1}$ and the phase angle ϕ_τ of the phase averaged variation, defined by $\langle\tau_w\rangle - \overline{\tau}_w = \hat{\tau}_{w1}\cos(T + \phi_\tau)$, are shown with respect to the reduced frequency $\omega\overline{\delta^*}/U_0$ together with some examples of the output signal of the shear stress meter. For the frequency range of the present experiment, τ_w always delays relative to the external flow oscillation, and both the phase shift and the amplitude decrease gradually as the reduced frequency increases. These trends are not in favour of the theoretical results reported so far, but seems to be acceptable if we take into account the negative phase shift of the velocity variation near the wall (see Fig.5).

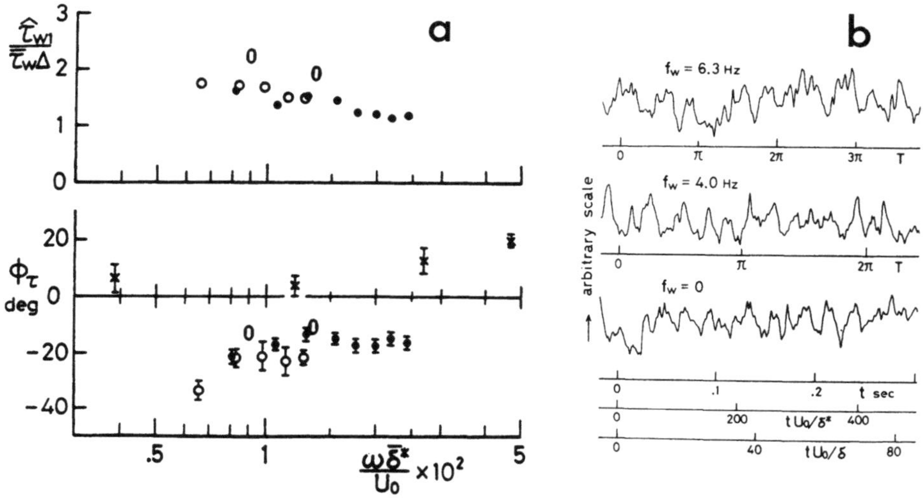

Fig.8. (a) Amplitude and phase angle of a wall shear stress
Karlsson: × (after Telionis and Thahalis)
Present results: ○ (r= 30mm), ● (r= 10mm),
0 (exterpolation of U near a wall)
(b) Out-put signals of a shear stress meter

Concluding Remarks

The oscillating boundary layers in the transition and the turbulent regimes have been investigated experimentally. Although the frequency range tested was not so high compared to the characteristic frequency of turbulence, some of the effects of the flow unsteadiness on the turbulence structure of the boundary layer have been revealed. Some of the results are not in favour of the theoretical evaluations reported so far. More extensive measurements, especially at higher frequency of the flow oscillation, are obviously needed.

References

1. Karlsson, S.K.F.: An unsteady turbulent boundary layer. J. Fluid Mech. 5 (1959) 622-636.

2. Telionis, D.P.: REVIEW—— Unsteady boundary layers, separated and attached. ASME J. Fluids Eng. 101 (1979) 29-43.

3. Houdeville, R.; Desopper, A.; Cousteix, J.: Experimental anslysis of average and turbulent characteristics of an oscillating boundary layer. Rech. Aerosp. 4 (1976) 183-191.

4. Cousteix, J.; Desopper, A; Houdeville, R.: Structure and development of a turbulent boundary layer in an oscillating external flow. Turbulent Shear Flows I (1979) 154-171.

5. Patel, M.H.: On turbulent boundary layers in oscillatory flow. Proc. Roy. Soc. Lond. A-353 (1977) 121-144.

6. Kobashi, Y.; Hayakawa, M.: Development of turbulence through non-steady boundary layer. Lecture Notes in Pysics 75 (1978) 277-288.

7. Obremski, H.J.; Fejer, A.A.: Transition in oscillating boundary layer flows. J. Fluid Mech. 29 (1967) 93-111.

8. Fiedler, H.; Head, M.R.: Intermittency measurements in the turbulent boundary layer. J. Fluid Mech. (1966) 719-735.

9. Blackwelder, R.F.; Kaplan, R.E.: On the wall structure of the turbulent boundary layer. J. Fluid Mech. (1976) 89-112.

Turbulence Modulated by a Coherent Shear Wave in a Wall Boundary Layer

K. RICHTER, D. RONNEBERGER

Drittes Physikalisches Institut, Universität Göttingen
Bürgerstr. 42-44, D-3400 Göttingen, F.R. Germany

Summary

The turbulence in a wall boundary layer is viewed as a dynamic system which is composed of various elements (coherent structures) which act together and constitute a dynamic equilibrium. The general aim of our investigations is to study this system by introducing small external perturbations and observing the resulting time-dependent departure from the equilibrium. The coherent shear wave used as perturbation throughout our experiments may either be excited by a sound wave propagating in a duct carrying turbulent flow, or may be directly excited by longitudinal oscillations of the wall. The results obtained by different measurement methods indicate that the shear wave causes the turbulence to deviate temporally from its state of dynamic equilibrium. So a modulation of the Reynolds' shear stress is observed. Our latest experiments carried out in a recirculating oil-tunnel yield that the modulation of the normal stress ($-\rho u'u'$), which has been measured by phase avering, results from a modulation of the probability of the occurrence of the so-called bursts. There are good arguments that the modulation of the shear stress ($-\rho u'v'$) originates from this effect, too.

Introduction

The initiative for these investigations was provided by a study of sound propagation in turbulent pipe flow. The attenuation of a plane sound wave propagating through a pipe had been found to increase strongly compared with that for the medium at rest if the Strouhal number ($Sr = \nu\omega_M/u_\tau^2$) was decreased below a critical value, i.e. where the Stokes' length $\delta_\omega = (\nu/\omega_M)^{1/2}$ exceeds the thickness of the viscous sublayer. The sound attenuation is caused by the diffusion of momentum and heat: the boundary

---

ν = kinematic viscosity, ω_M = angular frequency of the shear wave, $u_\tau = (\bar{\tau}_w/\rho)^{1/2}$ = friction velocity, $\bar{\tau}_w$ = mean wall shear stress, ρ = density conditions at the pipe wall (no slip, and no oscillation of the temperature) require that the sound wave excites a shear wave propagated by viscous and turbulent shear stresses and a heat conduction wave at the wall. Our experimental investigations are concerned with shear waves which are polarized in the mean-flow direction and propagate in the direction normal to the wall.

These waves can either be excited by small perturbations of the mean-flow velocity (sound) or by longitudinal oscillations of the pipe wall. Both methods have been used [1]. The wall impedance (shear stress / velocity) of the waves has been measured in the first case by evaluating the phase velocity and the attenuation of a plane sound wave which propagates in the turbulent pipe flow, and in the second case by evaluating the resonance frequency and the quality factor of a longitudinally vibrating glass pipe which carries turbulent flow. The results, which were obtained over a wide range of Strouhal numbers, exhibit very good agreement between the two measuring methods. The wall shear stress impedance is strongly affected by the turbulence. This indicates that the turbulent shear stress is modulated by the shear wave.

Unfortunately the shear stress as a function of the distance y from the wall could not be evaluated from these experiments, though these data are needed to answer the question, whether it is adequate to describe the propagation of the shear wave by an effective viscosity $\nu_{eff}(y^+, \delta_\omega^+)$.

<u>Oil-tunnel experiment</u>

In order to fill up this lack of information a new experimental set-up has been built, which allows the excitation of a pure shear wave and the measurement of its propagation into the thick turbulent boundary layer at the wall of a recirculating oil-tunnel (fig. 1). The shear wave is excited by longitudinal os-

---

+ denotes the non-dimensionalizing by ν and u_τ, $\delta_\omega^+ = u_\tau/(\nu\omega_M)^{1/2}$

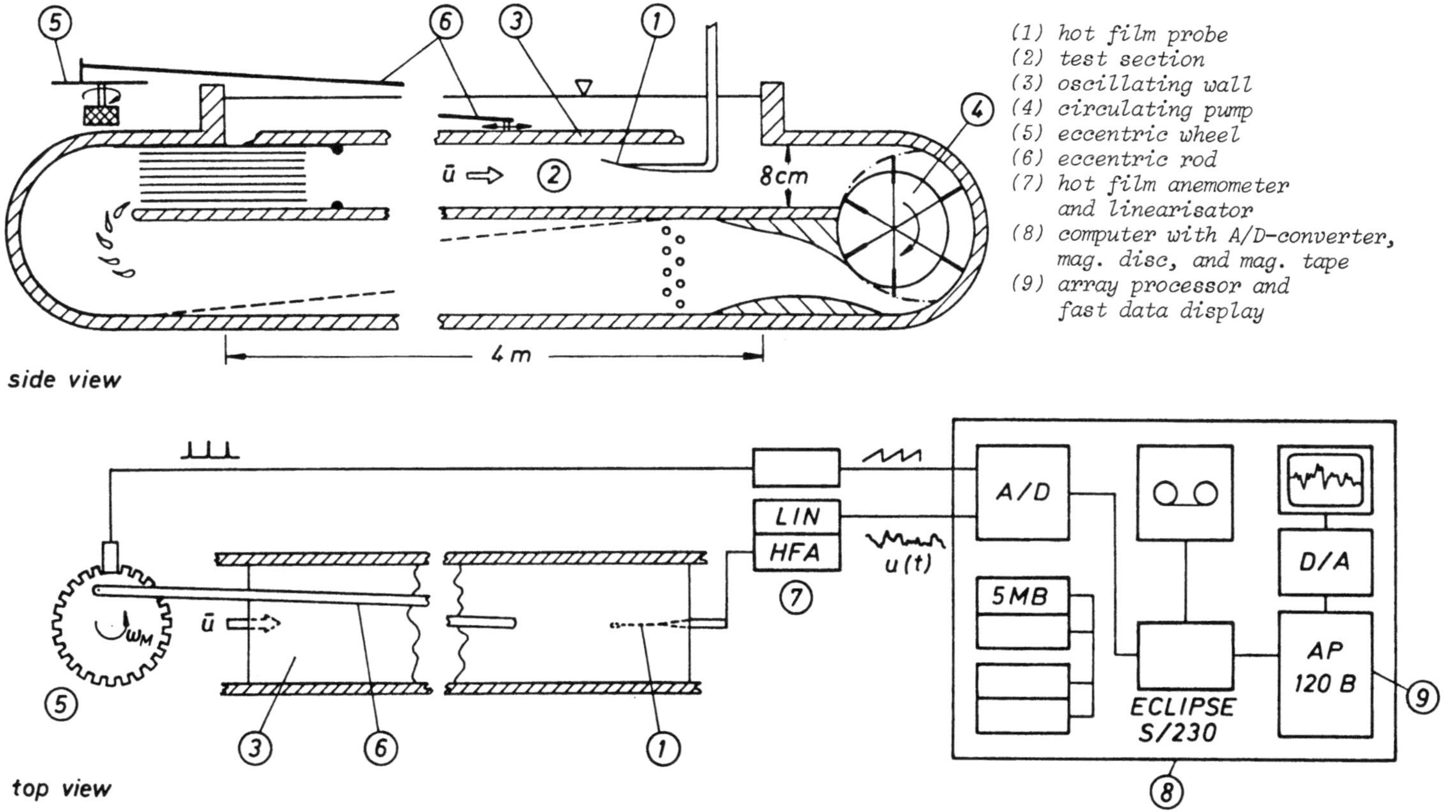

Fig.1. Sketch of the experimental set-up (oil-tunnel).

cillations of the upper wall of the test section. In order to obtain a coherent wave over a large streamwise and spanwise distance, the whole wall of about 4 m in length is moving driven by an eccentric wheel. The streamwise component of the velocity (u) is measured by means of a hot film probe, and the data are directly transmitted to a computer, where they are stored and evaluated. The angular position of the eccentric wheel is used as a reference signal for phase averaging.

Data evaluation

We assume that it is possible to decompose all the field quantities into three parts

$$f = \bar{f} + \tilde{f} + f' \;, \tag{1}$$

which denote the time average of the quantity, the fluctuation caused by the coherent perturbation, and the turbulent fluctuation, respectively. In the case of periodic perturbations, $\tilde{f}$ may be extracted from the total quantity f by phase averaging and subtraction of the mean part $\bar{f}$. This corresponds, in the case of small amplitude sinusoidal perturbations, to the determination of the frequency response of a very noisy system, the input and the output of which are the external perturbation and the quantity f, respectively.

The measurement of $\tilde{u}(y^+)$ offers an indirect method to determine the modulated part of the shear stress $-\rho\widetilde{u'v'}$. The transport of streamwise momentum towards the wall, i.e. the shear stress τ is composed of a diffusive and a convective part, so

$$\tilde{\tau} = \mu \frac{\partial \tilde{u}}{\partial y} + \tilde{\tau}_t ; \qquad \tilde{\tau}_t = -\rho\widetilde{u'v'} \tag{2}$$

($\mu = \nu\rho$ = dynamic viscosity), and the momentum equation

$$\rho \frac{\partial \tilde{u}}{\partial t} = \frac{\partial \tilde{\tau}}{\partial y} \tag{3}$$

yields

$$\widetilde{u'v'}(y) = -\int_{\infty}^{y} \frac{\partial \tilde{u}}{\partial t}(y')\, dy' + \nu \frac{\partial \tilde{u}}{\partial y} . \qquad (4)$$

This evaluation corresponding to equation (4) has been done by G. Höhler [2]. He could show that an effective viscosity or visco-elasticity is not appropriate for describing the turbulent shear stress $(-\rho\widetilde{u'v'})$ near the wall ($y \lesssim 30$).

Phenomenological model

A preliminary phenomenological model based on the findings of [2] has been proposed by one of the authors (D.R.). He assumes that near the wall the shear stress and the shear rate are totally uncoupled. The turbulent shear stress is produced at some characteristic distance from the wall, e.g. $y^+ = 30$, and propagates towards the wall like a wave.

$$\tilde{\tau}_t(y^+) = \begin{cases} \tilde{\tau}_t(d^+)\ \alpha(y^+/d^+)\ \exp\left[i\ \beta\ Sr\ (y^+ - d^+)\right] & \text{if } y^+ \leq d^+ \\ \text{visco-elastic (R.E. Davis [4])} & \text{if } y^+ > d^+ \end{cases} \qquad (5)$$

The parameters β and d^+ can be computed from the wall shear stress impedance by least square fit; the shear stress amplitude was preliminarily assumed to follow the law

$$\alpha(y^+/d^+) = 2\ (y^+/d^+) - (y^+/d^+)^2 . \qquad (6)$$

This very simple model is able to reproduce the actual shear stress as a function of y^+ far better than any of the models that were tested in [1].

An interpretation for the modulation of the Reynolds' stresses

A further experiment [3] conducted in the oil-tunnel was concerned with the modulation of the energy of the streamwise velocity fluctuations; i.e. the normal stress $-\rho\widetilde{u'u'}$. The purpose was to measure the modulation itself, and to collect data about the dependence on the modulation frequency (shear wave), and on the frequency of the velocity fluctuations being modulated.

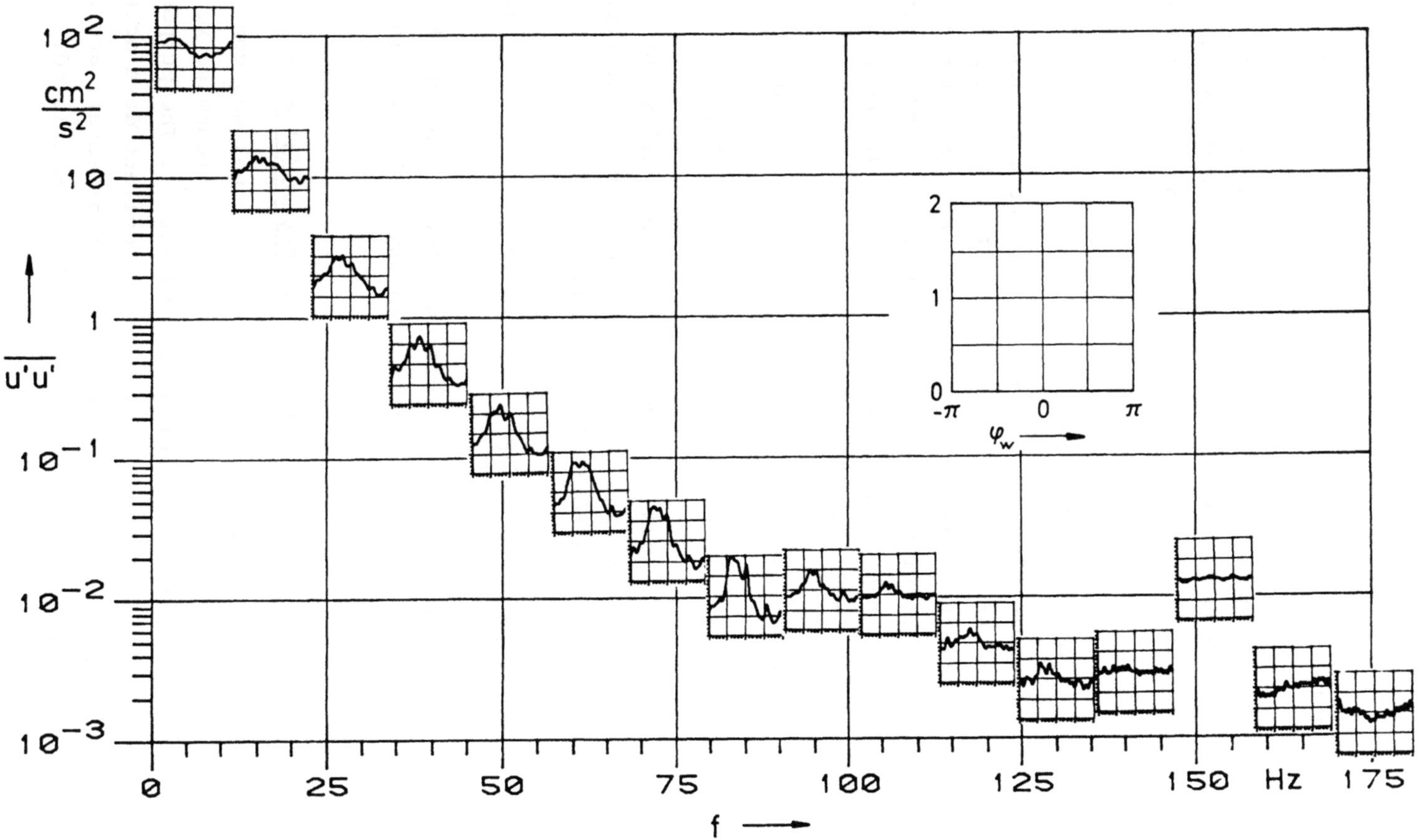

Fig.2. The energy of the streamwise velocity fluctuations. (See next page for detailed legend.)

Figure 3 shows some of the results obtained for $\omega_M/2\pi = .705$ Hz and $u_\tau = 4.89$ cm/s corresponding to $\delta_\omega^+ = u_\tau/(\nu\omega_M)^{1/2} = 8.8$; m is the absolute value of the amplitude modulation, and φ is the phase of it relative to the reference phase of the shear wave. The magnitude of the modulation varies strongly with the frequency of the velocity fluctuations and reaches values up to .8; note that the frequency corresponding to the maximum of m is two orders of magnitude higher than the modulation frequency.

Three different methods have been used to evaluate the modulation. First we tried to compute it from the correlation of the spectral components the frequencies of which differed by ω_M and $2\omega_M$. These correlation should be zero without a modulation. Using this method it is theoretically possible to obtain both the modulation of the amplitude and of the phase of the velocity fluctuations; but in practice the iterations carried out on our computer became steady only after a non-acceptable long time.

Figure 2

The energy of the streamwise velocity fluctuations ($\overline{u'u'}$, $\widetilde{u'u'}$) at $y^+ = 7$ (distance from the wall) and for $\delta_\omega^+ = 8.8$ ($\omega_M/2\pi = .705$ Hz, $u_\tau = 4.89$ cm/s).
S m a l l diagrams: Phase averages $\widetilde{u'u'}$ of the bandpass filtered signal. The horizontal dimensions of the small diagrams mark the lower and the upper limit frequency of the bandpass filter, respectively. The phase averages have been taken with respect to the reference phase φ_W, i.e. the phase of the velocity of the oscillating wall. The correspondent mean values $\overline{u'u'}$ have been used for normalizing.
B a c k g r o u n d diagram: The centre points of the small diagrams indicate the mid frequency of the spectral band and the corresponding mean value $\overline{u'u'}$.

Figure 3

The modulation of the energy of the streamwise velocity fluctuations (normal stress $-\rho\widetilde{u'u'}$) as a function of the mid frequency of the corresponding spectral band. y^+ is the distance from the wall (non-dimensionalized by ν and u_τ, $m = |m_{AM}|$, $\varphi = \arg(m_{AM})$. The uppermost sub-plot (for $y^+ = 7$) corresponds to the data presented in figure 2.

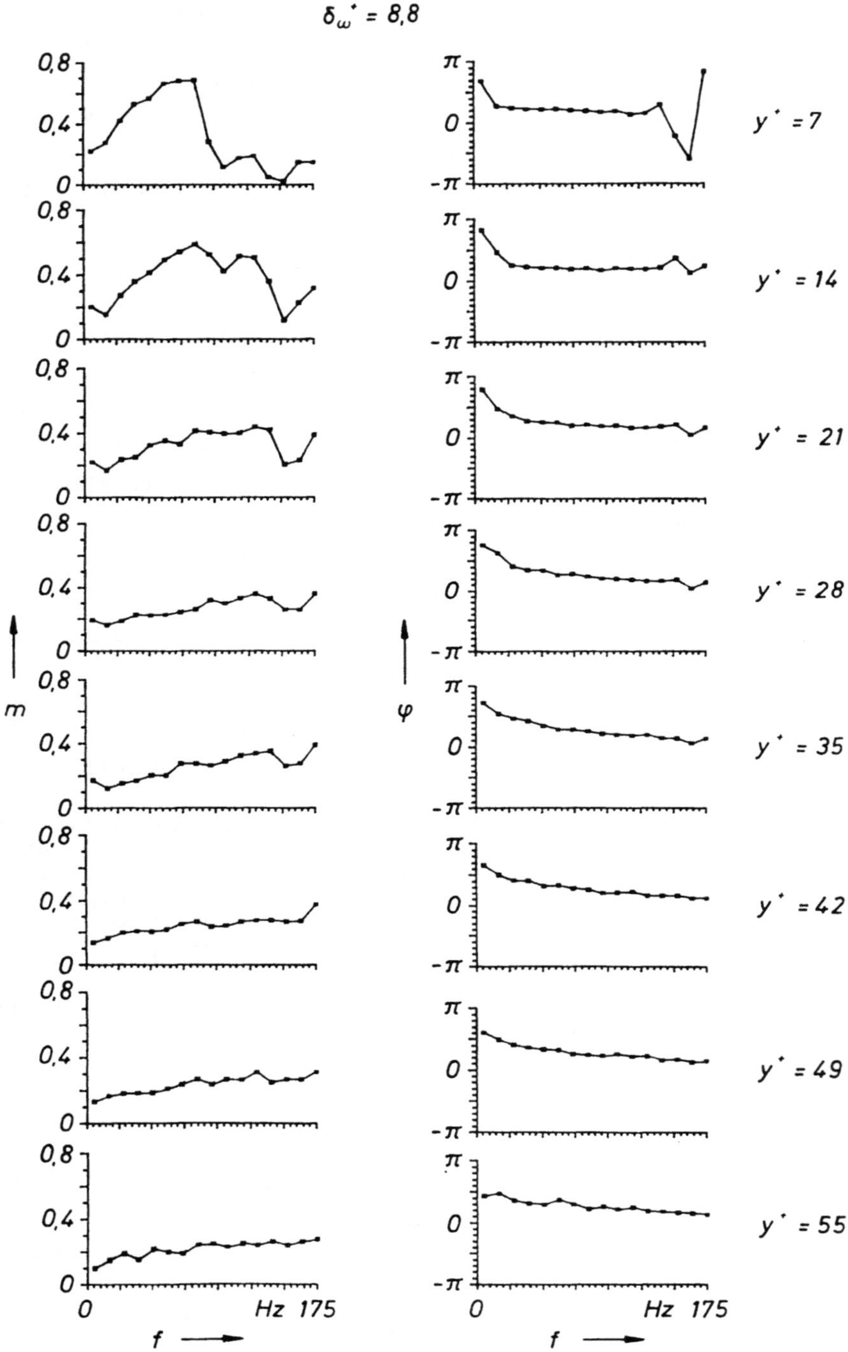

Fig.3. The modulation of the energy of the streamwise velocity fluctuations. (See page before for detailed legend.)

So we decided to compute the modulation from the phase averages (relative to the velocity of the wall) of the squared amplitude of the filtered analytical signal. The histograms of these phase averages are shown in figure 2 for $y^+ = 7$ and $\delta^+_\omega = 8.8$; they make the modulation of the turbulence energy quite evident. The amplitude and phase of the modulation coefficient as a function of the mid frequency of the regarded spectral band of the velocity fluctiations can then be extracted from these histograms. (Figure 3 corresponds to the uppermost sub-plot in figure 2.)

Yet the origin (or the originator) of this unexpectedly high modulation had not been found. It shall be emphasized that the modulation cannot be seen in the u(t)-trace on the oscilloscope: the velocity fluctuations do not behave like a sinusoidally increasing and decreasing broad band noise; the modulation appears only in the phase averages. The coherent structures (bursts) in the wall boundary layer were suspected to be responsible for this effect. Bursts occur at unpredictable time instants, and this fact is not changed in principle by the shear wave. But nevertheless the shear wave could be able to trigger the burst; i.e. if a burst is going to be initiated, if it is immanent but not yet released, the shear wave might promote its initiation.

In order to prove this suspect the third method was applied. The velocity data that had been recorded were digitally filtered as appropriate, and then displayed on the scope starting and ending at a time instant corresponding to the reference phase (shear wave) equal to zero. It is the task of the operator to detect a burst and to enter the corresponding reference phase. (An event was classified as a "burst", if the envelope of the filtered signal exceeds a specified magnitude.) The computer then calculates histograms of the frequency of occurrence of a burst as a function of the reference phase. These histograms were compared to the histograms of the phase averages shown in figure 2, and exhibit sufficiently good agreement. Obviously using this method for data evaluation takes a long time, and it is a rather stupid task for the operator. Therefore it could only be applied to a short number of data records.

Up to now only the normal stress data have been analysed in this detailed manner, but there is good reason for argueing that the shear stress data behave similarly, and that their modulation arises from the same origin, i.e. the modulation of the frequency of occurrence of the bursts.

References

1. Ronneberger, D.; Ahrens, C.D.: Wall shear stress caused by small amplitude perturbations of turbulent boundary-layer flow: an experimental investigation. J. Fluid Mech. 83 (1977), 433-464.

2. Höhler, G.: Ausbreitung von Scherwellen in einer turbulenten Wandgrenzschicht. Diss. Math. Nat. Fak., Univ. Göttingen, May 1978.

3. Richter, K.: Modulation der Turbulenz in einer schwach instationären Wandgrenzschicht. Diss. Math. Nat. Fak., Univ. Göttingen, Jan. 1981.

4. Davis, R.E.: Perturbed turbulent flow, eddy viscosity and the generation of turbulent stresses. J. Fluid Mech. 63 (1974), 673-693.

Experimental Study of Two- and Three-Dimensional Boundary Layer Separation

H. U. Meier, H.-P. Kreplin, L.W. Fang*

Deutsche Forschungs- und Versuchsanstalt für Luft- und Raumfahrt e.V., Institut für Experimentelle Strömungsmechanik, D-3400 Göttingen, Bunsenstraße 10

*) Visiting Scientist, on leave from Nanjing Aeronautical Institute, Peoples' Republic of China

Summary

The separation phenomena on a circular cylinder in cross-flow and on an inclined prolate spheroid are studied experimentally at subsonic velocities. On the circular cylinder the steady and unsteady surface pressures and shear stresses are measured in order to investigate the separation of a quasi two-dimensional, oscillating boundary layer. The boundary layer separation occurring on the inclined prolate spheroid is examined by means of similar experimental techniques, supplemented by velocity measurements in boundary layers and in the vortex flow. The emphasis in the following sections is on the presentation of the experimental facts applying different measuring techniques and on the description of two-dimensional, as well as three-dimensional, separation phenomena.

1. Introduction

1.1 Two-Dimensional Boundary Layer Separation on a Circular Cylinder

It is generally accepted that for a steady flow the separation of a two-dimensional boundary layer implies a vanishing wall shear stress. On a circular cylinder in a cross-flow the vortex shedding (von Kármán vortex street) creates an oscillating boundary layer flow and consequently the location of the separation is not constant. It is known that the vortex shedding itself is strongly dependent on the Reynolds number. Roshko [1] introduced three Reynolds number ranges which lead to characteristic flow patterns for the cross-flow over a circular cylinder**

a. Subcritical regime: $Re < 4 \times 10^5$, $S \approx 0.2$

b. Supercritical regime: $4 \times 10^5 < Re < 3.5 \times 10^6$

c. Transcritical regime: $Re > 3.5 \times 10^6$, $S \approx 0.3$.

**) Due to different test set-ups and flow conditions the Reynolds number ranges differ slightly in the reported literature.

In the subcritical regime (a) the boundary layer separates in a laminar fashion while in the transcritical regime (c) the boundary layer is turbulent before separating. In these Reynolds number regimes the vortex shedding is periodic and the measured, predominant frequencies lead to almost constant Strouhal numbers S. This implies that the interaction between the wake form and the vortex shedding frequencies is independent of the Reynolds number. As stated by Naumann and Quadflieg [2] an alternating vortex street can also exist in the range of supercritical Reynolds numbers (b). However, it was found by different investigations that this regime is characterized by quasi-random, unsteady pressure signals (e.g. Refs. [3], [4]). As reported by Roshko [1] and Achenbach [5] the upper transition (from supercritical to transcritical Reynolds numbers) is related to a vanishing laminar separation bubble on the rear side of the cylinder. For this reason the supercritical regime is extremely sensitive to the wind tunnel turbulence, side wall effects and the model surface roughness.

The purpose of our investigation was not to produce a new set of data concerning the cross flow over a circular cylinder. Instead the purpose was to examine, whether unsteady pressure and wall shear stress measurements can be applied for the investigation of the flow phenomena discussed above as well as for the investigation of three-dimensional boundary layers with separation.

1.2 Three-Dimensional Boundary Layer Separation on an Inclined Prolate Spheroid

In contrast to the two-dimensional separation, the three-dimensional separation is by far more complex and difficult to interpret physically. On an inclined body of revolution we observe two different kinds of flow separation:

a. An afterbody flow separation which is similar to the two-dimensional separation and is characterized by reverse flow and a vanishing wall shear stress.

b. A boundary layer separation resulting from the convergence of streamlines, (originating upstream) leading to a local thickening of the boundary layer and the possible formation of a vortex. This concept of an open separation was introduced by K.C. Wang [6]. This regime of flow separation, as discussed by Hirschel and Kordulla [7], is characterized by cross flow reversal and a wall shear stress minimum.

Some experimental results concerning these flow problems, will be discussed. They were obtained from pressure as well as wall shear stress measurements on a prolate spheroid.

2. Experiments Related to Two-Dimensional Separation

The experiments were performed in the open test section of the DFVLR-AVA 1-m Low Speed Wind Tunnel. A cylinder was placed spanwise between two side walls, which were carefully adjusted to the wind-tunnel flow in order to avoid a leading edge separation. The test cylinder, made of perspex, had a diameter of 0.1 m and span of 0.55 m, (Fig. 1). The cylinder was equipped with two parallel

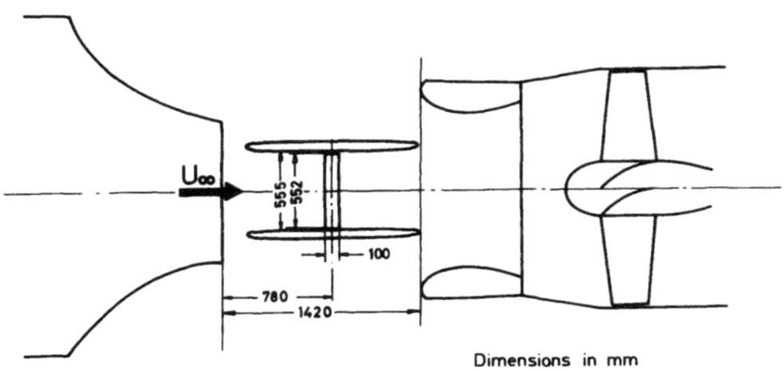

Fig. 1 Cylinder model in the DFVLR 1-m Low Speed Wind Tunnel

hot films which were flush mounted to the surface, and a pressure transducer which was fixed directly beneath the pressure orifice inside of the cylinder wall. Details of the probe arrangements and a schematic sketch of the electronic equipment is shown in Fig. 2. The cylinder can be rotated around its longitudinal axis, which allowed us to measure the circumferential distributions of the steady and unsteady pressures and wall shear stresses with the two sensors. Results of the steady pressure distribution are given in Fig. 3. For a Reynolds number Re = 10^5, the boundary layer flow is completely laminar and leads to a typical

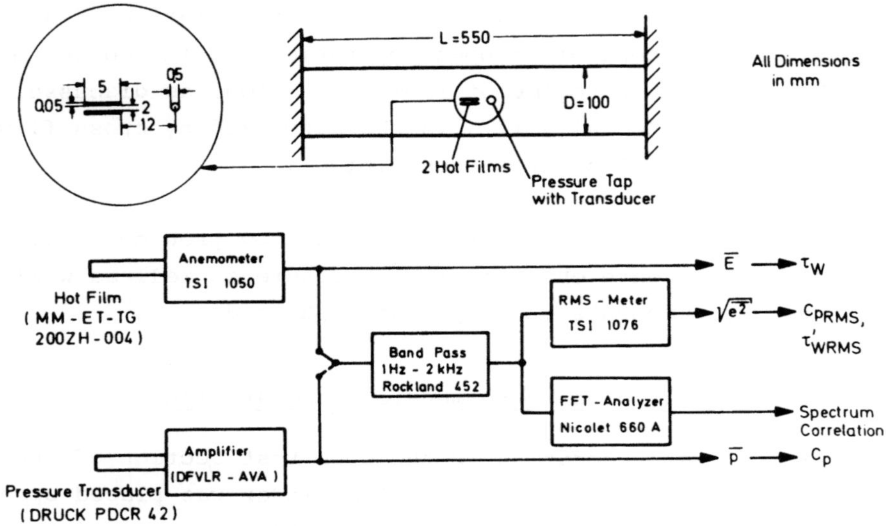

Fig. 2. Probe arrangement and electronic equipment

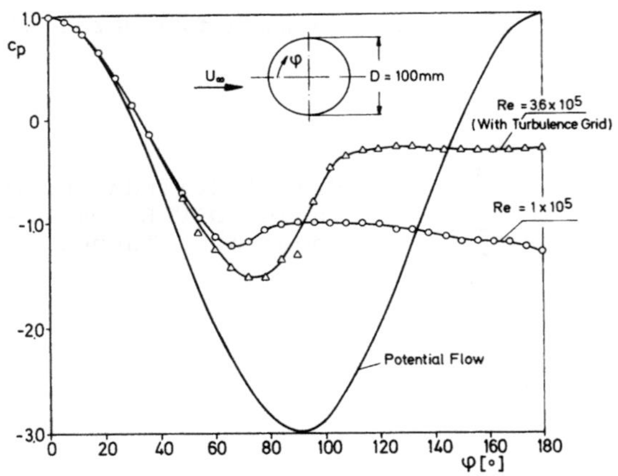

Fig. 3. Pressure distributions on a circular cylinder at sub- and supercritical flow conditions

"subcritical" pressure distribution. In order to create supercritical flow conditions, the wind tunnel turbulence intensity of $Tu \approx 0.2\%$ was increased by means of a turbulence grid to $Tu \approx 1.3\%$, at the location of the model. Thus obtained pressure distributions are qualitatively in agreement with those obtained at higher Reynolds numbers as reported in Refs. [1] - [5]. The corresponding RMS pressure distributions have much higher values for the laminar boundary flow than for the supercritical flow conditions (Fig.4). At subcritical Reynolds numbers, the separation is known to be

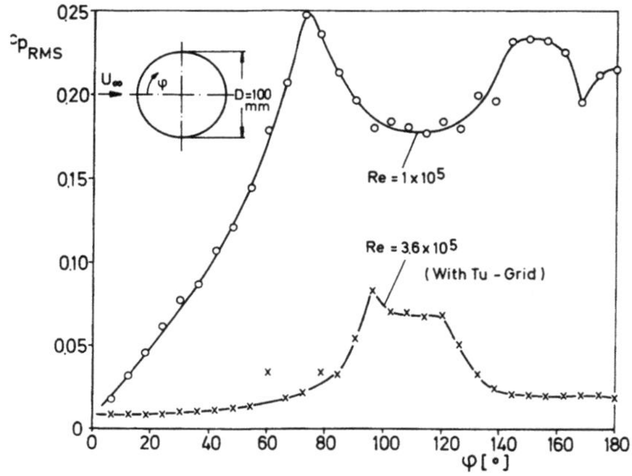

Fig. 4. RMS-values of pressure fluctuations at sub- and supercritical flow conditions

laminar and occurs early on the front part of the cylinder. Consequently this results in a large diameter of the wake flow compared to the cylinder diameter, and in large pressure fluctuations. With increasing Reynolds number in the supercritical regime, the wake diameter decreases from values larger than the cylinder diameter and the separation location moves from the front to the back of the cylinder. As found by van Numen [4] and James et al [8]. The decrease of the pressure fluctuations is an indirect measure of the decrease of the total drag* of the cylinder. The peaks in the unsteady pressure distributions coincide with the location of the flow separation. This is emphasized by the results of the wall shear stress measurements (Fig. 5). For $Re \approx 10^5$ the laminar boundary layer separation

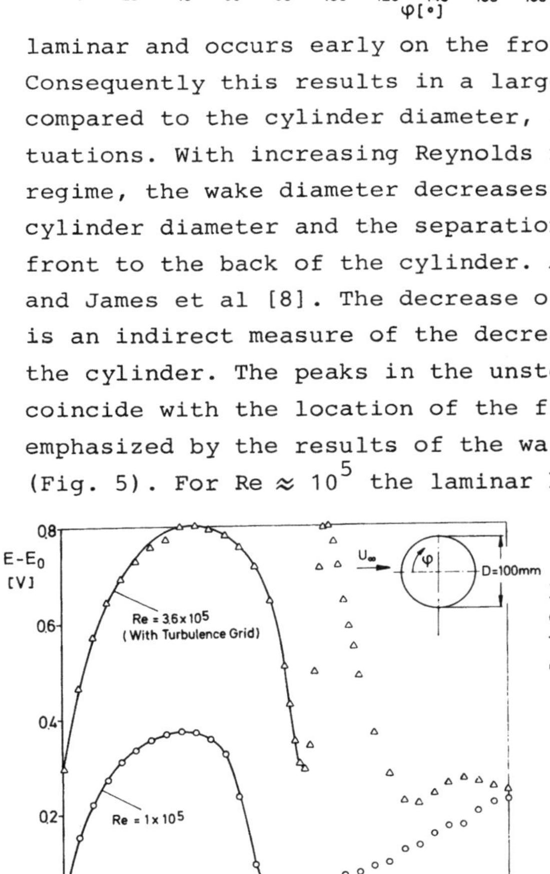

Fig. 5. Wall shear stress distributions (anemometer voltage) at sub- and supercritical flow conditions

*) Drag coefficients C_D: $Re \approx 10^5 \longrightarrow C_D \approx 1.2$; $Re \approx 10^6 \longrightarrow C_D \approx 0.3$

occurs at $\varphi \approx 75°$ as found from an oil flow pattern. For two reasons the corresponding measured wall shear stress in this regime does not vanish: firstly the hot film cannot distinguish between forward or reverse flow, and secondly the measuring technique becomes rather insensitive in the region of vanishing wall shear stress. This is of course also true for the region of separated flow. The shear stress distribution for supercritical flow conditions (Re = 3.6×10^5) indicates a minimum at $\varphi \approx 100°$. This was also found by Achenbach [5] who interpreted this phenomenon as the start of a separation "bubble". Downstream an intensive rise of the shear stress follows, indicating a transition to a turbulent boundary layer which separates finally at $\varphi \approx 140°$. These results are reinforced by measured unsteady wall shear stresses and pressure signals, shown in Fig. 6. The hot film signals in

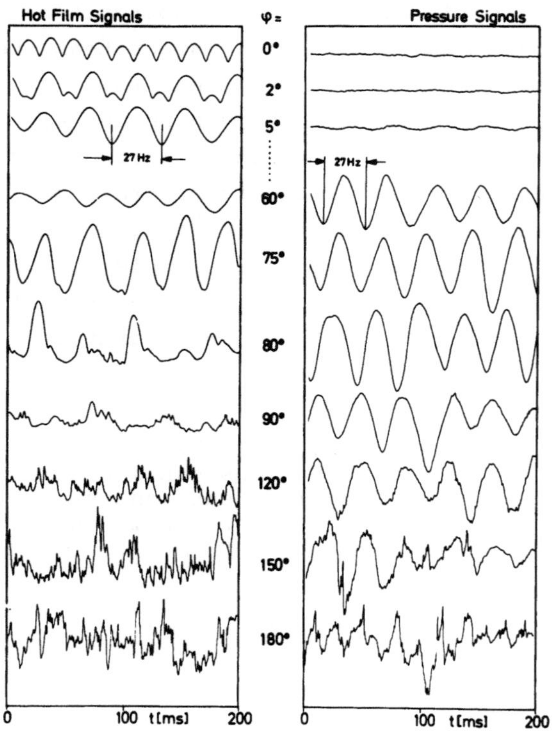

Fig. 6. Instantaneous wall shear stress and pressure signals for Re = 10^5. (Subcritical flow conditions)

Fig. 6 indicate clearly the oscillations of the laminar boundary layer. In the stagnation region the hot film cannot identify the flow direction and consequently rectifies the sinusoidal signals. This effect disappears for $\varphi > 3°$ and leads to a stagnation line oscillation of $\approx \pm 3°$. At $\varphi \approx 75°$ the first occurrence of reverse flow is indicated in a similar manner. Both the stagnation line oscillations and the start of the separation cannot be identified from the pressure signals. However, the two measuring techniques give identical Strouhal numbers S = 0.18. For supercritical flow conditions (Re = 3.6×10^5) the unsteady wall shear stress, as well as the pressure signals, become highly random. No clear periodic vortex formation can be detected from the measurements. Nevertheless, at φ = 100° - 110° the hot film signals indicate a high turbulence level, which underlines the interpretation of the corresponding mean shear stress measurements.

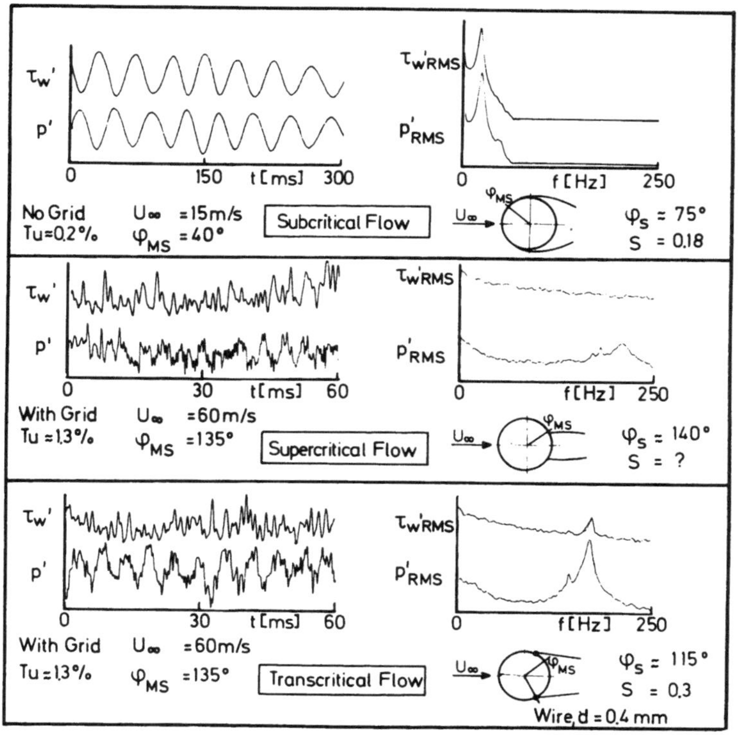

Fig. 7. Instantaneous wall shear stress signals and frequency spectra.
φ_{MS} location of the surface hot film
φ_S location of the separation

According to reported results (e.g. [1], [2]) the vortex shedding frequencies seem to be independent of the Reynolds number for the transcritical flow condition. Unfortunately we could not establish transcritical Reynolds numbers in our wind-tunnel.Therefore we applied Prandtl's idea of emanating disturbances in the boundary layer to effect an artificial transition or to fix the separation location. Within a certain limitation, it is possible to simulate the main features of the transcritical flow conditions. This was realized by fixing turbulence wires with a diameter of d = 0.4 mm at chosen angles φ on the cylinder. The obtained wall shear stress and pressure signals or the corresponding frequency spectra (Fig. 7) now clearly indicate a shedding frequency which leads to a Strouhal number of S = 0.3. It was found that in turbulent flow the unsteady pressure signals are more dominant when compared with the hot film signals, while the frequency spectra in both cases show a peak at f = 177 Hz.

3. Experiments Related to Three-Dimensional Boundary Layer Separation

These experiments were performed in the open test section of the DFVLR-AVA 3-m Low Speed Wind Tunnel. The wind tunnel model, a prolate spheroid 1 : 6, was especially designed for the investigation of three-dimensional boundary layers. The dimensions of the model and specifications of the wind tunnel are given in Fig. 8. A detailed description of the applied measuring technique is given in Ref. [9].

Fig. 8. Prolate spheroid (a:b=6) in the DFVLR 3m x 3m Low Speed Wind Tunnel

Model length L = 2a = 2.4 m

Maximum free stream velocity U_∞ = 62 m/s

Turbulence level Tu $\approx$ 0.2%

In Fig. 9 the circumferential distribution of wall shear stress vectors and representative oscilloscope traces for a central cross section of the inclined ellipsoid (α = 30°) are shown. As indicated in the schematic sketch, a laminar boundary layer develops up to $\varphi \approx$ 50°, which is followed by a transitional boundary layer

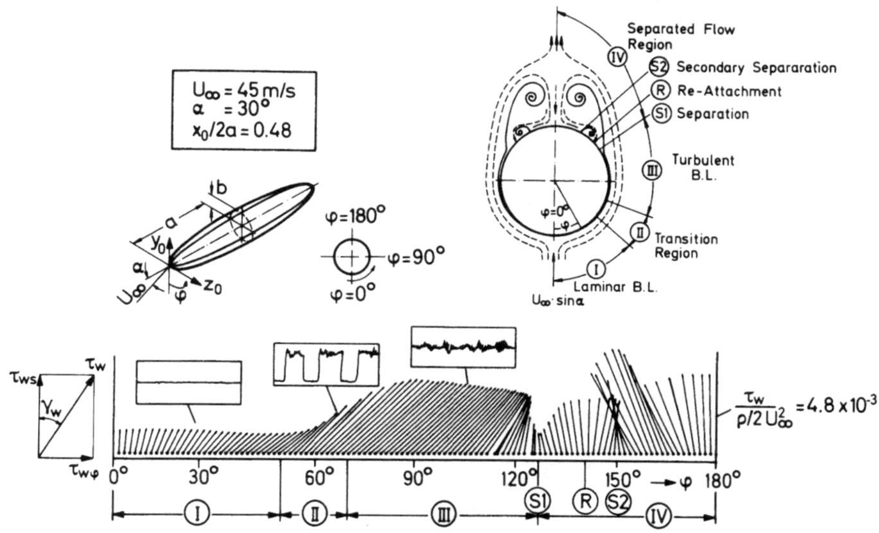

Fig. 9. Schematic sketch of the flow field based on measured wall shear stress vectors

becoming fully turbulent at $\varphi \approx$ 70°. This is characterized by a significant increase of the wall shear stress and the corresponding fluctuating components. In the regime of the three-dimensional boundary layer separation

- the circumferential wall shear stress component becomes zero (cross flow reversal), and
- the wall shear stress magnitude reaches a minimum.

The vortex flow pattern is characterized by the reattachment and secondary boundary layer separation, which is also clearly indicated by the wall shear stress vectors. The large τ_w-values on the lee-side result from the induced velocities due to the separated vortex flow. This kind of wall shear stress measurements were performed in 12 cross-sections for different angles of incidence and Reynolds numbers. The results were used to generate a B-spline approximation of the field of directions of the wall shear stress. From this approximation the direction can be evaluated at any point on the prolate spheroid between the first and last cross-section. In Ref. [10] these data were used to calculate the limiting streamlines applying a numerical integration of the differential equation for the shear stress direction. In Fig. 10 limiting streamlines, calculated for two different flow

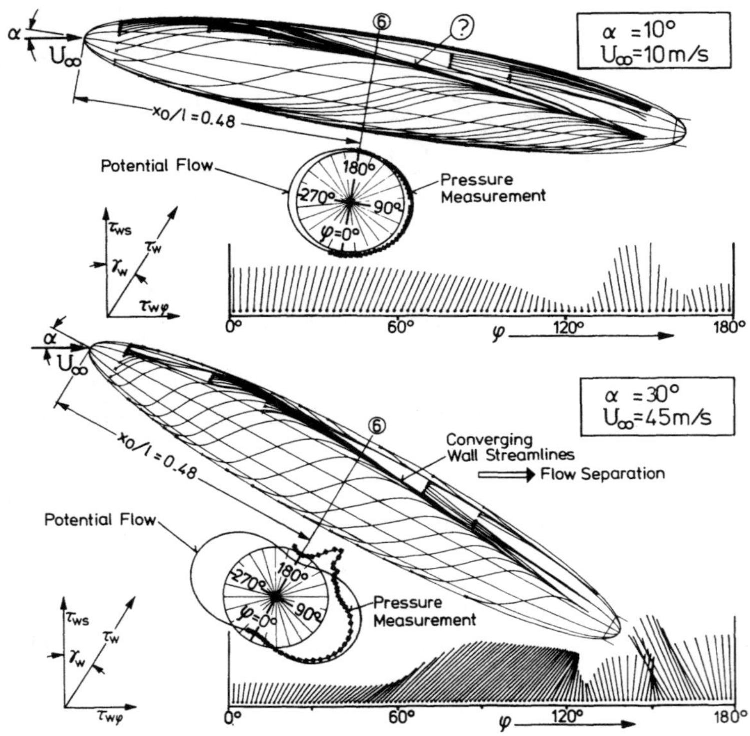

Fig. 10. Limiting streamlines calculated from wall shear stress measurements, with pressure and shear stress distributions in a central cross-section

cases, are presented together with wall shear stress vectors and pressure distributions in a central cross-section. For both flow conditions a shear stress minimum occurs in the regime of converging streamlines and the circumferential component vanishes. In Refs. [10] and [11] it is demonstrated that for $\alpha = 10°$ and $U_\infty = 10$ m/s the three-dimensional boundary layer is completely laminar. The differences between the measured and the potential pressure distributions are small, but differences between the corresponding pressure gradients in the circumferential and longitudinal direction are essential. These pressure gradients are responsible for the occurence of circumferential flow reversal in the three-dimensional boundary layer. The flow reversal exists only very close to the surface, and is combined with a strong thickening of the boundary layer, as shown by measurements of mean velocity distributions [11]. A vortex flow seems to be created, which is embedded in the boundary layer. At a higher angle of incidence (α = 30°) the shear layer separates from the surface and results in a vortex flow, as shown in Fig. 11. This is confirmed by a

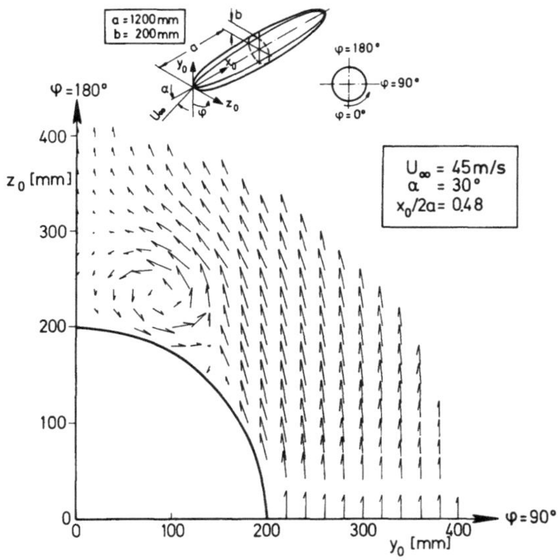

Fig. 11. Cross flow pattern in a vortex flow resulting from a three-dimensional boundary layer separation

strong suction peak in the pressure distribution. The separation "line" in Fig. 10 (α = 30°) indicates a deflection in the nose region ($x_0/2a$ < 0.2). In this regime the boundary layer is still laminar before separation, as found from both wall shear stress measurements and oil flow patterns [10].

4. Summary and Conclusions

The experimental determination of the two-dimensional boundary layer separation by means of wall shear stress measurements was found to be more reliable compared to wall pressure measurements. For subcritical flow conditions,the unsteady hot film signals clearly indicate the flow separation that could not be detected from unsteady pressure signals. For these flow conditions,both measuring techniques lead to identical Strouhal frequencies. At trancritical flow over the cylinder a regular vortex shedding can be identified by the hot film as well as by the pressure sensor in the attached flow regime. More distinct signals were obtained from the pressure sensor.

The investigation of three-dimensional boundary layer "separation" phenomena leads to the following conclusion: the existence of converging limiting streamlines, the occurrence of a wall shear stress minimum and its vanishing circumferential component need not be a definitive indicator of flow separation as manifested by the boundary shear layer departing from the surface and converting into a vortex.

References

1. Roshko, A.: Experiments on the flow past a circular cylinder at very high Reynolds number. J. Fluid Mech. 10 (1961) 345-356.

2. Naumann, A.; Quadflieg, H.: Vortex generation on cylindrical buildings and its simulation in wind tunnels. Proc. "Symposium on Flow-Induced Structural Vibration", Karlsruhe, W-Germany, (1972) 730-747.

3. Szechenyi, E.: Supercritical Reynolds number simulation for two-dimensional flow over circular cylinders. J. Fluid Mech. 70 (1975) 529-542.

4. Van Numen, J.W.G.: Pressure and forces on a circular cylinder in a cross-flow at high Reynolds numbers. Proc. "Symposium on Flow-Induced Structural Vibration", Karlsruhe, W-Germany, (1972), 748-754.

5. Achenbach, E.: Distribution of local pressure and skin friction around a circular cylinder in cross-flow up to $Re = 5x10^6$. J. Fluid Mech. 34 (1968) 625-639.

6. Wang, K.C.: Separation patterns of boundary layers over an inclined body of revolution. AIAA Journal 10 (1972) 1044-1050.

7. Hirschel, E.H.; Kordulla, W.: Local properties of three-dimensional separation lines. Z. Flugwiss. Weltraumforschung 4 (1980) 289-295.

8. James, D.W.; Paris, S.W.; Malcolm G.N.: Study of viscous cross flow effects on circular cylinders at high Reynolds numbers. AIAA Journal 18, No. 9 (1980) 1066-1072.

9. Meier, H.U.; Kreplin, H.-P.: Transition and separation phenomena on a body of revolution. Z. Flugwiss. Weltraumforschung 4 (1980) 65-71.

10. Kreplin, H.-P.; Vollmers, H.; Meier, H.U.; Experimental determination of wall shear stress vectors on an inclined prolate spheroid. Proc. 5th U.S. - FRG Data Exchange Agreement Meeting "Viscous and Interacting Flow Field Effects", AFFDL-TR-80-3088 (1980) 315-332

11. Meier, H.U.; Kreplin, H.-P.; Vollmers, H.: Velocity distributions in 3-d boundary layers and vortex flows developing on an inclined prolate spheroid. Proc. 6th U.S. - FRG Data Exchange Agreement Meeting "Viscous and Interacting Flow Field Effects", 1981, to be published.

Measurements of the Periodic Velocity Oscillations Near the Wall in Unsteady Turbulent Channel Flow

G. BINDER, J.L. KUENY

Institut de Mécanique de Grenoble, B.P. 53X, 38041. GRENOBLE-CEDEX, France

Summary

Measurements in turbulent channel flow subjected to small periodic velocity oscillations are reported. The mean flow and the mean turbulent intensity are not affected by the forced oscillations. The periodic oscillations follow closely the laminar Stokes solution near the wall, except a low frequency where the phase shift tends to zero. It is shown that the non dimensional Stokes thickness l_s^+ in terms of the wall thickness is the important parameter.

Experimental investigations of unsteady turbulent boundary layers have so far been rather scarce and have yielded little information on the wall layer. For instance, such vital information as the amplitude and phase shift of the wall shear stress is almost completely lacking. On the other hand, several turbulent closures developped for steady flows have been applied to unsteady flows but their predictions of the wall shear stress differ considerably from one another giving even different trends for its phase shift towards large non-dimensional frequencies (fig. 6 of ref. 6). Since there is little experimental data available with which the various predistions can be confronted, it is difficult to evaluate the merits and defects of the various models.

The need for measurements in unsteady turbulent boundary layers is also patent and has been stressed many times in recent years. The purpose of the present investigation is to provide some data in this field and an attempt to clarify the transport mechanisms especially in the wall region.

Apparatus. Instrumentation. Data acquisition

The measurements are performed in a water channel : length : 2600 mm, height, 2h = 100 mm, span : 1000 mm , u_c = 0-50 cm/s. The test section is 1000 mm long and located at the down stream end. A divergence, adjustable up to about 20°, can be imposed on the section walls in order to subject the flow to a mean unfavourable pressure gradient. The oscillating flow is driven by a reciprocating piston with adjustable speed and stroke mounted on the caisson upstream of the grids and honeycomb. The return flow is via a free surface in order to minimize the effects of unsteady pressure gradients.

The velocities were measured with a laser velocimeter operating in the fringe mode and the Doppler frequency was measured with a counter. Phase-locked ensemble averages are performed on a microprocessor. The averaged measurements are transfered to a computer where they are stored and where various calculations are performed in particular Fourier analysis (by a least square method). The ensemble averages are determined at N_p points equally spaced over the period (N_p = 00 in general) and at each point the sample size was at least 100, often 400 or more. A compromise had to be reached between the length of the measuring time and statistical convergence, especially for measurements close to the wall, where this time is long because of the low particle crossing rate due to the low mean velocities associated with the small size of the measuring volume and where simultaneously the convergence is slow because of the high relative turbulent intensity.

Numerous experimental difficulties had to solved such as :

- vibrations of the test section reduced by reenforcement
- size of the measuring volume, waist of the focused beam reduced to 0,08 mm by a beam expander (length 0,8 mm). Measurements were thus possible as close as 0,25 mm from the wall typically y^+= 2-3 and no correction for measuring volume truncation was necessary. This also helped minimizing shear noise which contributed at most 3% to the turbulent intensity at $y^+ = 2$.
- precise determination of the position of the measuring volume.
- bias towards larger velocities (TIEDERMAN effect) important near the wall because of large turbulent intensities. The correction is effected by averaging the Doppler periods (not the frequencies). The digital value of the Doppler period from the counter is fed directly to the microprocessor.
- phase errors when the frequency of validated measurements is smaller than fN_p, avoided by taking a validated Doppler period only once at the time of occurence.

The instrumentation and data acquisition were checked in two ways. Firstly in steady turbulent flow in the channel where good agreement was found for both $\bar{u}/u_\tau$ and $\sqrt{\overline{u'^2}}/u_\tau$ vs y^+ with previous measurements (fig. 1,2) and secondly in laminar oscillating flow in an oil channel where good agreement was found for the amplitude A_{1u} and the phase shift $\emptyset_{1u}$ vs y_s with the theoretical values of the Stokes flow.

Experimental conditions

$\bar{u}_c$ = 15.5 cm/s   X = 2.40 m u_τ = 0.80 cm/s R_{eh} = 8.8.10^3

$\frac{A_{1u_c}}{\bar{u}_c} \simeq 5\%$ T = 0.7 - 13.5 s Ω_* = 120 - 6

Reproducibility of oscillating flow : of amplitude A_{1u_c} $\pm$ 2,5%

of phase $\emptyset_{1u_c}$ $\pm$ 1.1°

Results and discussion

1) Mean flow

The mean velocity profiles measured in oscillating flow are indistinguishable from the steady flow curve at all frequencies tested. This confirms the previously published results (4-6) but is at variance with the latest data of RAMAPRIAN and SHUEN-WEI (7) whose profile seems to show a point of inflexion near the wall. These authors argue that the mean flow in their experiment has been affected by the periodic oscillations because the forcing frequency f is of the same order as the characteristic frequency f_t of the turbulence ($f \approx f_t$) and that an interaction between the two types of fluctuations takes place. Since our highest Strouhal number based on the half channel width $\Omega_h \simeq 2.5$ is even larger than the one of RS which was about 0.5 ($R\omega / u_m$ in their case ; pipe radius : R), this argument alone cannot explain the differences between these observations. It should be added that the highest frequency in KARLSSON's (5) experiment and the two frequencies used by ACHARIA (4) were also such that $f \approx f_t$ and no effect on the mean flow was observed.

The other parameter which, together with the frequency, would affect the turbulence and therefore the mean flow is the relative amplitude of the oscillation. This parameter is quite different in (7) and the present experiments, being respectively about 35% and 5%. In KARLSSON's measurements at high amplitude (33%) and at a medium or hight frequency, their seems to be a slight effect on the mean velocity but from the plotted data, it is difficult to conclude whether this effect is significant or not. In the

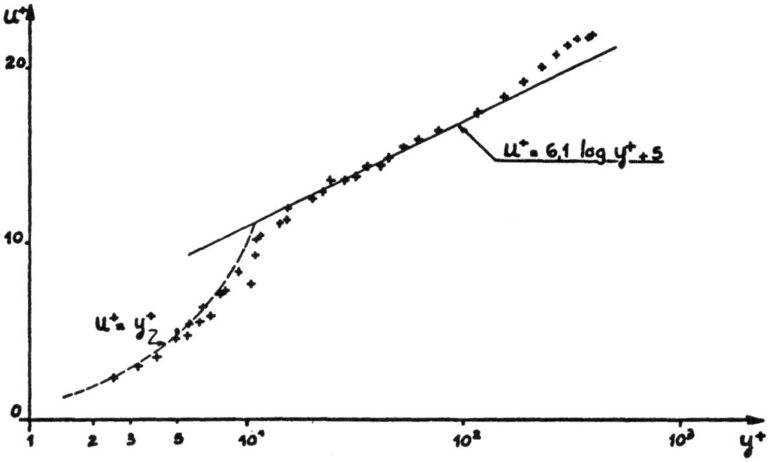

Fig. 1 - Mean velocity in steady and unsteady flow

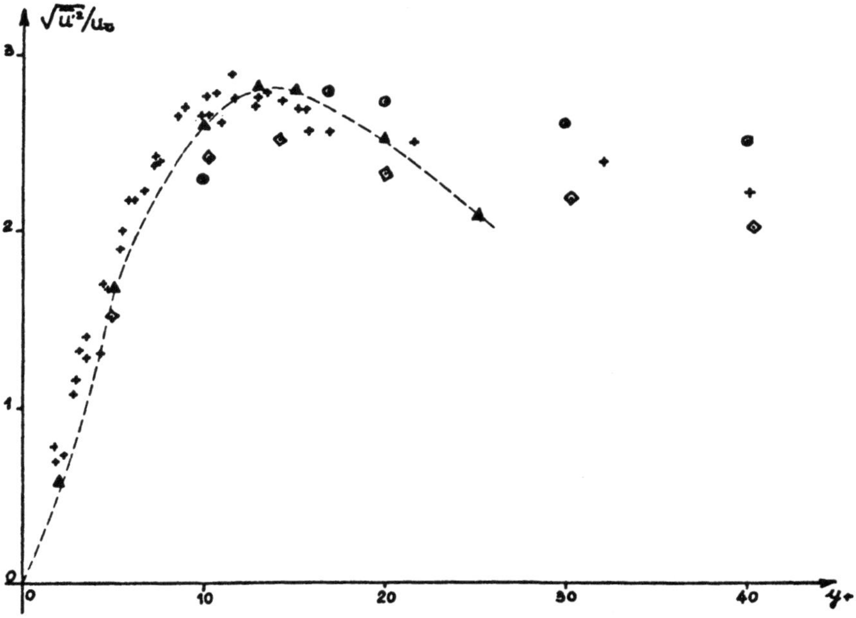

Fig. 2 - Turbulent intensity near the wall in channel flow

Steady flow : ⊙ Comte-Bellot , ▲ Eckelmann
Steady and unsteady flow : ◈ Acharia , ✚ present investigation

present work the amplitude had to be kept small in order to avoid flow reversals near the wall because the laser velocimeter was not equiped with a frequency shifting device.

2) Periodic velocity oscillation.

The evolutions of the fondamental mode of the oscillating velocity near the wall for two frequencies differing by one order of magnitude are shown on the figures 3-4. Even at the lower frequency, the variations of the amplitude and of the phase occur entirely in the wall region as already observed in (4) and (5).

The remarkable features of these experimental results are that : i) the amplitude near the wall follows very closely the laminar Stokes flow for both frequencies as far as $y^+ \simeq 12$. Similar results are reported in (7) and (4). It should be noticed that this value of y^+ corresponds roughly to the point where the turbulent and laminar shear stresses are equal (3). In the high frequency case (fig. 3) the amplitude ratio is almost one at this point and there is little left to be diffused outward by the turbulence. The overshoot on the other hand, despite the scatter of the data, is more important than in laminar flow and the dip down to about 0.85 does not exist in laminar flow. In the low frequency case a certain erosion of the gradient by the turbulence is noticeable (fig. 4).

ii) the phase of the velocity oscillation in the high frequency case is almost exactly that predicted by the Stokes solution. It appears that the extrapolation to the wall gives a value close to +45° which is characteristic of the laminar high frequency case. At the lower frequency on the contrary the phase shift near the wall is completely different from that of the Stokes flow. Extrapolation to the wall is not possible here because of the scatter of the data points, but nevertheless is seems that the value at the wall is close to zero. Regretably, there is no phase shift data in (7). There is qualitative agreement of the present results with KARLSSON's (4) such as smaller phase shifts at lower frequencies. This author also mentions that the largest phase lead measured was 35° at $y = 0.01''$ for $f = 7.65$ hz and $\Omega_x = 23$. The corresponding Stokes flow at the same distance gives a phase lead of 36° which confirms the present high frequency result.

ACHARIA's (4) results are completely different since he measured a phase lead of about 100° for $\Omega_k = 0.72$ and a phase lag of about 200° for $\Omega_k = 1.2$ at the closest point to the wall ($y^+ = 2$) with respect the center of the channel. No plausible explanation for these large descrepancies has been

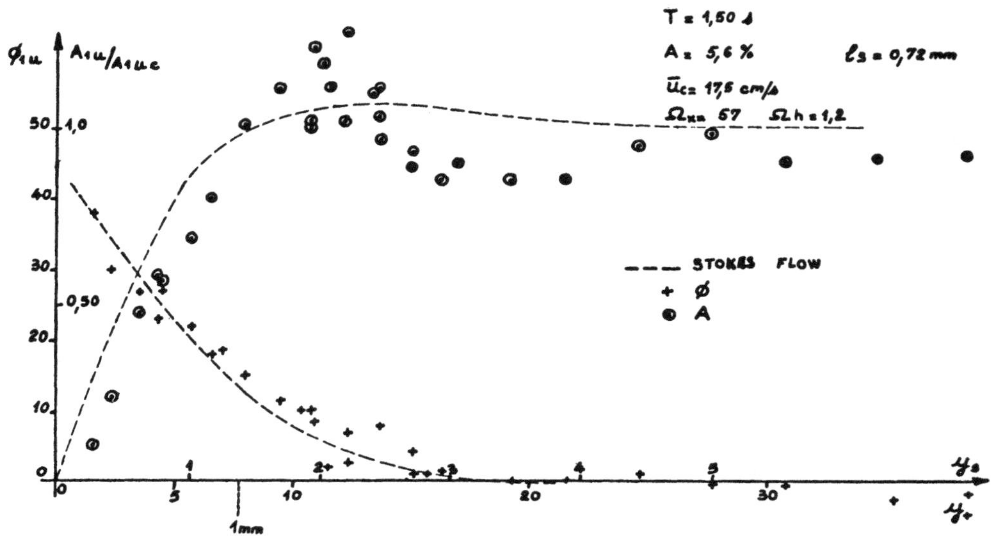

Fig. 3 - Amplitude and phase of velocity oscillations vs distance in the wall region

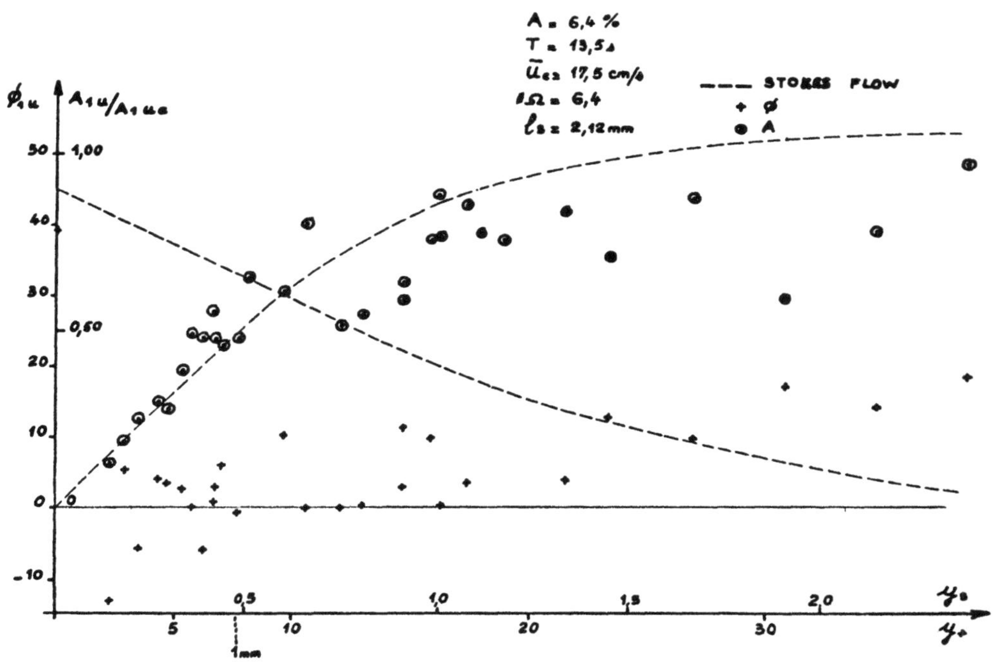

Fig. 4.- Amplitude and phase of velocity oscillations vs distance in the wall region

found yet.

3 Oscillating wall shear stress

The preceding results show that for $\Omega_x \gtrsim 5$ or $\Omega_h \gtrsim 0.13$ (values coresponding to the lowest fequency tested), the amplitude of the oscillating wall-shear stress is given by the laminar Stokes solution as if there were no turbulence present quite contrary to the mean wall shear which is imposed by the turbulent shear stress. Therefore :

$$|\tilde{\tau}| \simeq \rho\nu\left(\frac{\partial|\tilde{u}|}{\partial y}\right)_{y=0} = \sqrt{2}\,\rho\nu\,\frac{|\tilde{u}_e|}{l_s}$$

so that :

$$a_\tau = |\tilde{\tau}|/\bar{\tau} = \sqrt{2}\,\frac{|\tilde{u}_e|}{u_\tau}\cdot\frac{l_\nu}{l_s} = \sqrt{2}\,\frac{|\tilde{u}_e|}{u_\tau}\cdot\frac{1}{l_s^+}$$

or

$$a_\tau/a_{u_e} \simeq \sqrt{2}\,\frac{\bar{u}_e}{u_\tau}\cdot\frac{1}{l_s^+}$$

Let us recall that in a laminar boundary layer this relative amplitude ratio equals $3\sqrt{\Omega_x}$ in the high frequency limit and $h/\sqrt{2}\,l_s$ in laminar channel flow. The above relation shows the importance of the non-dimensional Stokes thickness expressed in wall parameters :

$$l_s^+ = \frac{l_s}{l_\nu} = \frac{\sqrt{2}\,u_\tau}{\sqrt{\omega\nu}}$$

The values of l_s^+ in the experiments which have been quoted are as follows

Experiment	KARLSSON	ACHARIA	COUSTEIX	RAMARRIAN	Present
l_s^+	4.5-5.4	7.1-9.4	35	6.2	3.8-16.

If Ω_x has a clear physical meaning in laminar oscillating boundary layers where $\Omega_x = \delta^2/l_s^2$ this is not so in the turbulent regime. In this case the parameter l_s^+ seems more relevant and it is suggested to use it to correlate the data.

It is also seen that the shear stress ratio a_τ depends upon $|\tilde{u}_e|/u_\tau$. This may explain some of the differences between the present results and those of (7) where the amplitude was large.

Finally, the proportionality $a_\tau \propto |\tilde{u}_e|/u_\tau$ points towards the importance of a mean unfavorable pressure gradient which leads to a decrease in u_τ especially in conditions near separation since there $u_\tau \to 0$.

Periodic flow reversal requires $a_\tau > 1$, i.e. $a_{u_e} > \frac{u_e}{\sqrt{2}\,\bar{u}_\tau} \cdot l_s^+$ which shows that in the absence of an unfavorable pressure gradient the forcing amplitude has to be large unless the frequency is very high (small l_s^+).

4) Turbulence

The measured mean longitudinal turbulent intensity $\sqrt{\overline{u_1^2}}$ was the same in the unsteady as in the steady case as already observed in (4) showing that small amplitude oscillations do not affect the mean turbulence.

Since turbulence is affected only by velocity gradients -and not by velocity oscillations whatever the frequency- the uniform velocity oscillations in the core which may cover most of the channel or the boundary layer have no influence the turbulence. Moreover, the turbulence in the wall region will only be modified by the oscillations if $\frac{\partial \tilde{u}}{\partial y} / \frac{\partial \bar{u}}{\partial y} \propto a_\tau$ is of the ordre of one since otherwise the oscillating flow contributes little to the instantaneous velocity gradient. In addition it may be conjectured that a substancial $\frac{\partial \tilde{u}}{\partial y}$ must subsist beyond $y^+ = 12$ since below this distance the flow is dominated by viscous forces. It is then simultaneously required that $a_\tau \gtrsim 1$ and $l_s^+ > 12$. This leads for a value of $\bar{u}_c / u_\tau \simeq 25$ to $a_{u_e} > 50\%$, also a very large forcing amplitude.

It seems, therefore, the turbulence and hence the mean flow can only be affected by very severe oscillations of high frequency. This could be a sufficient explanation for most observations which show that the mean flow is quite insensitive to forced oscillations in the absence of an unfavorable pressure gradient.

Acknowledgement

Financial support from the Direction des Recherches, Etudes et Techniques from the Minisètre des Armées is greatfully acknowledged.

References

1. Tellonis, D.P., Unsteady boundary layers separated and attacked, J. Fluids Engineering, 101, 29-43, 1979.

2. Comte-Bellot, G., Ecoulement turbulent entre deux parois parallèles. Publications Scientifiques et Techniques du Ministère de l'Air, 160 p., 1965.

3. Eckelmann, H., The structure of the viscous sublayerand the adjacent wall region in a turbulent channel flow, J.F.M., 65, 439-459, 1974.

4. ACHARYA, M., Measurements and predictions of a fully developed turbulent channel flow with imposed controlled oscillations. Stanford Univ., Ph. D., Xeros University Microfilms, Ann Arbor, Mi., réf. 75-25, 492, 1975.

5. Karlsson, S.K., An unsteady turbulent boundary layer, J.F.M., 5, 622-636n and Ph. D. Dissertation, The Johns Hopkins University,1969 (1958)

6. Cousteix, J.. Desopper, A., Houdeville, R., Structure and Development of a turbulent boundary layer in oscillatory external flow, Proc. Symp. on Turbulent Shear Flows, Penn. State Univ., Univ. Park, Pa., 1977

7. Ramaprian, B.R., Shven-Wei Tu, An experimental study of oscillatory pipe flow at transitional Reynolds numbers, J.F.M., 100, 513-544, 1980.

Nomenclature

any quantity q : $q = \bar{q} + \tilde{q} + q'$, $\langle q \rangle = \bar{q} + \tilde{q}$ ensemble average

$\bar{q}$ = time mean , $\tilde{q}$: periodic oscillation , q' = turbulent fluctuation,

$|\tilde{q}|$: amplitude

$\langle q \rangle = \bar{q} + \sum_{n=1} A_{nq} \cos(n\omega t + \phi_{nq})$, $a_q = |\tilde{q}|/\bar{q}$ = relative amplitude

$\bar{\tau} = \rho u_\tau^2$: wall shear stress , $l_\nu = \nu/u_\tau$ = wall length scale

$l_s = \sqrt{\frac{2\nu}{\omega}}$ Stokes length , $\omega = 2\pi f$ f = frequency of periodic flow

2h = height of channel

$y^+ = y/l_\nu$ $y_s = y/l_s$ $(.)_c$: center line

$\Omega_x = \frac{\omega x}{\bar{u}_c}$ $\Omega_h = \frac{\omega h}{\bar{u}_c}$

Some Features of Unsteady Separating Turbulent Boundary Layers

R. L. Simpson, B. G. Shivaprasad, and Y.-T. Chew

Department of Civil and Mechanical Engineering
Southern Methodist University
Dallas, Texas 75275 USA

Abstract

A survey of the physical features of steady and unsteady separating turbulent boundary layers is presented for practical Reynolds numbers and reduced frequencies for helicopter and turbomachinery flows. Well upstream of separation there is little interaction between the periodic motions, so the flow away from the wall has little phase variation from the freestream. Near the wall between the viscous sublayer and the semi-logarithmic region, unexpected phase shifts of the velocity and turbulence oscillations occur.

Near separation and downstream more interaction occurs between the periodic and turbulent motions since the characteristic frequency of the large scale structures is much lower than upstream. Significant phase variations between the velocity and turbulence exist in the detached and back flows. For moderate oscillation amplitudes there is no effect of oscillation waveform on the mean flow features. Large amplitude oscillations affect the flow structure significantly.

1. Introduction

Unsteady turbulent boundary layers have become the subject of much recent interest because of unsteady aerodynamic phenomena associated with blades in compressors and turbines and with helicopter rotors in translating motion. While all turbulent flows are inherently unsteady, the term "unsteady" will mean here an organized time dependent motion, in contrast to the relatively aperiodic motion of turbulence. The boundary layers cannot be ignored in unsteady flow analyses of these devices because there is considerable interaction between the boundary layer and the inviscid flow during high lift operating conditions. In such cases the relatively thick boundary layer on the suction side of the lifting body is near separation. "Separation" must mean the entire process of "departure" or "breakaway" or the breakdown of the boundary-layer concept. An abrupt thickening of the rotational flow region next to a wall and significant values of the normal-to-wall velocity component must accompany breakaway, else this region will not have any significant interaction with the freestream inviscid flow [1].

"Detachment" is the locus of points where the limiting streamlines of the flow leave the surface.

This paper presents a survey of experimental results obtained in the wind tunnel shown in Fig. 1 at a practical Reynolds number of 5.1 x 10^6, based on the time-mean entrance velocity $\overline{U}_{ee}$ of 16.5 mps and the 4.9 m length C of the converging-diverging section. The time-averaged freestream velocity distribution for a steady flow and a sinusoidal waveform unsteady flow are shown in Fig. 2 to be essentially the same. A brief summary is given of the behavior of the steady freestream separating boundary layer so that the effects of periodic freestream unsteadiness can be distinguished. Results obtained at a practical reduced frequency $\omega C/2\overline{U}_{ee}$ of 0.55 are presented, and some effects of unsteadiness waveform and amplitude are discussed.

2. Experimental Analysis

All of the equipment and instrumentation details are given by Simpson *et al.* [2]. A programmable rotating-blade damper [3] was used to produce different amplitude and shape unsteadiness waveforms at a frequency of 0.596 Hz. Active suction and wall jet boundary layer controls were used to prevent separation on the non-test walls and to permit separation on the wind tunnel floor. Velocity measurements were made with a hot-wire anemometer and a two-velocity component laser velocimeter [4]. A Rubesin *et al.* [5] surface hot-wire anemometer was used to measure the surface shear stress τ_w while an Eaton *et al.* [6] wall-flow-direction probe [7] was used to measure the fraction of time that the flow moves downstream, γ_{pu}. The data were phase-sampled and ensemble-averaged for 200 cycles by a digital computer activated by the clock for the programmable rotating-blade damper.

3. The Nature of a Steady Freestream Mean-Two-Dimensional Separating Turbulent Boundary Layer

For steady freestream separating turbulent boundary layers, a set of quantitative definitions on the detachment state near the wall has been proposed [1]: *incipient detachment* (ID) occurs with instantaneous backflow 1% of the time; *intermittent transitory detachment* (ITD) occurs with instantaneous backflow 20% of the time; *transitory detachment* (TD) occurs with instantaneous backflow 50% of the time; and *detachment* (D) occurs where the time-averaged wall shearing stress $\overline{\tau}_w = 0$. Thus, γ_{pu} is a descriptive parameter for identifying these stages and should be documented in all separated flow experiments [1,7].

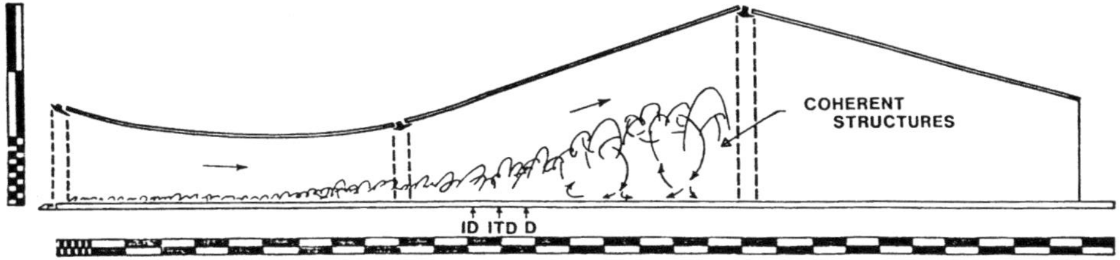

Fig. 1. Sideview schematic of the test section with the steady freestream separating turbulent boundary layer [8,9] on the bottom wall. Major division on scales: 10 inches.

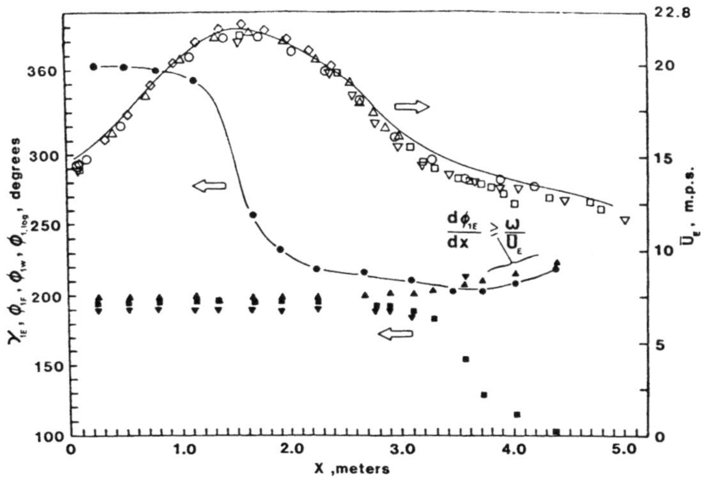

Fig. 2. $\overline{U}_e$: — steady flow; open symbols, sinusoidal unsteady flow on different days. Phase angles of first harmonics: ▲, ϕ_{1e}, free-stream velocity; ■, ϕ_{1log}, semi-log region velocity; ●, ϕ_{1e} + 180° + γ_{1e}, pressure gradient (dashed line for visual aid); ▼, ϕ_{1w}, wall shearing stress.

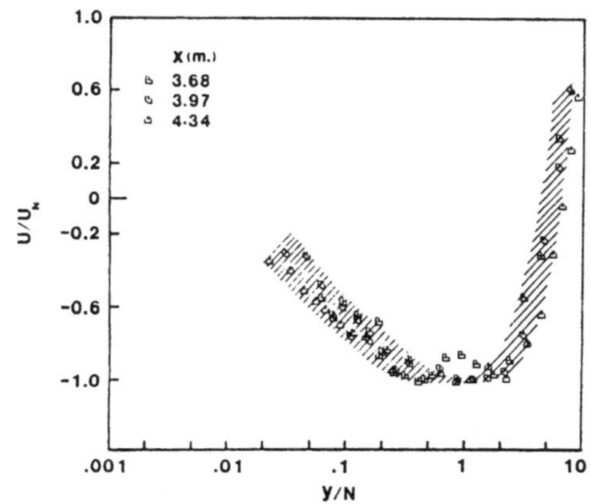

Fig. 3. Normalized backflow mean velocity profiles for sinusoidal unsteady flow. Shaded region, steady freestream case [8].

Figure 1 is a side view schematic of the 8 m long, 0.91 m wide test section of the SMU Wind Tunnel. Also shown are a qualitative sketch of the steady freestream bottom wall turbulent shear flow studied with a LDV at SMU [8,9] and the locations of ID, ITD, and D when determined 1 mm from the wall. The mean flow upstream of ID obeys the "law of the wall" and the "law of the wake" as long as the maximum shearing stress $-\rho\overline{uv}_{max}$ is less than $1.5\ \tau_w$. The qualitative turbulence structure is not markedly different from the zero-pressure-gradient case. The "bursting" frequency n of the most energetic eddies near the wall is correlated by $U_e/n\delta = 10$, where U_e is the mean velocity outside the boundary layer and δ is the boundary layer thickness.

When $-\rho\overline{uv}_{max} > 1.5\ \tau_w$, the Perry and Schofield [10] mean velocity profile correlation and the law of the wall apply upstream of ITD. Up to one-third of the turbulence energy production in the outer region is due to normal stresses effects, which modifies the relations between dissipation rate, turbulence energy, and turbulent shearing stress that are observed farther upstream. The spanwise integral length scale of the turbulence increases with δ^2 and the bursting frequency n continues to be equal to $U_e/10\delta$. Pressure gradient relaxation begins near ITD and continues until D.

Downstream of detachment, the mean backflow profile scales on the maximum negative mean velocity U_N and its distance from the wall N (Fig. 3). A U^+ vs. y^+ law of the wall is not consistent with this result since U_N and N increase with streamwise distance while $\nu/(\tau/\rho)_w^{1/2}$ varies inversely with $(\tau/\rho)_w^{1/2}$. High turbulence levels exist in the backflow, with u and v fluctuations of the same order as $|U|$. $-\overline{uv}/u'v'$ is very low in the backflow and is about 25% lower in the outer region than for the upstream attached flow. Mixing length and eddy viscosity models are adequate upstream of detachment and in the outer region, but are physically meaningless in the backflow. γ_{pu} never reaches zero, indicating that there is no location with backflow all of the time. Normal and shear stresses turbulence energy production in the outer region supply turbulence energy to the backflow by turbulence diffusion where it is dissipated [8,9]. Negligible turbulence energy production occurs in the backflow.

This turbulence energy diffusion and the small mean backflow are supplied intermittently by large-scale structures as they pass through the detached flow as suggested by Fig. 1. The backflow does not come from far downstream. The frequency of passage n of these large-scale structures also varies as U_e/δ and is about an order of magnitude smaller than the frequency far upstream of detachment. Reynolds shearing stresses in the backflow must be modeled by relating them to the turbulence structure and not to local mean velocity gradients. The mean velocity profiles in the backflow are a result of time-averaging the large fluctuations and are not related to the cause of the turbulence.

4. Experimental Results for a Sinusoidal Waveform

Fig. 2 shows the streamwise distributions of several parameters associated with the sinusoidal waveform flow [2] discussed here. The ensemble-averaged free-stream velocity outside the boundary layer can be expressed in terms of its Fourier components as

$$U_e = \overline{U}_e \left\{ 1 + \sum_{n=1}^{\infty} \left[\frac{U_{ne}}{\overline{U}_e} \cos\left(\omega n t - \phi_{ne}\right) \right] \right\} \tag{1}$$

$U_{1e}/\overline{U}_e$ was about 1/3 and U_{2e} and U_{3e} were about 2% of U_{1e}.

Using this streamwise distribution, the free-stream streamwise pressure gradient distribution can be calculated from the inviscid equation of motion

$$\begin{aligned} \frac{1}{\rho}\frac{dP}{dx} = -\left[\overline{U}_e\overline{U}_e' + \frac{U_{1e}U_{1e}'}{2}\right] + \Big[\left(\overline{U}_e U_{1e}' + U_{1e}\overline{U}_e'\right)^2 \\ + U_{1e}^2\left(\overline{U}_e\phi_{1e}' - \omega\right)^2\Big] \cos\left(\omega t - \phi_{1e} - 180° - \gamma_{1e}\right) \end{aligned} \tag{2}$$

where $$\gamma_{1e} = \arctan\left\{\frac{U_{1e}(\overline{U}_e\phi_{1e}' - \omega)}{\overline{U}_e U_{1e}' + U_{1e}\overline{U}_e'}\right\} \tag{3}$$

As in the steady free-stream case, there is pressure gradient relaxation after detachment so that dP/dx is *never* negative.

Fig. 2 shows ϕ_{1e} and $(\phi_{1e} + 180° + \gamma_{1e})$ and that the first harmonic pressure gradient strongly lags the local free-stream velocity in the converging section of the tunnel. The lag is considerably lower in the diverging section. After detachment the oscillatory pressure gradient only slightly leads the velocity oscillation with the onset of pressure gradient relaxation.

Ensemble-averaged values at each phase of the unsteadiness period were obtained for U, $\overline{u^2}$, $\overline{v^2}$, and $-\overline{uv}$. Only typical results are presented here.

Complete data are on magnetic tape filed with Dr. L. W. Carr* in the format used for the 1980-81 Stanford Conferences on Complex Turbulent Flows.**

Well upstream of separation where the periodic unsteadiness frequency is much lower than the turbulence frequencies, the flow in the outer region of the boundary layer is in phase (Figs. 2 and 4). Outside the viscous dominated region, but closer to the wall than the logarithmic region, the ensemble-averaged unsteady oscillatory velocity leads the logarithmic region oscillatory velocity by as much as 60° as shown in Fig. 4. The wall shear stress leads the velocity in the logarithmic region by about 3° (Fig. 2). Spectral measurements in this region indicate that the turbulence frequencies vary drastically during the cycle period. Higher frequency turbulent oscillations are observed during the higher velocity part of the cycle while frequencies an order of magnitude lower are observed during the low velocity part. The oscillatory turbulent fluctuations are not in phase with the periodic ensemble-averaged U oscillations in this region. This leads us to believe that there is some sort of "critical layer." On the side of this critical layer nearer the wall the turbulence magnitude leads the periodic ensemble-averaged oscillatory velocity, while the ensemble-averaged oscillatory velocity leads the turbulence magnitude on the free-stream side of the critical layer, as shown in Fig. 4. The amplitude of the first harmonic $U_1/\overline{U}$ is about constant in the logarithmic region and increases along the flow. The phase angle of the first harmonic U_1 is independent of y in the logarithmic region, as it must be when an inner wall layer and outer wake-like layer overlap [11].

The mean flow well upstream of detachment possesses a law-of-the-wall logarithmic region with approximately the same constants as for steady flow. The mean skin friction values are closely the same as those for the steady flow. Just upstream of detachment (upstream of $\gamma_{pu} = 0.8$ in Fig. 5), the mean velocity profiles obey Perry and Schofield's [10] correlation approximately, but not as well as the steady flow velocity profiles do.

Near and beyond detachment, the frequencies of the turbulence and the large-scale structures are much lower so that more interaction of the periodic and turbulent motions occurs. U and $-\overline{uv}$ appear to be in phase while

* U.S. Army Aeromechanics Lab., NASA-Ames Research Center, Moffett Field, CA. 94035.

** Contact Professor S. J. Kline, Thermosciences Division, Department of Mechanical Engineering, Stanford University, Stanford, CA. 94305.

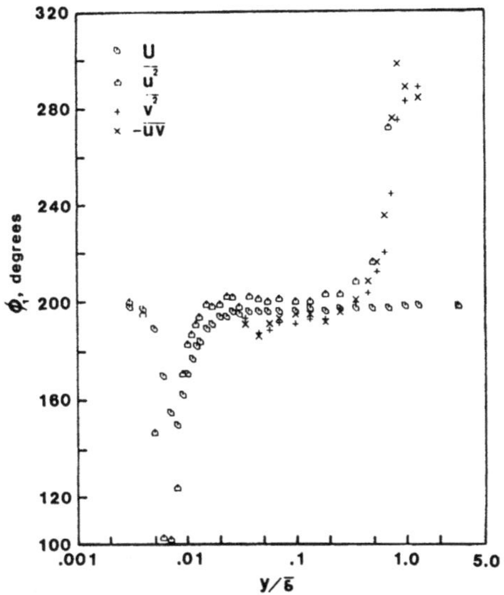

Fig. 4. Phase angle of first harmonic ϕ_1 of U, u^2, v^2, and $-\overline{uv}$ for the sinusoidal flow at 1.34 m.

Fig. 5. $\bar{\gamma}_{pu}$ (cycle-averaged γ_{pu}) at 1 mm from the wall. Solid line, LDV data, steady flow. Wall-flow-direction probe: dashed line, steady flow; □ , ■ , sinusoidal flow; △ , ▲ , compressor waveform; ○ , ● , large amplitude flow; solid symbols, data with probe orientation reversed.

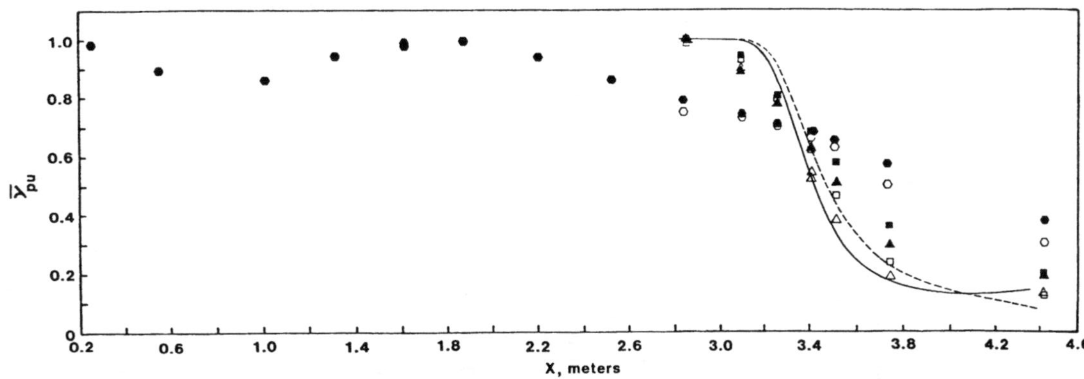

$\overline{u^2}$ and $\overline{v^2}$ progressively lag $-\overline{uv}$ and U more in the downstream direction (Fig. 6). Fig. 7 shows the ensemble-averaged velocity waveforms at several y positions from the test wall at the same location downstream of detachment. A significant phase lead occurs in the near wall region. Fig. 8 shows ensemble-averaged velocity profiles at several phases of the oscillation period. Fig. 9 shows the amplitude of the first two harmonics of the ensemble-averaged oscillation normalized on the local *average* velocity $\overline{U}$.

Downstream of separation approximate mean velocity profile similarity is observed (Fig. 3) for the reverse flow region when the absolute value of the minimum mean velocity U_N and the distance from the wall to its location N are used as velocity and length scales. This profile shape is the same as for the comparable steady free-stream case [8], even though the periodic ensemble-averaged velocity leads the freestream velocity oscillation by a large amount. As in the steady freestream velocity case, a law-of-the-wall profile based on the mean wall shear cannot hold since both U_N and N increase with streamwise distance, while the law-of-the-wall length scale $\nu/(\tau/\rho)_w^{1/2}$ varies inversely with the velocity scale $(\tau/\rho)_w^{1/2}$.

The turbulence structure is not greatly influenced by the unsteadiness upstream of separation. Eddy viscosity and mixing length distributions for the mean flow are closely the same as for the comparable steady free-stream flow. The eddy viscosity for the first harmonic is twice that of the mean flow, which is consistent with the fact that $(-\overline{uv})_1/(-\overline{uv})$ is twice $U_1/\overline{U}$ at a given location in the flow.

The Reynolds shearing stresses in the backflow region must be modeled by relating them to the turbulence structure and not to local mean velocity gradients, which is the same result for the comparable steady free-stream case. The ensemble-averaged velocity profiles in the backflow are a result of ensemble-averaging the large turbulent fluctuations and are not related to the cause of the turbulence. Naturally, eddy viscosity and mixing length models are physically meaningless in the backflow, the eddy viscosity being negative and the mixing length being imaginary.

5. The Effect of Amplitude and Waveform on Separation

Fig. 10 shows the sinusoidal waveform used in the experiments discussed above, an axial-compressor-type waveform, and a large amplitude oscillation. The parameter $R = (U_{emax} - U_{emin})/2\overline{U}_e$ is 0.287, 0.238, and 0.752 for these cases, respectively. In each case the mean free-stream velocity $\overline{U}_e$ was the

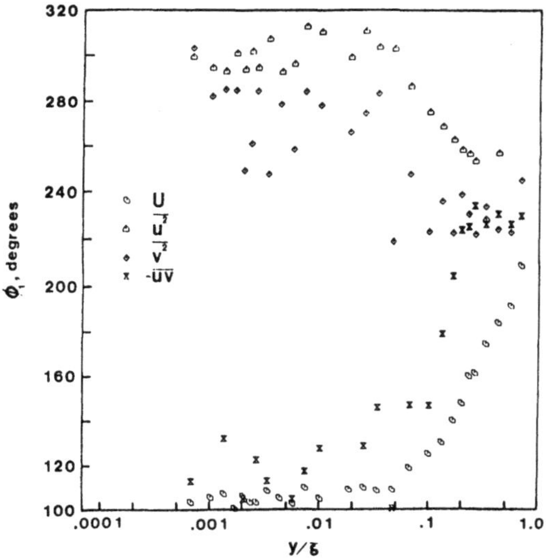

Fig. 6. ϕ_1 for sinusoidal flow at the 4.34 m location.

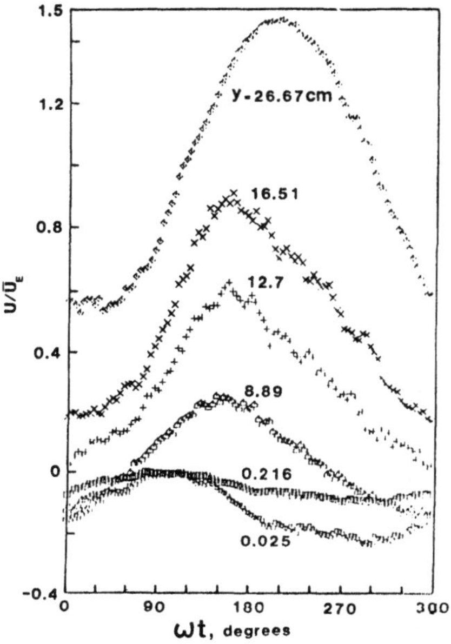

Fig. 7. $U/\overline{U}_e$ vs. ωt at various y at the 4.34 m location.

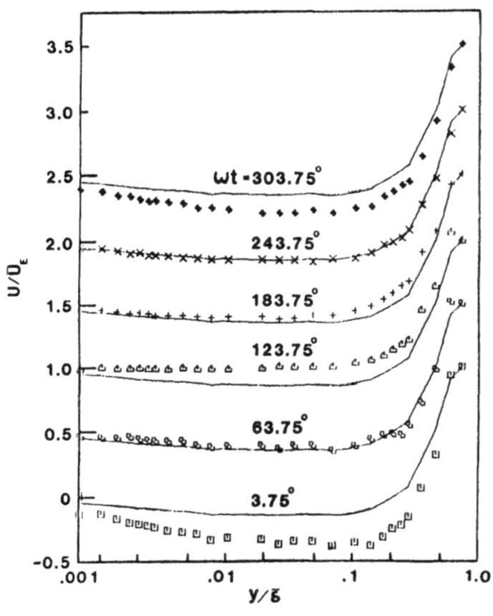

Fig. 8. $U/\overline{U}_e$ vs. y/δ at various ωt at the 4.34 m location.

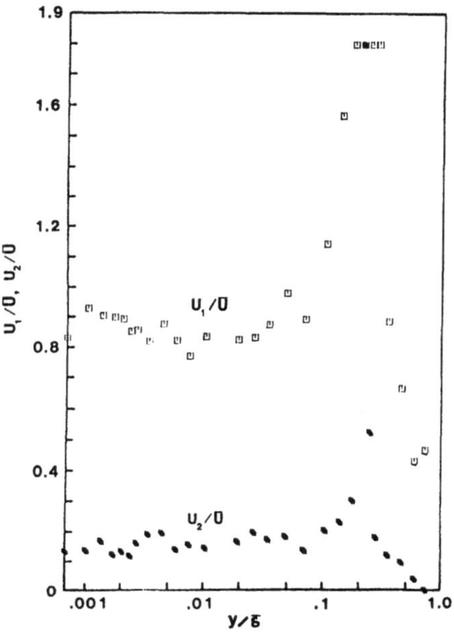

Fig. 9. Amplitude to mean ratios for the first two harmonics at the 4.34 m location.

same near the wind tunnel throat (1.62 m). For moderate values of R less than about 1/3, there is no effect of velocity waveform on the $\overline{\gamma}_{pu}$ distribution near the wall, as shown in Fig. 5.

For the high amplitude case, the $\overline{\gamma}_{pu}$ distribution differs significantly from the low amplitude cases. Note that even upstream of the wind tunnel throat $\overline{\gamma}_{pu} < 1$, so that reversed flow exists some of the time near the wall. Detailed laser and hot-wire anemometer measurements are currently being made on this flow.

6. Conclusions

A moderate amplitude ($R < 0.3$) oscillating turbulent boundary layer at practical reduced frequencies and Reynolds numbers has a mean flow structure that is close to that of a steady free-stream turbulent boundary layer. Mean velocity profiles, skin friction, and $\overline{\gamma}_{pu}$ are closely the same for the two cases. However, a significant phase lead occurs near the wall upstream of separation and in the backflow downstream. Since there is more interaction between the periodic and turbulent oscillations in the detached flow, U and $-\overline{uv}$ oscillations phase lead $\overline{u^2}$ and $\overline{v^2}$ oscillations. Large amplitude oscillations have a substantial effect on the mean flow behavior.

7. Acknowledgments

This work was sponsored by the U.S. Army Research Office under Grant DAAG29-76-G-0187 and by the National Aeronautics and Space Administration, Ames Research Center, under Grant NSG-2354.

References

1. Simpson, R.L., "A Review of Some Phenomena in Turbulent Flow Separation," in press, J. Fluids Engrg., 1981.

2. Simpson, R.L., Shivaprasad, B.G., and Chew, Y.-T., "Measurements of Unsteady Turbulent Boundary Layers with Pressure Gradients," Rept. WT-6, SMU Dept. Civil/Mech. Engrg., 1980; submitted to J. Fluid Mechanics.

3. Simpson, R.L., Sallas, J.J., and Nasburg, R.E., "Tailoring the Waveform of a Periodic Flow with a Programmable Damper," J. Fluids Engrg., 100, pp. 287-290, 1978.

4. Simpson, R.L. and Chew, Y.-T., "Measurements in Highly Turbulent Flows: Steady and Unsteady Separated Turbulent Boundary Layers," Proceedings of Third Int. Workshop on Laser Velocimetry, pp. 179-196, Hemisphere, 1979.

5. Rubesin, M.W., Okuno, A.F., Mateer, G.G., and Brosh, A., "A Hot-Wire Surface Gage for Skin Friction and Separation Detection Measurements," NASA TM X-62, 465, 1975.

6. Eaton, J.K., Jeans, A.H., Ashjaee, J., and Johnston, J.P., "A Wall-Flow-Direction Probe for Use in Separating and Reattaching Flows," J. Fluids Engrg., 101, pp. 364-366, 1979.

7. Shivaprasad, B.G. and Simpson, R.L., "Evaluation of an Improved Wall-Flow-Direction Probe for Measurements in Separated Flows," submitted to J. Fluids Engrg.

8. Simpson, R.L., Chew, Y.-T., and Shivaprasad, B.G., "The Structure of a Separating Turbulent Boundary Layer: Part I, Mean Flow and Reynolds Stresses; Part II, Higher Order Turbulence Results," both in press, J. Fluid Mechanics.

9. Shiloh, K., Shivaprasad, B.G., and Simpson, R.L., "The Structure of a Separating Turbulent Boundary Layer III: Transverse Velocity Measurements," in press, J. Fluid Mechanics.

10. Perry, A.E. and Schofield, W.H., "Mean Velocity and Shear Stress Distribution in Turbulent Boundary Layers," Physics Fluids, 16, pp. 2068-2074, 1973.

11. Simpson, R.L., "Features of Unsteady Turbulent Boundary Layers as Revealed from Experiments," paper 19, Symposium on Unsteady Aerodynamics, NATO-AGARD, Ottawa, Canada, Sept. 1977, AGARD-CP-227.

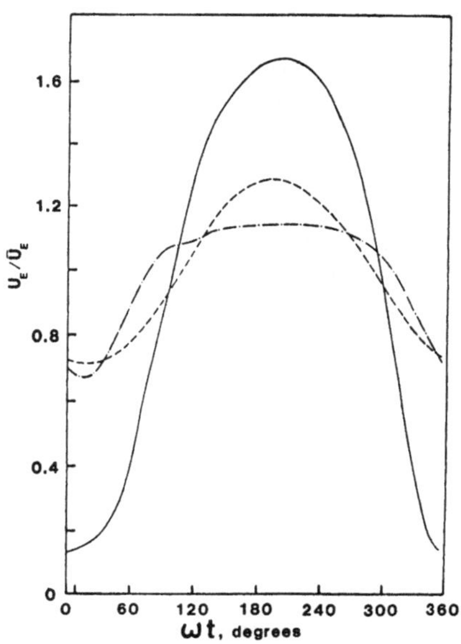

Fig. 10. Velocity waveform shapes: dashed line, sinusoidal case; — • —, compressor-type case; solid line, large amplitude case.

Response of a Turbulent Boundary Layer to a Pulsation of the External Flow With and Without Adverse Pressure Gradient

J. COUSTEIX - R. HOUDEVILLE - J. JAVELLE

ONERA/CERT
2 Av. E. Belin 31055 Toulouse (France)

Summary

The development of an oscillating turbulent boundary layer along a flat plate is studied here. Systematic measurements are carried out for Strouhal number from 1.5 to 18. The harmonic analysis of the main parameters is presented,and compared with the calculation results obtained from the integration of the global or local boundary layer equations.The occurrence of singularities in boundary layer equations solved with an imposed adverse pressure distribution,is discussed.

Introduction

Prediction of unsteady turbulent flows has developed greatly during the last few years, due to the improvement of calculation methods and closure models. Unfortunately, the knowledge of the physics of the phenomenon has not progressed as much, and there exists still a lack of experimental data to validate the theoretical work. This is one of the reasons for undertaking the study of the response of a turbulent boundary layer over a flat plate to an oscillating external velocity field. In this experiment, the Strouhal number varies systematically from 1.5 to 18 and all the important boundary layer parameters, such as the integral thicknesses and skin friction, are studied in terms of their harmonic analysis.

A similar study has been carried out for a turbulent boundary layer in presence of a strong mean adverse pressure gradient. Such a flow has important applications as, for example, the dynamic stall of helicopter blades. The problem of the occurrence of singularities in the boundary layer equations integrated with a prescribed pressure gradient, is analytically studied from the global equations, and it is shown that the use of an inverse mode of calculation, in which the pressure distribution is not imposed, avoids such difficulties.

I - Oscillating turbulent boundary layer on a flat plate

I.1. - Experimental set-up and flow parameters

The boundary layer develops on a 800 mm long, 220 mm wide and 15 mm thick flat plate. The aspect ratio of the elliptic leading edge is 6. A transverse wire, 0.8 mm in diameter and located 70 mm from the leading edge, trips the transition of the boundary layer. The X-notation in the figures corresponds to the distance from an arbitrary origin, 20 mm downstream the leading edge.

A rotating butterfly valve, located 1.3 meter downstream the trailing edge of the plate, induces a periodic pressure loss in the facility. Provided that only the resonance frequencies of the facility are used, the harmonic distortions of the velocity remain within a few per cent of the fundamental.

The large range of variation of the Strouhal number is obtained from two different experiments, the parameters of which are given in table 1. The mean velocity is constant in the X direction, but the relative amplitude of the fluctuation decreases linearly by a factor 1.5 at f=38 Hz and 3 at f= 62 Hz over 700 mm. The phase difference reaches 45° at 62 Hz between the two utmost measurement stations, which corresponds to a travelling wave with a celerity equal to 450 m/s.

The Strouhal number $\frac{\omega X_o}{U_e}$ is defined using a fictitious origin of the mean turbulent boundary layer. This origin is calculated, from the mean momentum thickness measured at a station X = 0.097 m, by a classic flat plate law of the form : $\Theta/\Delta X = 0.021\, R_{\Delta X}^{1/6}$. At the other stations, one simply has : $X_o = \Delta X + X$.

The quantity $U'max/\delta$ in table 1 gives an idea of the characteristic frequency of turbulence. Taking into account the fact that $U'max/\delta$ underestimates the turbulence frequency, it can be said that it remains larger than the periodic frequency, even for f = 62 Hz at the last station.

$U_e = \overline{U_e} + \Delta U_e \sin(\omega t + \varphi_{ue})$		
F (Hz)	38	62
$\overline{U_e}$ (m/s)	21.9	16.8
$\Delta U_e / \overline{U_e}$	0.152 - 0.0743 $(X_o - 0.047)$	0.118 - 0.114 $(X_o - 0.047)$
φ°_{Ue}	7.6 $(X_o - 0.047)$	88.9 $(X_o - 0.047)^2$ + 6.66 $(X_o - 0.047)$
X_o (m)	0.14 to 0.77	0.137 to 0.762
$\omega X_o / \overline{U_e}$	1.52 to 8.35	3.2 to 17.8
R_{X_o}	$0.2\ 10^6$ to $1.1\ 10^6$	$0.15\ 10^6$ to $0.87\ 10^6$
U'_{max}/δ (s^{-1})	350 to 120	270 to 90

Table 1 : External velocity and main parameters of the flows

I.2. - Measurements and data reduction

So far, only the longitudinal component of the velocity has been measured by constant temperature hot wire anemometry. The random fluctuation of the velocity is separated from the periodic component by a conditional sampling technique. U being the instantaneous velocity, φ the phase angle in the cycle and n the number of the cycle, the average velocity is defined as :

$$<U> = \sum_{n=1}^{N} U(nT + \varphi) / N \quad \text{with } U = <U> + u' \text{ and } <u'> = 0$$

The u-component of the turbulence energy is by definition : $<u'^2>$

In practice, a synchronisation signal, generated by the rotation of the butterfly valve, gives the phase information and the velocity is averaged over 600 cycles / 1 /.

I.3. - Experimental results

I.3.1. - Harmonic analysis of the velocity

The boundary layer profiles have been measured at different X-locations along the plate and for the two frequencies, in order to vary systematically the corresponding Strouhal number. The mean velocity profiles are identical to those obtained in a steady flow at the same mean velocity. The shape parameter decreases slightly from 1.52 to 1.46 at $\overline{U_e}$ = 17 m/s, and from 1.48 to 1.43 at $\overline{U_e}$ = 22 m/s. This indicates that even at f = 62 Hz, there is no influence of the unsteadiness on the mean flow development.

Profiles of the relative amplitude of fluctuation and of phase shift with the external velocity, are presented in figures 1 and 2 for different stations along the plate. Profile of $\Delta U/\Delta U_e$ shows a slight overshoot at the first station. This overshoot increases to reach 20% at X_o= 0.34. Farther downstream, it decreases and disappears completely at X_o= .54. Then, this evolution repeats. If the boundary layer is considered as a resonant mechanical system, the observed periodicity along X can be interpreted as a response of the boundary layer to the perturbations induced by the pulsation of the external flow.

By assuming the existence of a travelling wave which affects the successive profiles, the convective celerity of this wave is of the order $.75\overline{U_e}$. The results obtained at 62 Hz are identical. Further, the evolution with Strouhal number of the profiles of the phase shift of the velocity displays the same behaviour.

All the profiles of $\varphi_U - \varphi_{Ue}$ display a maximum of phase lead at $y^+ \simeq 50$, which lies in the logarithmic region of the velocity profile. This property also observed by Simpson / 2 / as well as the decrease of the phase lead near the wall, will be interpreted later.

As the Strouhal number increases, the phase shift decreases in the outer part of the boundary layer, which means that this region behaves as a quasi-steady flow, and that the unsteady effects lie in a thin layer near the wall.

I.3.2. - Integral boundary layer thicknesses

A synthetic representation of the response of the boundary layer to the pulsation of the flow can be obtained from the harmonic analysis of the displacement thickness, as a function of the Strouhal number (Fig. 3). For the sake of homogeneity, the relative amplitude of fluctuation of δ_1 is reduced by the relative amplitude of the external velocity. The striking result is the periodic oscillation of the quantities with a small damping as the Strouhal number increases. Results of Karlsson / 3 / along with those obtained at Onera, are also plotted.

The general agreement between the various experiments is good provided that a small shift in Strouhal number is effected. This can result from the differences in the initial conditions for the turbulent boundary layer. In the present experiments, the locations of the tripping wire correspond respectively to $S = 1.1$ and 2.3, whereas in the previous measurements the transition was natural.

It must be noted that to obtain a detailed description of the boundary layer response, it is necessary to measure the boundary layer at close intervals of Strouhal number. The pioneering experiments of Karlsson do not reveal any fluctuations because of the limited number of Strouhal numbers investigated by him. Therefore, great care must be taken in interpreting these results, particularly when comparisons with calculations are presented.

The qualitative behaviour of the response of the boundary layer can be easily described from a small perturbation development of the Von Karman equation. To do this, it is assumed that all the quantities which appear are of the form : $f = f_0 + f_1 \, e^{i\omega t}$. Subscripts 0 and 1 correspond respectively to the time average and the complex amplitude of f. With the assumption $|f_1| \ll f_0$, the Von Karman equation reads :

$$\frac{d\theta_1}{dX} = \frac{C_{f1}}{2} - \frac{i\omega}{U_{eo}} \left(\delta_{11} + \delta_{10} \frac{U_{e1}}{U_{eo}} \right) \tag{1}$$

If the variation of the shape parameter is neglected, as indicated by experimental results, and if the skin friction is supposed to be a function of θ and U_e, the equation reduces to :

$$\frac{d\theta_1}{dX} + \theta_1 \left[i\omega \frac{H_o}{U_{e1}} - \frac{1}{2} \left(\frac{\partial C_f}{\partial \theta} \right)_o \right] = \frac{U_{e1}}{2} \left(\frac{\partial C_f}{\partial U_e} \right)_o - i\omega H_o \theta_o \frac{U_{e1}}{U_{eo}} \tag{2}$$

Using the Strouhal number $S = \dfrac{\omega X}{U_{eo}}$ instead of X, the solution is of the form:

$$\theta_1 = \Theta \exp\left[\left(-iH_o + \frac{U_{eo}}{2\omega} \left(\frac{\partial C_f}{\partial \theta} \right)_o \right) S \right] \tag{3}$$

The period of the solution is equal to $\dfrac{2\pi}{H_o}$ which is close to the experimentally observed periodicity ($S_p = 5$ instead of $\dfrac{2\pi}{H_o} \simeq 4{,}5$). Moreover, the calculation shows that the damping factor is strongly dependent on the chosen skin friction law.

In the above calculation, no hypothesis is made regarding the nature of the boundary layer and therefore, it applies to laminar flow also.

Figure 4 presents the actual response of such a flow / 5 /. The oscillation in Strouhal number, although of small amplitude due to a large damping factor, is clearly visible.

I.3.3. -Skin friction coefficient - Phase shift near the wall

From steady flow measurements, the skin friction coefficient is often determined by using the technique of "Clauser plot", which consists in fitting the experimental velocity profile by the law of the wall. The same technique has been used in the present experiments on unsteady flow, by assuming that the law of the wall remains valid at every instant of the cycle. We have used :

$$U^{+} = \frac{1}{\chi} \operatorname{Ln} Y^{+} + C \qquad (4)$$

with $U^{+} = \frac{U}{U_{\tau}}$; $Y^{+} = \frac{Y U_{\tau}}{\nu}$; $\chi = 0{,}41$; $C = 5.25$.

Figure 5 gives a typical example of the velocity profiles plotted in semi-logarithmic coordinates. The profiles shown were measured at a given station, at different instants in the cycle. From each of these profiles, the value of U_{τ} which gives the best fit to the law of the wall, was selected. A good agreement is achieved between the experimental values and the universal formula (4) (represented by the solid line), over a rather wide range of y^{+} .

The evolution of the amplitude and of the phase shift of the wall shear stress with Strouhal number, are plotted in figure 6. The general trend is that the amplitude increases with the increase in Strouhal number, while the phase shift, defined with reference to the phase of the external velocity, tends to go to zero. As regards the evolution of the displacement thickness, an oscillation is superimposed on this general trend ; the periodicity in Strouhal number is about 5.5. Thus, one notices again the undamped response of the boundary layer which seems to be characteristic of the turbulent regime.

As noted by Simpson / 2 /, the existence of a logarithmic region of overlap, between the law of the wall and the law of the wake, implies that the velocity phase shift must be constant in such a region. In effect, the experiment shows that the velocity phase shift is nearly constant in the region where a logarithmic behaviour is observed. Actually, a local maximum of the velocity phase shift is measured for values of y^+ around 50. In addition, it will be shown that this phase shift must be equal to the shift in the wall shear stress phase.

This shift in the wall shear stress phase, determined by assuming the validity of the law of the wall, is represented by triangles in figure 2. These values compare well with the maximum velocity phase shift. The agreement so-obtained reinforces the hypothesis of the validity of the law of the wall.

Now, let us assume that, near the wall, the velocity profile at each instant in the cycle obeys not only the law of the wall, but more generally a universal law of the type :

$$U^+ = f(Y^+) \qquad (5)$$

which is assumed to be independent of time.

The above law is expanded in small perturbations by using, for U, U_e and U_τ the same notations as in paragraph I.3.2. By assuming : $|U_1| \ll U_o$, $|U_{e1}| \ll U_{eo}$ and $|U_{\tau 1}| \ll U_{\tau o}$, we get :

$$\frac{U_o}{U_{\tau o}} = f\left(\frac{Y U_{\tau o}}{\nu}\right) \qquad (6)$$

$$U_1 = U_{\tau_1}\left[f\left(\frac{Y U_{\tau o}}{\nu}\right) + \frac{Y U_{\tau o}}{\nu}\frac{df}{dY^+}\right] \qquad (7)$$

Equation (6) shows that the time-mean velocity profile obeys the steady universal law, while the second equation shows that the velocity phase shift is constant, and equal to the phase shift of the shear velocity or equivalently, to the wall shear stress phase shift ($\tau_w = \rho\, U_\tau^2$) / 2/.

This last conclusion is at variance with experimental results, which indicate that the velocity phase shift is not constant in the region from the wall to the logarithmic region. Therefore, we are led to conclude that a universal law such that (5) does not cover the whole wall region from y = 0 up to the logarithmic region. However, a relationship like $u^+ = f\ (y^+)$ does exist in the immediate vicinity of the wall, since we have $u^+ = y^+$ in this region.

On the other hand, figure 5 tends to prove that a $U^{+} = f\ (y^{+})$ law exists for higher values of y under the form of a logarithmic law. Then, it could be that the buffer layer does not obey the universal law $u^{+} = f(y^{+})$. This would mean that the buffer layer is affected by the unsteadiness of the flow.

Nevertheless, if the law of the wall is valid, the velocity phase shift in the logarithmic region must be equal to the wall shear stress phase shift. In this case, one must imagine that there is a rapid variation of the velocity phase shift for the smallest values of y. Indeed, such a hypothesis can be supported by the following considerations:

Let us assume that, in the sublayer, we have :

$$-\frac{1}{\rho}\frac{\partial P}{\partial X} + \frac{\partial}{\partial Y}\left(\mu\frac{\partial U}{\partial Y}\right) = 0 \tag{8}$$

By setting $-\frac{1}{\rho}\frac{\partial P}{\partial X} = \frac{\partial U_e}{\partial t}$ and using a small perturbation development with complex notations, the above sublayer equation gives :

$$\tan\varphi_U = \tan\varphi_{\tau w} - \frac{\omega \nu}{2U_{eo}^2}\frac{1}{(C_f/2)^{3/2}}\frac{|U_{e1}|/U_{eo}}{|\tau_{w1}|/\tau_{wo}}\frac{Y U_{\tau o}}{\nu}\frac{1}{\cos\varphi_{\tau w}} \tag{9}$$

where φ_U and $\varphi_{\tau w}$ are the phase shifts of the velocity and of the wall shear stress, with respect to the external velocity.

This equation shows that the velocity phase shift is not constant near the wall ; it decreases as y increases. For the frequency f = 38 Hz, substitution of typical values gives :

$$\tan\varphi_U = \tan\varphi_{\tau w} - 2\ 10^{-2}\ Y^{+}$$

and for frequency f = 62 Hz, one obtains :

$$\tan\varphi_U = \tan\varphi_{\tau w} - 4.5\ 10^{-2}\ Y^{+}$$

Thus, a substantial decrease of φ_U can take place between $y^{+}=0$ and $y^{+} = 5$. This decrease is about 6° for f = 38 Hz,and about 13° for f = 62 Hz.

Therefore, the behaviour of the velocity phase shift near the wall can be thought of, as given in figure below.

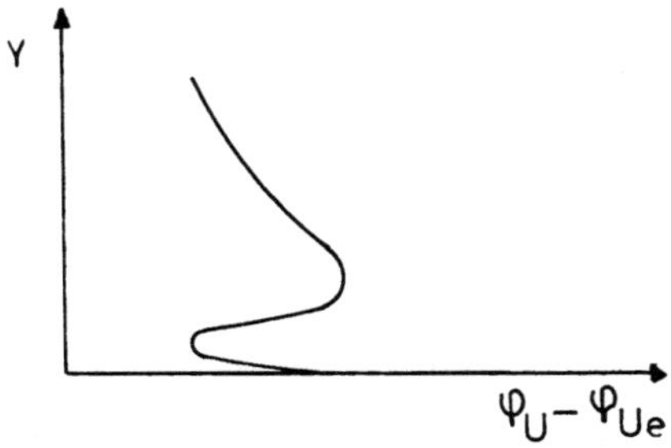

As shown above, φ_U decreases for the low values of y, and then increases to reach in the logarithmic region a local extremum which is equal to the wall shear stress phase shift. Such a behaviour seems to be supported by measurements reported by Simpson / 11 /.

Another matter of discussion could be the law of the wall. Figure 5 shows that the velocity profiles follow such a law over a rather extended region. However, it is not demonstrated that the scaling parameter U_τ is equal to $\sqrt{\tau_w/\rho}$. If U_τ was not equal to $\sqrt{\tau_w/\rho}$, the velocity phase shift would be constant, but not equal to the wall shear phase shift.

I.3.4. - A relationship for the velocity phase shift in the logarithmic region

Let us assume that, at each instant, the skin friction coefficient obeys the following steady flow relationship :

$$\left(\frac{2}{C_f}\right)^{1/2} = \frac{1}{\chi} \, Ln \, \frac{U_e \delta_1}{\nu} + D^* \tag{9}$$

with $\quad D^* = 2G - 4.25\,G^{1/2} + 2.12 \qquad G = \dfrac{(H-1)}{H\,(C_f/2)^{1/2}}$

The above relationship is now expanded in small perturbations. By using index 0 for a continuous component and index 1 for a complex amplitude, one gets :

$$\tan\varphi_{\tau w} = \frac{(C+D)A_{\delta 1}\sin\varphi_{\delta 1} - D\,A_\theta \sin\varphi_\theta}{(C+1)\,A_{U_e} + (C+D)A_{\delta_1}\cos\varphi_{\delta 1} - DA_\theta\cos\varphi_\theta} \tag{10}$$

$$\frac{1}{2}\frac{\tau_{w1}}{\tau_{wo}} = \left[\left((C+D)A_{\delta_1}\sin\varphi_{\delta_1} - DA_\theta\sin\varphi_\theta\right)^2 + \left((C+1)A_{U_e} + (C+D)A_{\delta_1}\cos\varphi_{\delta_1} - DA_\theta\cos\varphi_\theta\right)^2\right]^{1/2} \tag{11}$$

$$A=\left(2-\frac{4.25}{2G_o^{1/2}}\right)\frac{H_o-1}{H_o}-1 \qquad B=\left(2-\frac{4.25}{2G_o^{1/2}}\right)\frac{1}{H_o}$$

$$C=\sqrt{C_{f_o}/2}/\chi A \qquad D=B/A$$

A_{δ_1}, A_Θ, A_{U_e} are the moduli of the reduced amplitudes of δ_1, Θ, and U_e:

$$A_{\delta_1}=\frac{|\delta_{11}|}{\delta_{1o}} \qquad A_\Theta=\frac{|\Theta_1|}{\Theta_o} \qquad A_{U_e}=\frac{|U_{e1}|}{U_{eo}}$$

φ_{δ_1} and φ_Θ are the phase shifts of δ_1 and Θ, with respect to the external velocity.

By calculating the right hand side of formula (10) from experimental results, the value of the wall shear stress phase shift is deduced. Now, it has been shown in paragraph I.3.3 that the wall shear stress phase shift is equal to the phase shift in the log-region. So, these experimental values are compared with formula (10) in figure 7.

This figure also shows the results of Patel / 6 /, who studied the influence of a progressive wave on a turbulent boundary layer, and our results obtained in a study of an oscillating boundary layer with adverse pressure gradient / 1 /. An overall satisfactory agreement is seen except when regions of reverse flow exist. But in the latter case, the law of the wall, as well as the skin friction law are not valid.

A more detailed check of formulae (10) and (11) is presented in figure 6. In this figure, the values of wall shear stress phase shifts and amplitudes determined from the law of the wall, are compared with values calculated from formulae 10 and 11. The good agreement which is obtained tends to support the validity of the quasi-steady skin friction law (9).

I.3.5. - Measurements of turbulence quantities

Fig. 8 represents the evolution in time of the u-component of the kinetic energy of turbulence, divided by its mean walue. Two cycles are plotted every 5 cm along the plate, at a distance from the wall corresponding to a constant value of $Y/\overline{\delta}$ equal to 0.8, which lies in the intermittent region. The stars indicate the instants when $\langle u'^2 \rangle$ is maximum. The fact that they are along the dotted straight lines, the slope of which corresponds to .85 Ue, leads to a result similar to that obtained in steady flow from space-time correlation.

An other interpretation of the results of figure 8 is that the free edge of the boundary layer oscillates at the frequency of the external flow, and can be represented by a travelling wave, the celerity of which being equal to .85 Ue.

Among the various turbulence parameters, the correlation coefficient is important because in the closure models for calculation methods, it is more or less explicitely assumed constant.

The evolution in time of this coefficient is presented in figure 9 at various distances from the wall. Over greater part of the boundary layer, the correlation coefficient remains equal to .45, a value which is commonly found in steady turbulent boundary layers.

This result tends to prove that, up to Strouhal number of 5,the development of the turbulence is not strongly affected by the unsteadiness of the flow. This result can be used to calculate unsteady flow using quasi-steady closure assumptions.

II - Boundary layer calculations

II.1. - Calculation methods

These methods can be divided into two groups. In the first one, the global boundary layers equations are considered and in the second group, the local equations are numerically integrated in the x, y and t directions.

The global methods developed here are based on the integration of the Von Karman and entrainment equations. Closure relationships are derived from the study of unsteady similarity solutions, determined by using a mixing length formulation / Ref. 10 /.

In the local methods, transport equations for the kinetic energy of turbulence, dissipation rate and the Reynolds stress $\langle u'v' \rangle$ are used. The closure relations are identical to those used in steady flow calculations. The sublayer region is calculated using a mixing length formulation, and the continuity of U, K and $<u'v'>$ insures the matching with the transport equations at a value of y^+ equal to 30.

II.2. - Calculation results for flat plate boundary layer

Figure 10 shows the harmonic analysis of the displacement thickness calculated using the full global equations (solid line), and a linearised development of these equations (dotted line).

The experimental points correspond to F = 62 Hz. The initial conditions for calculations come from the experiment. The periodicity in Strouhal number is correctly predicted but the damping is not correct, especially for the results corresponding to the complete equations. It has been observed that this factor is strongly dependent on the skin friction evolution. The same disagreement is observed in the experimental and calculated evolution of the momentum thickness.

II.3. - Boundary layer with adverse pressure gradient

II.3.1. - Description of the experiment

The boundary layer develops on the floor of the wind tunnel. The central airfoil induces the mean adverse pressure gradient, and suction of a certain part of the upper mass flow allows a constant mean velocity along X in the entrance section (fig. 11).

The device to produce pulsating flow is the same as for the flat plate experiment. Table 2 gives the main parameters of the flow, and figure 14 shows the harmonic analysis of the external velocity.

$X_0 = X + \Delta X$

$\Delta X = 0.4$ (m)

X (mm)	$\Delta U/\overline{U_e}$	$10^{-6} R_e$	$\omega X_o/\overline{U_e}$	u'/δ (Hz)
100	0,19	0,74	3,0	350
240	0,16	0,91	4,6	300
390	0,16	1,0	7,0	230
604	0,13	1,15	10,0	70

Table 2 : Main parameters

Various calculation methods have been applied to the above experiment, and figure 12 represents typical results. The harmonic analysis of the displacement thickness is considered here. The complete or linearised form of the global equations give similar results. They correctly predict the oscillations of the reduced amplitude and of the phase difference, as the Strouhal number increases. Unfortunately, the calculated amplitude of $\varphi_{\delta_1} - \varphi_{U_e}$ is too large. To improve the agreement, the integration of the local equations with three transport equations is needed.

Experimentally, the most forward location of the region of negative velocities corresponds to x = 435 mm. To calculate such a region, it will be seen in section III that downstream boundary conditions have to be imposed to keep a well posed problem. It is not easy to do this in practice and therefore, it has been prefered to stop calculations before the occurrence of reverse flow.

To compare various closure models, the harmonic analysis of the velocity is presented in figure 13, at a station located in the adverse pressure gradient region. Experimentally, the overshoot of the relative amplitude of fluctuation of U reaches 40%. Neither the simplest model which used a mixing length formulation, nor a $k-\varepsilon$ model can predict such an overshoot.

A decisive improvement is achieved by considering a five transport equation model (with k, ε, $\langle u'v'\rangle$, $\langle u'^2\rangle$ and $\langle v'^2\rangle$), and incorporating term of the form $-(\langle u'^2\rangle - \langle v'^2\rangle)\frac{\partial U}{\partial x}$ in the k-equation. This quantity, commonly neglected in boundary layer assumptions, can reveal not negligible near separation in steady flow, as shown by Simpson / 7 /.

Moreover, the sensitivity of the results to the choice of the numerical constant appearing in the production and diffusion terms of the $\langle u'v'\rangle$ equation, is an important matter which needs to be carefully examined.

III - Singularities of the direct mode - Inverse mode

A boundary layer calculation, in which the external pressure distribution is prescribed, is called a direct method. In two-dimensional steady flow, it is now well established that singularities occur in this mode when the skin friction vanishes. In contrast, the unsteady case has not been thoroughly investigated.

The approach we have proposed is based upon analytical considerations of the properties of a set of global boundary layer equations. The derivation and the consequences of this analysis are given in details in references / 8 /, / 9 /. Here, only the main results are presented.

III.1- Properties of global equations - Singularities

The analysis is based on the entrainment and global momentum equations, closed by relationships deduced from self-similar solutions /10/. These equations read :

$$C_E = \frac{1}{U_e}\frac{\partial}{\partial X}[U_e(\delta-\delta_1)] \qquad C_E = \frac{\partial \delta}{\partial X} - \frac{V_e}{U_e} \tag{12}$$

$$\frac{C_f}{2} = \frac{\partial \theta}{\partial X} + \theta\frac{H+2}{U_e}\frac{\partial U_e}{\partial X} + \frac{1}{U_e^2}\frac{\partial}{\partial t}(U_e\delta_1) \tag{13}$$

In fact, to facilitate the analysis, it is advantageous to consider the parameter $H^* = (\delta - \delta_1)/\theta$ as a function of the shape parameter $H = \delta_1/\theta$ alone. A good approximation of this function is :

$$H^* = \frac{\beta H^2 + H}{H-1} \qquad \beta = 0.631 \tag{14}$$

Then, the global equations can be written as :

$$H^{*\prime}\frac{\partial \delta_1}{\partial X} + \frac{1+H^{*\prime}}{U_e}\frac{\partial \delta_1}{\partial t} + (H^* - HH^{*\prime})\frac{\partial \theta}{\partial X} + \frac{H^* - HH^{*\prime}}{U_e}\frac{\partial \theta}{\partial t} = A \tag{15}$$

$$\frac{1}{U_e}\frac{\partial \delta_1}{\partial t} + \frac{\partial \theta}{\partial X} = B \tag{16}$$

with $A = C_{Es} - \frac{\delta-\delta_1}{U_e}\frac{\partial U_e}{\partial X}$ $\quad B = \frac{C_f}{2} - \theta\frac{H+2}{U_e}\frac{\partial U_e}{\partial X} - \frac{\delta_1}{U_e^2}\frac{\partial U_e}{\partial t}$ $\quad H^{*\prime} = \frac{dH^*}{dH}$

Above equations constitute a first order partial derivative system for δ_1 and θ, the nature of which is investigated by means of its characteristic equation. Let $\lambda = dX/U_e dt$ be the reduced characteristic direction. Then, λ satisfies the following equation :

$$\lambda^2(H^* - HH^{*\prime}) + \lambda(1 - H^* + H^{*\prime}(1+H)) - H^{*\prime} = 0 \tag{17}$$

The roots of which are :

$$\lambda_1=[1+(H-1)(\beta/(1+\beta))^{1/2}]/H$$
$$\lambda_2=[1\ (H-1)(\beta/(1+\beta))^{1/2}]/H$$

In the domain $H > 1$, the following conclusions can be drawn:

- Since there are always two distinct real characteristic directions, the system is hyperbolic.
- One of the direction is such that $0 < \lambda_1 < 1$.
- The second direction λ_2 is positive for $H < H_c$, and it is negative for $H > H_c$. The point $H = H_c \simeq 2.6$ is defined by $H^{*'} = 0$, very close to the point where the skin friction vanishes.

In steady flow, the point $H = H_c$ (i.e. $H^{*'} = 0$) raises a difficulty. In effect, the steady flow equations are :

$$H^{*'}\frac{d\delta_1}{dX} + (H^* - HH^{*'})\frac{d\theta}{dX} = C_{Es} - \frac{\delta-\delta_1}{U_e}\frac{dU_e}{dX} \tag{18}$$

$$\frac{d\theta}{dX} = \frac{C_f}{2} - \frac{H+2}{U_e}\frac{dU_e}{dX} \tag{19}$$

In general, $\frac{d\delta_1}{dX}$ and $\frac{d\theta}{dX}$ can be calculated from these equations, except when $H^{*'} = 0$, which leads generally the derivative $\frac{d\delta_1}{dX}$ to be infinite.

In contrast, the unsteady equations do not exhibit, in general, such a behaviour when $H^{*'} = 0$. In the unsteady case, the point $H = H_c$ is associated with $\lambda_2 = 0$, which does not imply singularities. The negative values of λ_2 for $H > H_c$ are simply the result of the influence of the downstream region on the upstream region when there is a reverse flow.Thus, the point $H = H_c$ is a limit between upstream influence and downstream influence.

These properties of λ_2 have influence on the boundary conditions.Let us consider, for example, a calculation domain (X_o , X_1) such that in the upstream region the velocity is positive everywhere in the boundary layer (two positive characteristic directions), and that in the downstream region there is a reverse flow (one positive and one negative direction). Considering that, to obtain a well posed problem, it is necessary to prescribe as many boundary conditions as entering characteristic directions , it is seen that it is necessary to add to the two upstream boundary conditions (which set for example $\delta_1(X_0,t)$ and $\theta(X_0,t)$) an additional downstream condition. This data can be constituted by the function $\delta_1(X_0,t)$ or $\theta(X_0,t)$, or by a relation linking these functions.

In unsteady flow, singularities different from those of the steady case, are likely to exist. The system of equations actually has weak solutions involving discontinuity lines. By means of numerical examples, we were able to show cases in which the λ_2-characteristic lines accumulate and even focus to form a line in the plane XOt ,across which the displacement thickness is discontinuous / 8 /.

Physical meaning for this phenomenon should not be looked for, no more than in the case of the singular solutions in steady flow. We feel that such anomalies are attributable to the classic boundary layer approach associated with a theory,whereby the coupling with the external flow is a weak coupling, while an approach involving a strong coupling is most often required by the physics of the phenomena. These techniques involve the use of the so-called inverse mode for calculating the boundary layer which enables one to avoid the singularities.

III.2 - Inverse mode

In the inverse mode, the pressure distribution is not a data of the boundary layer equations, but becomes an unknown and is a result of the solution. These data are replaced by the prescription of the evolution of another parameter. Several inverse procedures can be developed, depending on the choice of this parameter. In fact, this choice depends on the aim for which the inverse method is designed.

Two procedures have been examined with the objective to include the method in a coupling technique, which involves simultaneous calculation of the viscous and inviscid flow. In both cases, the prescribed quantity is a condition applied to the direction of the flow on the boundary line of the inviscid flow.

This flow is extended to within the boundary layer, and its boundary is placed at a distance Y from the wall. Let $\alpha(Y)$ be the angle of inclination of the velocity vector, with respect to the wall in the inviscid fluid at a distance Y from the wall. The coupling condition between the boundary layer and the external flow allows $\alpha(Y)$ to be related to boundary layer characteristics, which results in an equation for α depending on the choice of Y. The two possibilities which have been examined are : 1) $Y = 0$, 2) $Y = \delta_1$. Accordingly, the coupling equations are respectively :

$$\alpha_{(Y=0)} = \frac{\partial \delta_1}{\partial X} + \frac{\delta_1}{U_e}\frac{\partial U_e}{\partial X} \tag{20}$$

$$\alpha_{(Y=\delta_1)} = \frac{\partial \delta_1}{\partial X} \tag{21}$$

The corresponding inverse boundary layer calculations can be used either independently, or in combination with an inviscid flow calculation. In the second case, the quantity α is a result of the solution while in the first, α is a prescribed data which is determined from boundary layer measurements.

The properties of the two inverse methods have been analyzed by considering them as independent boundary layer methods. Thus, for example, if $\alpha(Y = 0)$ is prescribed, the set of equations examined is :

$$(H^* - HH^{*'})\frac{\partial \theta}{\partial X} + \frac{(H^* - HH^{*'})}{U_e}\frac{\partial \theta}{\partial t} + H^{*'}\frac{\partial \delta_1}{\partial X} + \frac{(1 + H^{*'})}{U_e}\frac{\partial \delta_1}{\partial t} + \frac{H^*\theta}{U_e}\frac{\partial U_e}{\partial X} = C_{Es} \tag{22}$$

$$\frac{\partial \theta}{\partial X} + \frac{1}{U_e}\frac{\partial \delta_1}{\partial t} + \theta\frac{H+2}{U_e}\frac{\partial U_e}{\partial X} + \frac{\delta_1}{U_e^2}\frac{\partial U_e}{\partial t} = \frac{C_f}{2} \tag{23}$$

$$\frac{\partial \delta_1}{\partial X} + \frac{\delta_1}{U_e}\frac{\partial U_e}{\partial X} = \alpha_{(Y=0)} \tag{24}$$

in which the simplified relationship $H^*(H)$ has been used to write the entrainment and global momentum equations.

The above system has three characteristic directions. Two of them are identical and parallel to the X-axis. The third direction is given by :

$$\lambda = \frac{H+1}{2H}$$

which is always positive for $H > 1$. Moreover, it has been demonstrated / 8 / that discontinuous singular solutions are impossible.

If $\alpha_{(Y=\delta_1)} = \frac{\partial \delta_1}{\partial X}$ is prescribed, or equivalently if δ_1 is prescribed, the characteristic roots of the resulting set of equations are real distinct and always positive for $H > 1$, even in regions with reverse flow. Thus, whatever be the state of the boundary layer in a given domain of computation, the boundary conditions to be prescribed are only the distributions of the main unknowns along the upstream boundary, i.e. $\theta\,(X_0\,,t)$ and $U_e(X_0,t)$. However, if reverse flow exists at the downstream end of the domain, the influence of the downstream region on the upstream region is not ignored, because it is felt through the prescription of δ_1.

This situation is identical to that observed in steady flow when the partial differential equations are solved in the inverse mode, by neglecting the longitudinal convective term in the reverse flow zones.

In the inverse mode for which δ_1 is prescribed, the impossibility of discontinuous singular solution is not guaranteed. However, singularities are most improbable because the variation of the slopes of the characteristic directions as a function of H, is quite small.

III.3. - Application - Comparison with experimental results

An inverse boundary layer calculation has been implemented and applied to our own experiments, involving an adverse pressure gradient. As discussed above, reverse flow exists periodically in the downstream region. So, the calculation in the direct mode of the whole experimentally investigated field would require the prescription of a downstream boundary condition. Moreover, this would not ensure the absence of a singularity. For these reasons, it seemed to us easier to apply the inverse method.

In this calculation, the distribution of the displacement thickness $\delta_1\,(X\,,t\,)$ is prescribed from the experimental results, while the external velocity and the momentum thickness, or the shape factor, are the results of calculation. They are compared with experimental values in figures 14 and 15. Rather good agreement is observed, even in the region located downstream of S = 435 mm, where flow reversals occur periodically.

Conclusions

The experimental study of the response of a turbulent boundary layer over a flat plate to an oscillating external velocity field has revealed an interesting feature, that the main parameters like δ_1 , θ ... oscillate with Strouhal number. The damping of this oscillation, in the range S = 1.5 to 18, is much smaller than for a laminar flow. Despite this phenomenon, the logarithmic law of the wall remains valid, which allows estimation of the phase shift of the skin friction from the velocity profiles. This phase shift with respect to the external velocity,remains small in the entire domain studied, even at Strouhal number as high as 18. These conclusions have to be confirmed by direct hot wire measurements at the wall.

A comparison of various prediction methods, applied to the case of flat plate with and without mean pressure gradient, shows that the best results are obtained by the integration of local equations closed by transport equations for k , ε and $<u'v'>$. However, integral methods predict correctly the overall response of the boundary layer, and seem very useful for practical applications due to their simplicity and low computation time.

Integral methods are also well adapted to analytically discuss the problem of the occurrence of singularities in the boundary layer equations,solved with an imposed pressure gradient distribution. It is shown that an inverse mode of calculation, in which the pressure distribution is an unknown of the boundary layer problem, avoids such difficulties. This is particularly interesting in practical applications, where a coupling between the inviscid flow field and the viscous boundary layer has to be achieved.

References

1. Houdeville, R. and Cousteix J.: Couches limites turbulentes en écoulement pulsé avec gradient de pression moyen défavorable. La Recherche Aérospatiale 1979-1 Nasa Tr. TM 75799 - N 8017400.

2. Simpson R.L.: Features of unsteady turbulent boundary layers as revealed from experiments AGARD-CP-227

3. Karlsson S.K.F: An unsteady turbulent boundary layer J.F.M. Vol. 5 pp. 622-636. 1959.

4. Cousteix J.; Desopper A. and Houdeville R.: Structure and development of a turbulent boundary layer in an oscillatory external flow. Turbulent shear flows I Springer-Verlag. 1977.

5.Mc Croskey and Yaggy: Laminar boundary layers on helicopter rotors in forward flight AIAA Journal Vol. 6 N° 10 Oct. 1968.

6.Patel M.H.: On turbulent boundary layers in oscillatory flow Proc.R.Soc Lond A 353 121-144 (1977).

7.Simpson R.L.; Strickland J.H. and Barr P.W.: Features of a separating turb ulent boundary layer in the vicinity of separation J.F.M. Vol.79 part 3 1977.

8.Cousteix J.; Le Balleur J.C. and Houdeville R.: Calculation of unsteady turbulent boundary layers in direct or inverse mode, including reverse flow. Analysis of singularities La Recherche Aérospatiale N° 1980-3 in English.

9.Cousteix J. and Houdeville R.: Singularities in three-dimensional turbulent boundary layer To be published in AIAA Journal. 1981.

10.Cousteix J. and Houdeville R.: Turbulent boundary layer calculations in unsteady flow. Numerical Methods in Applied Fluid Dynamics. Academic Press. 1978.

11.Simpson R.L.; Shivaprasad B.G. and Chew Y.T: Some features of unsteady separating turbulent boundary layers. IUTAM Symposium on Unsteady Turbulent Shear Flows Toulouse 5-8 May 1981.

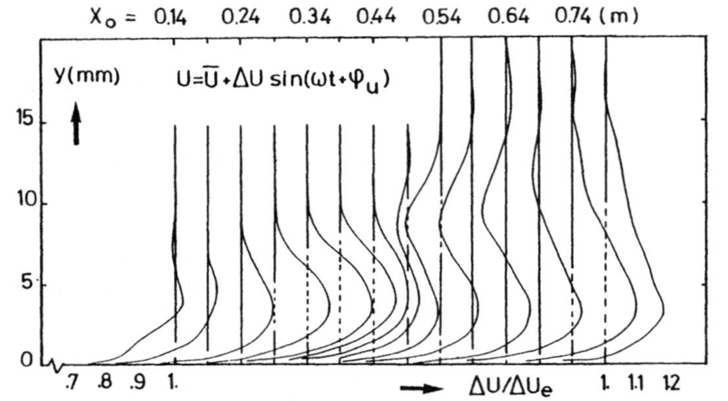

Fig. 1. Harmonic analysis of the velocity - Relative amplitude of fluctuation - f = 38 Hz

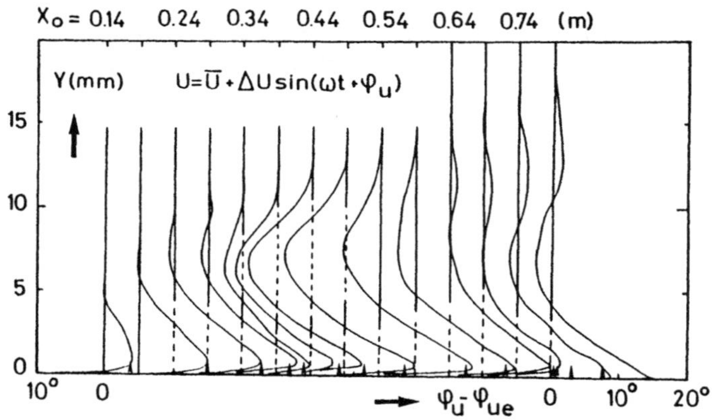

Fig. 2. Harmonic analysis of the velocity - Phase shift - ▲ : $\varphi_{\tau w}$ calculated from the law of the wall - f=38 Hz

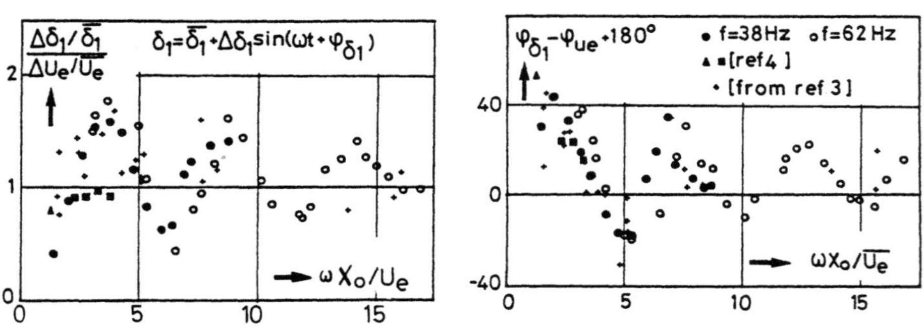

Fig. 3 Harmonic analysis of the displacement thickness versus Strouhal number

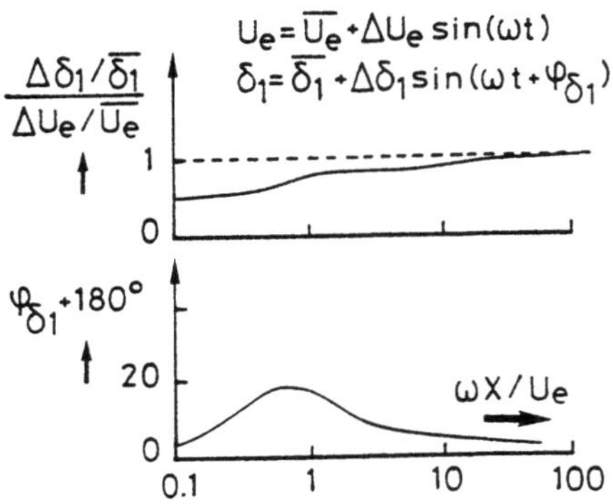

Fig. 4 .Harmonic analysis of the displacement thickness versus Strouhal number - Laminar boundary layer

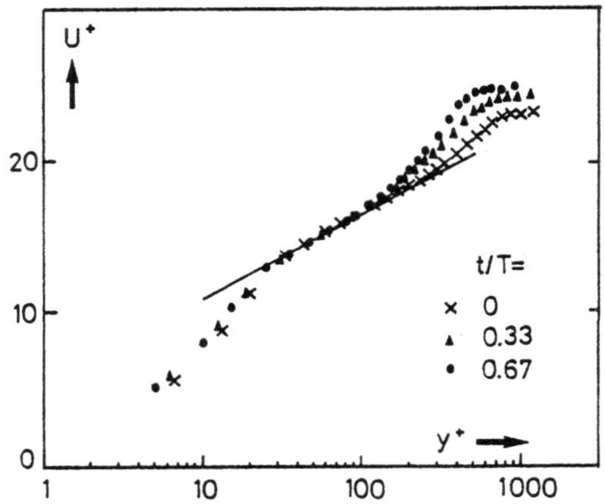

Fig. 5. Velocity profiles at different instants; semi-logarithmic coordinate.
S = 5.9; —— Law of the wall (eq. 4)

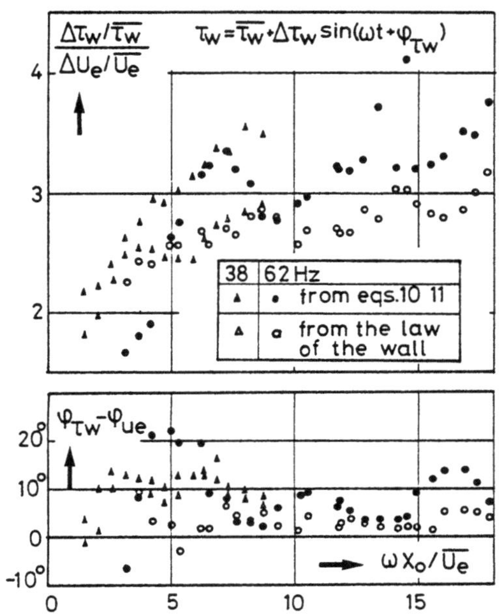

Fig. 6. Harmonic analysis of the wall shear stress versus Strouhal number

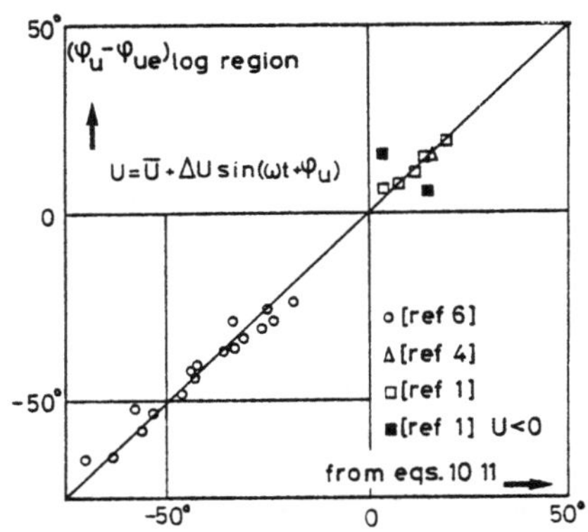

Fig. 7. Comparison between the velocity phase shift in the wall region and a small perturbation estimation of the wall shear stress phase shift

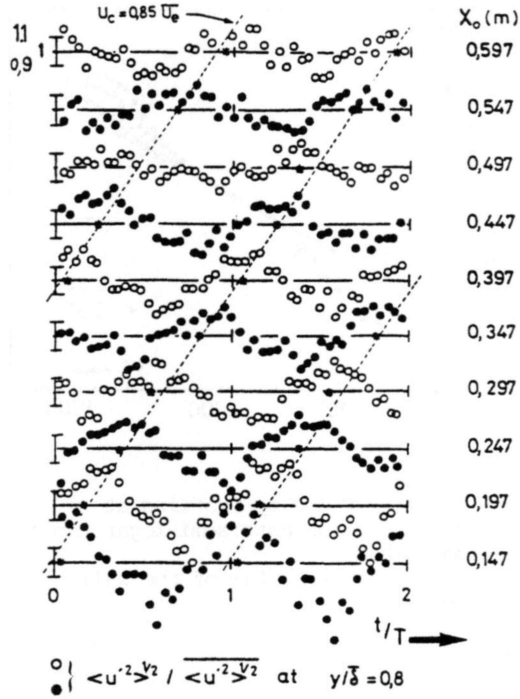

Fig. 8 . Evolution in time of the turbulence intensity at $Y/\delta = 0.8$ for different stations along X.
f = 62 Hz

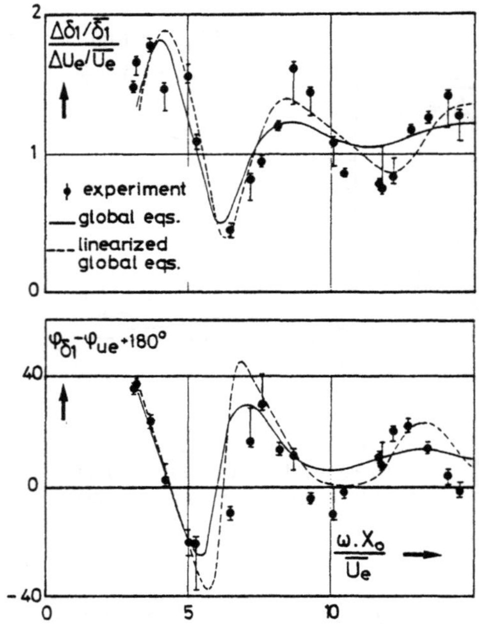

Fig. 10. Harmonic analysis of the calculated displacement thickness. Comparison with experiment. f = 62 Hz

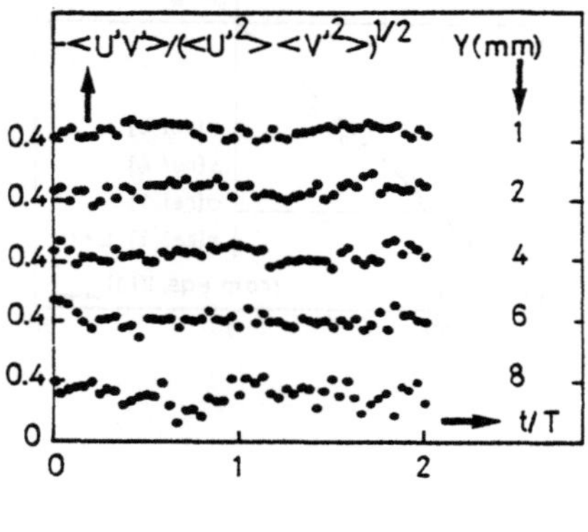

Fig. 9. Evolution in time of the correlation coefficient at different distances from the wall.
S = 5

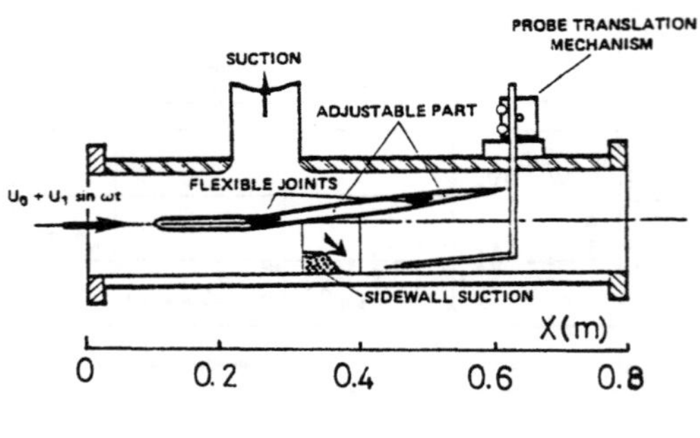

Fig. 11. Experiment with adverse pressure gradient. Schematic view of the test section

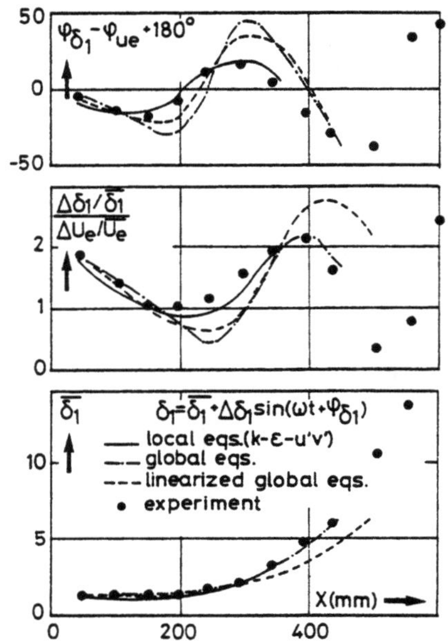

Fig. 12. Adverse pressure gradient. Harmonic analysis of the calculated displacement thickness. Comparison with experiment

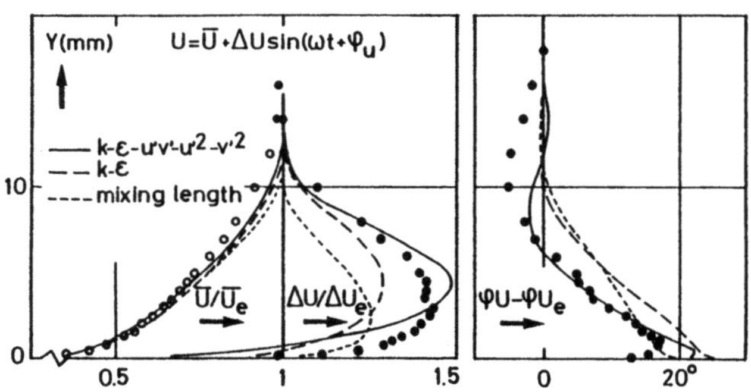

Fig. 13. Adverse pressure gradient. Harmonic analysis of the calculated velocity. Comparison with experiment.
X = 390 mm.

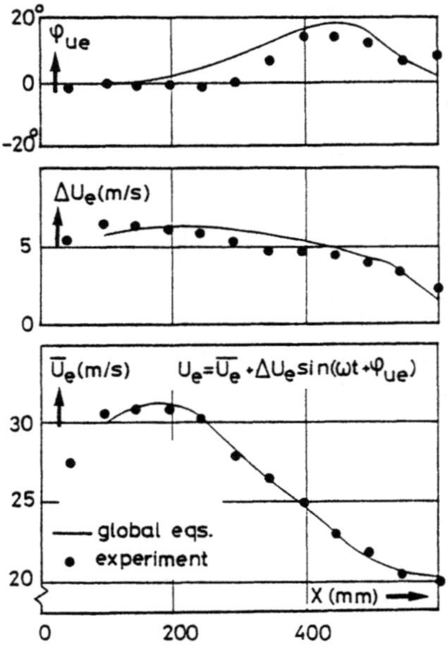

Fig. 14. Adverse pressure gradient. Inverse mode calculation of the external velocity. Comparison with experiment

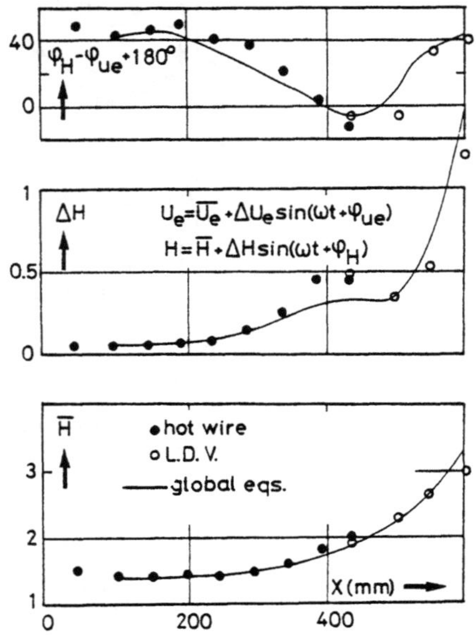

Fig. 15. Adverse pressure gradient. Inverse mode calculation of the shape parameter. Comparison with experiment

Prediction of Boundary-Layer Characteristics of an Oscillating airfoil

Tuncer Cebeci and Lawrence W. Carr

Mechanical Engineering Dept.
California State University
Long Beach, California

U.S. Army Aeromechanics Lab.
NASA Ames Research Center
Moffett Field, California

Abstract

The evolution of unsteady boundary layers on oscillating airfoils is investigated by solving the governing equations by the Characteristic Box scheme. The difficulties associated with computing the first profile on a given time-line, and the velocity profiles with partial flow reversal are solved. A sample calculation has been performed for an external velocity distribution typical of those found near the leading edge of thin airfoils. The results demonstrate the viability of the calculation procedure.

Introduction

The effect of unsteady motion of an airfoil on its stall behavior is a problem of great importance in many types of fluid motion, for example, helicopter rotor blades and jet engine compressors. A recent detailed study has been published by Carr, McAlister and McCroskey(1); this study reveals a very complicated phenomenon which depends in a subtle way on a large number of parameters. There is one important characteristic of the data they have compiled which serves as a focal point of the approach which we are planning to apply sometime later in our studies; namely that at some stage of the cycle, a large vortex is formed near the surface of the airfoil and very shortly afterwards stall occurs. It seems also that the stall is associated with flow reversals in the unsteady boundary layer which may spread downstream or upstream depending on the leading-edge radius of the airfoil.

The present paper concerns itself with the calculation of boundary-layer characteristics of an oscillating airfoil in order to investigate the evolution of unsteady boundary layers on such airfoils. It is one phase of a study which will be extended later in the hope of throwing light on the dynamic stall problem. In our present study we focus our attention on the calculation of time-dependent boundary layers for a given pressure distribution.

There are three difficulties associated with unsteady boundary layers that requires careful attention. First of all appropriate initial values at $t=0$ must be chosen for the velocity distribution. Strictly speaking they can be arbitrary but in that event, the values of $\partial u/\partial t$ at $t = 0$ is non-zero and this implies an inviscid acceleration of the fluid in the boundary layer and in consequence a velocity of slip begins to grow at the wall. This is smoothed out by an inner boundary layer initially of thickness $(\nu t)^{1/2}$ in which viscous forces are of importance. Thus a double structure develops in the boundary layer which may be treated by a generalization of the Keller-box scheme(2). However, if interest is centered on the solution at large times, this feature may be reduced in importance by requiring that the initial velocity distribution satisfies the steady-state equation with the instantaneous external velocity. In addition it is necessary to smooth out the external velocity $u_e(x,t)$ so that $\partial u_e/\partial t = 0$ at $t = 0$ and then standard methods may be used and are stable. The use of the smoothing function makes for some loss of accuracy at small values of t but the error soon decays to zero once the required value of u_e is specified. In the present problem the choice of parameters in the specified external velocity distribution is such that the smoothing function is actually unnecessary but it can easily be incorporated into the scheme.

The second difficulty arises when u changes sign over part of the profile at some x-station where the x-axis is parallel to the streamwise direction and u is the corresponding velocity component. Normally this does not occur and one can integrate away from the profile in the direction of positive u without any difficulty by using a standard numerical scheme. However, if the change in sign of u does occur, we encounter numerical instabilities since in the negative u-region we would be integrating against the stream. The instability can be avoided by changing the scheme either to the zig-zag box or the characteristics box. These new schemes have already been shown to be effective in such circumstances when the flow is unsteady(3) and in three-dimensional flows(4-6). The essence of these schemes is that, to an increasing extent, they take into account the fact that small disturbances are carried along with the local fluid velocity.

The third difficulty arises when it is desired to compute the first velocity profile at the new time line. Given, as we are, the complete velocity profile distribution on the previous time line, there is in principle no difficulty in computing values on the next time-line by an explicit method, but if we wish to avoid the stability problems associated with such a method by using an implicit method, we are immediately faced with the problem of generating a starting profile on the new time-line.

In order to explain the problem further, it is instructive to see what happens to the stagnation point as a function of time. For this purpose let us assume that the external velocity distribution for an oscillating airfoil is given by the following function,

$$u_e = \frac{x + \xi_0 (1 + A \sin \omega t)}{(1 + x^2)^{\frac{1}{2}}} \tag{1}$$

where A and ξ_0 denote parameters that need to be specified. This equation is a good approximation to the external velocity distribution near the leading edge of a class of thin airfoils at variable angles of attack and, when $A = 0$, has recently been used by Cebeci, Stewartson, and Williams[7] to study leading-edge separation in steady flow.

Since by definition $u_e = 0$ at the stagnation point, its location, x_s, is given by

$$x_s = -\xi_0 (1 + A \sin \omega t) \tag{2}$$

and so the upper and lower surfaces of the airfoil as functions of time are defined in particular by $x > x_s$ and $x < x_s$. For example, let us take $A = 1$, $\omega = \pi/4$ and plot x_s/ξ_0 in the (t,x) plane, as shown in Figure 1 for one cycle $(0 \leq t \leq 8)$. When $t = 2$, the stagnation point x_s is at $-2\xi_0$, when $t = 6$ it is at 0, etc. If x_s were fixed we could assume that $u = 0$ at $x = x_s$ for all time and all y. Further profiles at this time-line then follow by use of one of the box schemes that have been developed. However x_s is not fixed and it is clearly unjustified to assume a priori that $u = 0$ there. Instead therefore we use the characteristic box, with an

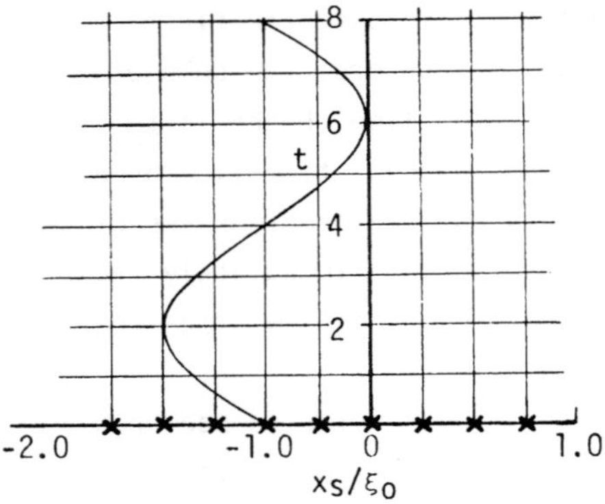

Fig. 1. Variation of stagnation point with time for one cycle according to Eq. (2), with $\omega = \pi/4$.

extrapolated normal velocity, to complete the profile at $x = x_1$, the nearest x station to x_s on the new time-line. Thus we avoid the need to use two x-stations at the new time-line to dimensionalize the governing equations. Once this step is completed, the determination of further profiles on the new time line proceeds normally and moreover we can improve our initial estimate of the normal velocity at $x = x_1$ by iteration.

Governing Equations

The governing boundary-layer equations for an incompressible laminar or turbulent flow past an oscillating airfoil are well known and, with the eddy viscosity (ε_m) concept, they can be written as

$$\frac{\partial u}{\partial x} + \frac{\partial v}{\partial y} = 0 \tag{3}$$

$$\frac{\partial u}{\partial t} + u\frac{\partial u}{\partial x} + v\frac{\partial u}{\partial y} = \frac{\partial u_e}{\partial t} + u_e\frac{\partial u_e}{\partial x} + \frac{\partial \tau}{\partial y} \tag{4}$$

where $\tau = \nu(\partial u/\partial y) - \overline{u'v'}$. In the absence of mass transfer, these equations are subject to the boundary conditions given by

$$y = 0, \quad u = v = 0; \quad y \to \delta, \quad u \to u_e(x,t) \tag{5}$$

The presence of the Reynolds shear stress term, $-\overline{u'v'}$, requires a closure assumption; in our study we use the algebraic eddy-viscosity formulation developed by Cebeci and Smith. For details, see ref. 8.

To complete the formulation of the problem, initial conditions must be specified in the (t,y) plane at some $x = x_0$ either on the lower or upper surface of the airfoil (see Fig. 1) as well as initial conditions in the (x,y) plane on both surfaces of the airfoil. In the latter case, if we assume that steady-flow conditions prevail at $t = 0$, then the initial conditions in the (x,y)-plane can easily be generated for both surfaces by solving the governing equations for steady flow, which in this case, are given by Eq. (3) and by

$$u \frac{\partial u}{\partial x} + v \frac{\partial u}{\partial y} = u_e \frac{du_e}{dx} + \frac{\partial}{\partial y} (b \frac{\partial u}{\partial y}) \tag{6}$$

where $b = \nu + \varepsilon_m$. There is no problem with the initial conditions for Eqs. (1) and (6) since the calculations start at the stagnation point, $x = x_s$.

Generation of the initial conditions in the (t,y) plane at $x = x_0$, which is one of the purposes of our study, is not so easy as was discussed in the previous section. The following section describes our solution procedure at $t = 0$ and $t > 0$.

Solution Procedure

We use Keller's Box method to solve the governing equations of the previous section. This is a two-point finite-difference method which has been used to solve a wide range of parabolic partial-differential equations as discussed by Bradshaw, Cebeci and Whitelaw[9]. The solution procedure for $t = 0$ (Eqs. (3) and (6)) and $t > 0$ (Eqs. (3) and (4)) are described separately below.

Solution Procedure for $t = 0$

As explained in the introduction any velocity profile may be chosen at $t = 0$ to initiate the computation but it is convenient to select one which obviates the need for double-structured numerical schemes and joins smoothly on to the solution for $t > 0$. We insist that these profiles satisfy the steady state equations with u_e given by (1) and $t = 0$. The details of the procedure for computing these profiles differs slightly from previous procedures used in steady two-dimensional flows where we have used the definition of the stream function and reduced Eqs. (3) and (6) to two first-order ordinary differential equations and to one partial-differential equation. Here we consider the solution of Eqs. (3) and (6) without the use of the stream function. For this purpose, with primes denoting differentiation with respect to y, we let

$$u' = f \tag{7a}$$

and write Eqs. (3) and (6) as

$$v' = -\frac{\partial u}{\partial x} \tag{7b}$$

$$(bf)' - fv = -\frac{d}{dx}\left(\frac{u_e^2}{2}\right) + \frac{\partial}{\partial x}\left(\frac{u^2}{2}\right) \tag{7c}$$

The finite-difference approximations of Eqs. (7) are also somewhat different than those reported in our previous studies dealing with two-dimensional flows[10]. All quantities except for the normal velocity component v, are centered at the center of the box $(y_{j-1/2}, x_{i-1/2})$, see Fig. 2, by taking the values of each parameter, say q, at the four corners of the Box, that is,

$$q_{j-1/2}^{i-1/2} = \frac{1}{2}(q_j^{i-1/2} + q_{j-1}^{i-1/2}) = \frac{1}{4}(q_j^i + q_j^{i-1} + q_{j-1}^i + q_{j-1}^{i-1}) \tag{8a}$$

However, the centering of the y-velocity component v is done by writing it as

$$v_{j-1/2}^{i-1/2} = \frac{1}{2}(v_j^{i-1/2} + v_{j-1}^{i-1/2}) \tag{8b}$$

The unknown parameters in Eqs. (8) correspond to q_j^i and $v_j^{i-1/2}$ so that when a solution of the system given by Eqs. (7) is obtained, f and u are computed at (i,j) and v at $(i-1/2, j)$. This modified centering procedure is necessary in order to avoid oscillations due to the use of the continuity equation in the form given by Eq. (7b) rather than the use of the stream function. The centering of Eqs. (7) and the subsequent linearization procedure by Newton's method allows the resulting linear system to be written in the form[10]

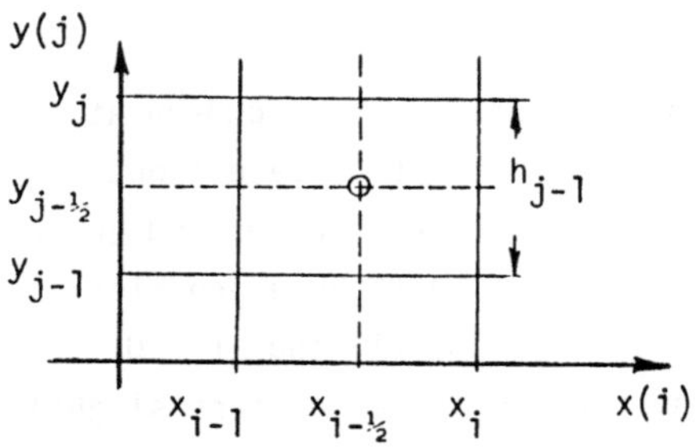

Fig. 2. Net rectangle for finite-difference approximations.

$$\delta u_j - \delta u_{j-1} - \frac{h_{j-1}}{2}(\delta f_j + \delta f_{j-1}) = (r_3)_{j-1} \tag{9a}$$

$$\frac{1}{h_{j-1}}(\delta v_j - \delta v_{j-1}) + (s_7)_j\,(\delta u_j + \delta u_{j-1}) = (r_1)_j \tag{9b}$$

$$(s_1)_j\delta f_j + (s_2)_j\delta f_{j-1} + (s_3)_j\delta v_j + (s_4)_j\delta v_{j-1} + (s_5)_j\delta u_j + (s_6)_j\delta u_{j-1} = (r_2)_j \tag{9c}$$

After linearization, the wall boundary conditions

$$u_o = 0, \qquad v_o = 0 \tag{10a}$$

and the edge boundary condition

$$u_J = u_e \tag{10b}$$

become

$$\delta u_o = 0, \qquad \delta v_o = 0, \qquad \delta u_J = 0 \tag{11}$$

The resulting linear system consisting of those given in Eqs. (9) and by those in Eq. (11) can be solved by the block elimination method discussed by Cebeci and Bradshaw[10].

Solution Procedure for $t > 0$

We have already pointed out that if $\partial u_e/\partial t \neq 0$ when $t = 0$, a double structure scheme should strictly be used to advance the solution from $t = 0$. However the choice of parameters in our study is such that $\partial u_e/\partial t$ is small and the difficulties that arise from using a standard method are of a sufficiently minor nature that no further refinement is necessary. For larger values of the relevant parameters it is easy to incorporate a smoothing function into u_e and one can always use the general method[2].

Nevertheless there is still the difficulty about obtaining the velocity profile on the first x-station at any new time-line. It can be resolved with the use of the characteristic box method developed by Cebeci and Stewartson[14]. Defining the streamline by

$$\frac{dt}{1} = \frac{dx}{u} \tag{12}$$

and using the definition of f', and with s denoting the local streamline, we write Eq. (4) as

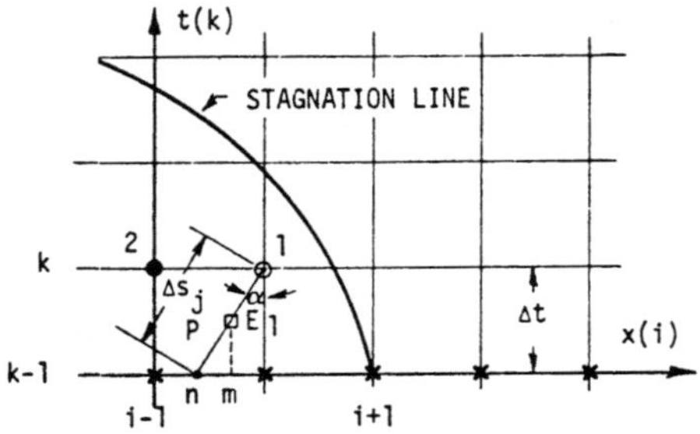

Fig. 3. Notation and finite-difference molecule for the Characteristic Box 2.

$$(bf') - fv + \sqrt{1 + u_e^2}\,\frac{\partial u_e}{\partial s} = \sqrt{1 + u^2}\,\frac{\partial u}{\partial s} \tag{13}$$

To obtain the solution of the unsteady boundary-layer equations given by Eqs. (7a,b) and (13) at the first x-station on either side of the stagnation line, let us consider the grid of Fig. 1 and direct our attention to the point denoted by 1 (see Fig. 3).

To write the difference approximations of Eq. (13) we define

$$\Delta s_j = \Delta t/\cos \alpha_j \tag{14}$$

where, with u_j denoting an average velocity, we compute α_j from

$$\alpha_j = \tan^{-1} u_j \tag{15}$$

assuming that at first v at point P is known and is equal to its value at $v^{i-1/2,k-1}$. This assumption decouples the continuity equation, Eq. (7b) from Eqs. (7a) and (13) and reduces the problem to a "two-dimensional" one with f and u being the only unknowns. The finite-difference approximations of Eq. (7a) are written in the usual way and the finite-difference approximations to Eq. (13) are written by centering it at point P. This procedure leads to

$$h_{j-1}^{-1}(f_j^{i,k} - f_{j-1}^{i,k}) - u_{j-1/2}^{i,k} = 0 \tag{16a}$$

$$\frac{(bv)_j^{i,k} - (bv)_{j-1}^{i,k}}{2h_{j-1}} + \frac{(bv)_j^{n,k-1} - (bv)_{j-1}^{n,k-1}}{2h_{j-1}} - \frac{1}{2}[f_{j-1/2}^{i,k} + f_{j-1/2}^{n,k-1}][v_{j-1/2}^{P}]$$

$$+ \frac{1}{2}[(\sqrt{1 + u_e^2})^{i,k} + (\sqrt{1 + u_e^2})^{n,k-1}]\left(\frac{u_e^{i,k} - u_e^{n,k-1}}{\Delta s_j}\right)$$

$$= \frac{1}{2}[(\sqrt{1 + u^2})_{j-1/2}^{i,k} + (\sqrt{1 + u^2})_{j-1/2}^{n,k-1}]\,\frac{u_{j-1/2}^{i,k} - u_{j-1/2}^{n,k-1}}{\Delta s_j} \tag{16b}$$

The profiles b_j, f_j and u_j as well as u_e at $(n,k-1)$ are obtained by interpolating the profiles at $(i,k-1)$ and $(i-1,k-1)$. To find the angle α_j, we define u_j in Eq. (15) by

$$u_j = \frac{1}{2}(u_j^{i,k} + u_j^{n,k-1}) \tag{17}$$

Since the system given by Eqs. (16) is linear, there is no need for linearization and we solve it subject to the two boundary conditions, namely,

$$u_o = 0, \qquad u_J = u_e \tag{18}$$

by using the block-elimination method in which case the matrices are 2 x 2. We shall refer to this scheme as Characteristic Box 2.

Once a solution of Eqs. (7a) and (13) has been obtained, we compute v from Eq. (7b) which, in finite-difference form, for the center of the net rectangle, point E, can be written as (see Fig. 3)

$$\frac{V_j - V_{j-1}}{h_{j-1}} = -\frac{u_j^{i,k-1/2} - u_j}{x_i - x_m} \tag{19}$$

Here V_j denotes the value of v_j at E and u_j is given by Eq. (17). Since the right-hand side of Eq. (19) is known, we can solve this equation for V_j and with $V_o = 0$, find V_j for $1 \leq j \leq J$. We then substitute this new value of V_j into Eq. (16) for v^P and solve the system again to compute new values of V_j. This procedure is repeated until convergence.

For convenience we use the same procedure to compute point 2 to the left of point 1. Once two points on a given t-line are computed by this procedure, we then use the values of V_j at E_2 and E_1, compute a new value v^P and repeat the solution procedure for Eqs. (7a) and (13), and later Eq. (19). After that the stations to the left of point 2 and the stations to the right of point 1 are computed by using the Regular Box scheme if there is no flow reversal across the layer and by the Characteristic Box scheme if there is flow reversal. The "new" Characteristic Box scheme is now slightly different than the Characteristic Box 2 so we shall refer to it as Characteristic Box 3.

To describe the Characteristic Box 3 scheme which solves Eqs. (7a,b) and (13) without decoupling the continuity equation from Eqs. (7a) and (13), we consider the sketch shown in Figure 4.

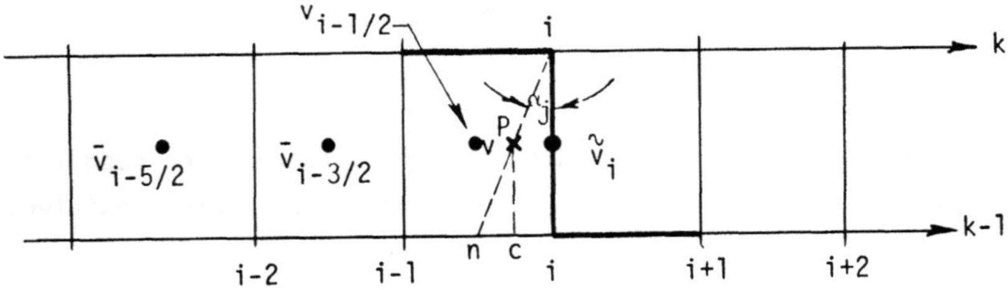

Fig. 4. Notation for Characteristic Box 3.

Using the Zig-Zag Box scheme discussed in detail in Ref. 5, we write Eq. (7b) in the following finite-difference form

$$h_{j-1}^{-1}\,(\tilde{v}_j^i - \tilde{v}_{j-1}^i) + \tilde{\theta} u_{j-1/2}^{i,k} = -\beta_1 + \tilde{\theta} u_{j-1/2}^{i-1,k} \tag{20}$$

where

$$\tilde{\theta} = \frac{(x_{i+1} - x_i)}{(x_{i+1} - x_{i-1})}\,\frac{1}{(x_i - x_{i-1})},$$

$$\beta_1 = \frac{x_i - x_{i-1}}{x_{i+1} - x_{i-1}}\,\frac{1}{(x_i - x_{i+1})}\,(u_{j-1/2}^{i,k-1} - u_{j-1/2}^{i+1,k-1}) \tag{21}$$

Since

$$\bar{v}_{i-1/2} = \bar{v}_{i-3/2} + \frac{(x_{i-1/2} - x_{i-3/2})}{(x_{i-3/2} - x_{i-5/2})}\,(\bar{v}_{i-3/2} - \bar{v}_{i-5/2}),$$

The relation between $\bar{v}_{i-1/2}$ and $\tilde{v}_i$ can be written as

$$\tilde{v}_i = (1 + \beta_2)\bar{v}_{i-1/2} + \beta_3 \tag{22}$$

where β_2 and β_3 are given by

$$\beta_2 = -\frac{(x_i - x_{i-1/2})}{(x_{i-3/2} - x_{i-1/2})}, \qquad \beta_3 = -\beta_2\bar{v}_{i-3/2} \tag{23}$$

Introducing Eq. (22) into Eq. (20) and rearranging, we get

$$h_{j-1}^{-1}\,(1 + \beta_2)(\bar{v}_j - \bar{v}_{j-1}) + \tilde{\theta} u_{j-1/2} = \beta_4 \tag{24}$$

where

$$\beta_4 = -\beta_1 + \tilde{\theta} u_{j-1/2}^{i-1,k} + h_{j-1}^{-1}\,(\beta_3^{j-1} - \beta_3^{i}) \tag{25}$$

As in Characteristic Box 2, we center Eq. (13) at the midpoint of (i,k) and (i,n) to get the finite-difference equations given by Eq. (16b) with v^P being obtained by linear interpolation of $\bar{v}_{i-1/2}$ and $\tilde{v}_i$, which is

$$v^P = \tilde{v}_i + (x_c - x_i)\,\frac{v_{i-1/2} - \tilde{v}_i}{x_n - x_i} \qquad (26)$$

Equations (16a), (24) and (16b) are then linearized by Newton's method, and again are solved by the block elimination method.

When there is no flow reversal across the layer, we use the Regular Box scheme described in detail in Ref. 5.

Results and Discussion

To date calculations have been carried out in only one test-case, namely when $\xi_0 = 0.10$, $A = 1$, $\omega = \pi/4$, and for a limited range of x ($|x| < 0.3$). With the use of the various procedures described in this paper the calculations were quite straightforward and the formal validity and efficacy of the numerical schemes were established. The results are summarized in Figs. 5-7. In Fig. 5 we display the variation of wall-shear with time at different x-stations and in Fig. 6 the variation with x at different times. These graphs are entirely in line with expectations and we note that the flow reversal at the wall is quite smooth. A similar remark applies to the velocity profiles on either side of the stagnation line displayed in Fig. 7.

The next phase in our studies is to extend the computations to larger values of A, ξ_0 and smaller values of ω so as to more closely approach the conditions of dynamic stall. It is of interest to comment on the fluid mechanical problems that may then arise. First, if ξ_0 is increased beyond 1.155, the steady-state solution at $t = 0$ separates on the upper side of the airfoil and the calculation terminates. This is not a serious drawback unless the unsteady boundary develops a singularity because the smoothing function mentioned earlier may be adapted to ensure that ξ_0 is initially less than 1.155 and rises to a value greater than that after a finite time. The unsteady boundary layer then includes regions of reversed flow which may well become extensive if A is also allowed to increase to mimic more closely the conditions of dynamic stall. Even if the boundary layer remains smooth, the displacement thickness may then become much thicker and have a significant modifying effect on the external flow. It would be useful then to consider an interactive problem in which the external stream depends in part on the displacement thickness thus generalizing the studies reported for steady flow[7]. Consideration then has to be given to the variation of circulation with time which may lead to a more complicated expression for the dependence of u_e on the displacement thickness than was used in Ref. 7 but the computation should not be any more complicated as a result. Finally, in order to mimic the dynamic stall problem most effectively[1,11], ω should be reduced to very small values (as typical of dynamic stall problem). So long as $\omega > 0$, the difficulties reported in Ref. 3 at separation in uninteracted flows and in post-separation flows otherwise, should not be present. On the other hand, van Dommelen and Shen[12] have provided quite strong evidence that a singularity can occur in an unsteady boundary layer for which the external velocity is steady. This phenomenon is still somewhat controversial[13] but there seems no doubt that the boundary layer will exhibit dramatic properties for small enough values of ω and it is possible that these may give further insight into dynamic stall.

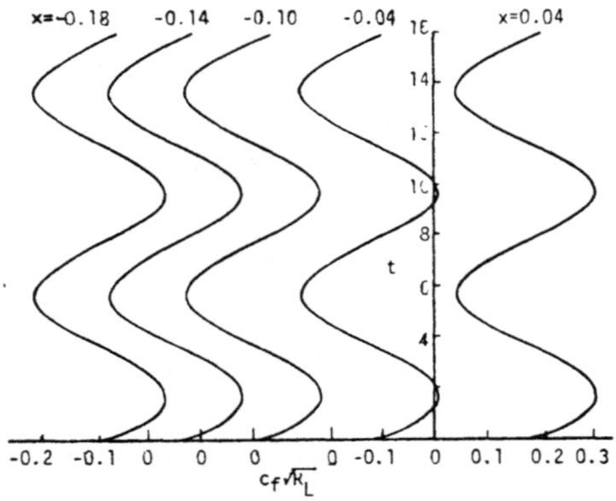

Fig. 5. Variation of wall shear parameters with time at different x-stations.

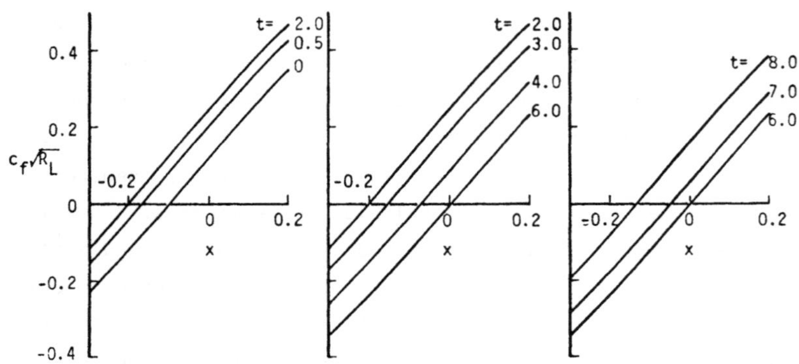

Fig. 6. Variation of skin-friction coefficient with x at different t-intervals. Note $t = 8$ corresponds to one cycle.

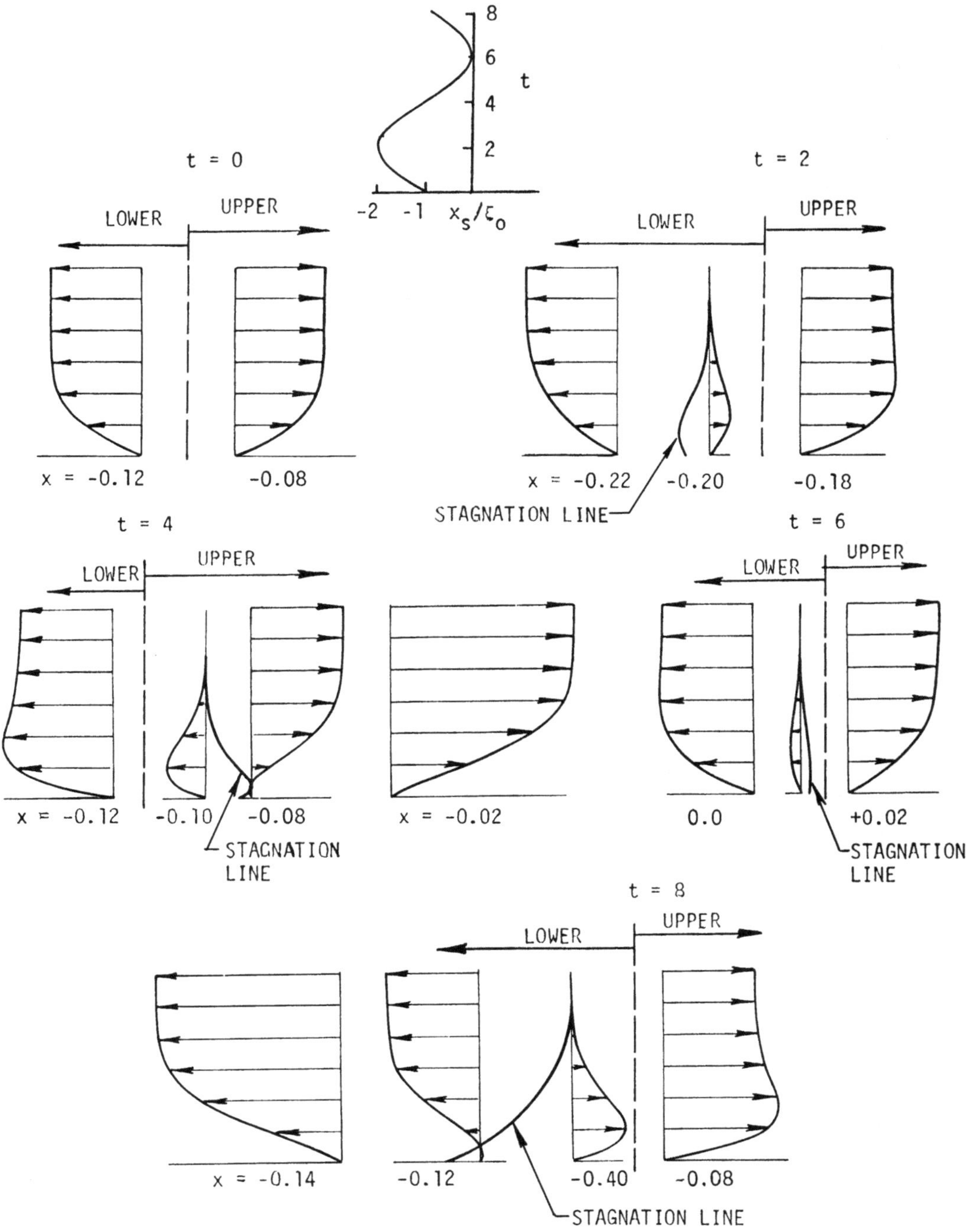

Fig. 7. Velocity profiles in the immediate neighborhood of the stagnation lines at different times. $u_e = 0$ on the dashed line at the specified time and serves to "define" upper or lower surfaces.

References

1. Carr, L.W.; McAlister; K.W., and McCroskey, W.J.: Analysis of the development of dynamic stall based on oscillating airfoil experiments. NASA TN D-8382 (1977).

2. Cebeci, T.; Thiele, F.; Williams, P.G. and Stewartson, K.: On the calculation of symmetric wakes. I. Two-dimensional flows. Num. Heat Trans. 2, (1979) 35-60.

3. Cebeci, T.: The laminar boundary layer on a circular cylinder started impulsively from rest. J. of Comp. Phys., 3, No. 2, (1979).

4. Cebeci, T. and Stewartson, K.: Unpublished work (1978).

5. Cebeci, T. and Carr, L.W.: Computation of unsteady turbulent boundary layers with flow reversal and evaluation of two separate turbulence models. NASA TM 81259 (1981).

6. Cebeci, T.; Khattab, A.A. and Stewartson, K.: Three-dimensional laminar boundary layers and the ok of accessibility. To appear in J. Fluid Mech. (1981).

7. Cebeci, T.; Stewartson, K. and Williams, P.G.: Separation and reattachment near the leading edge of a thin airfoil at incidence. AGARD Symp. on Computation of Viscous-Inviscid Interacting Flows, Colorado Springs, Colo. (1980).

8. Cebeci, T. and Smith, A.M.O.: Analysis of Turbulent Boundary Layers. Academic Press, N.Y. (1974).

9. Bradshaw, P.; Cebeci, T. and Whitelaw, J.H.: Engineering Calculation Methods for Turbulent Flow. Academic Press, London (1981).

10. Cebeci, T. and Bradshaw, P.: Momentum Transfer in Boundary Layers. Hemisphere/McGraw-Hill, Washington, D.C. (1977).

11. McCroskey, W.J. and Pucci, S.L.: Viscous-inviscid interaction on oscillating airfoils. AIAA Paper No 81-0051, Jan. 1981.

12. VanDommelen, L.L. and Shen, S.F.: The genesis of separation. Proc. of Numerical and Physical Aspects of Aerodynamic Flows, California State University, Long Beach, January 19-21, 1981.

13. Cebeci, T.: Unsteady separation. Proc. of Numerical and Physical Aspects of Aerodynamic Flows, California State University, Long Beach, January 19-21, 1981.

Acknowledgment. This work was supported by NASA Ames under contract NAS2-10799. The authors would like to thank Professor Keith Stewartson for many helpful discussions on this problem and to Janet Chiu for her help in programming.

Unsteady Adverse Pressure Gradient Turbulent Boundary Layers

P. ORLANDI

Istituto di Aerodinamica, Università di Roma, Roma

SUMMARY

A one-equation turbulence model, taking into consideration the viscous and buffer regions and including some dynamics of the turbulence, has been used to solve unsteady adverse pressure gradient boundary layers. A fully developed steady flat-plate turbulent boundary layer has been assumed at the inlet of the unsteady region. The numerical results have been compared with the experimental ones obtained by Reynolds and Parikh [7], showing a very good agreement for the displacement thickness behavior. Discrepancies are obtained for the skin friction at high frequencies.

INTRODUCTION

In a previous study one-equation [1] and two equation [2] turbulence model, which take into account the viscous layer were applied to calculate unsteady flate-plate boundary layers with unsteady free-stream and initial conditions. Both models, were able to predict the qualitative trend, of the displacement thickness phase angle of the Karlsson's [3] experiments interpreted by Mc Croskey [4], that is a change of phase at intermediate reduced frequency. But were not able to predict the, experimentally observed [3], overshoot of the inphase velocity. However a comparison with the experimental results it is difficult, because for such a flow there is a lack of "validated" experimental results. This lack of experiments brought about a large number of numerical results [5] widely differing one from each other, probably due to the different initial conditions, assumed by each modeller. In the round table discussion of the Agard Conference on Unsteady Aerodynamics [6], came out the necessity, by the theoreticians, to have experiments with well defined initial conditions, expecially for unsteady boundary layers with adverse pressure gradient.

W.C. Reynolds and P.G. Parikh at Stanford proposed an experimental study, the results of which are presented at this Symposium [7],where at the inlet of the test section a fully developed steady flat-plate turbulent boundary layer is achieved. This condition is very usefull for tubulence modellers, in order to check if their models are able to simulate unsteady boundary layers. The Stanford's experimental study is carried out with an adverse pressure gradient , it is one of the main characteristics required to understand the behavior of helicopter rotor blades. The unsteadiness of the free-stream velocity should move the separation point, if it occurs, in the streamwise direction increasing the frequency.

Usually it is very difficult, even for the steady case, to obtain measurements of mean quantities in the near wall region , and the measurements of each term of the turbulence energy balance across the entire boundary layer. A greater difficulty is to be expected in unsteady conditions. Thus a numerical simulation by a turbulence model taking into consideration the viscous region, could be a usefull toll to explain those behaviors of unsteady layers, the experimental analysis is not able to investigate.

In this paper the one-equation turbulence model has been used, this model has been prefered to the two-equation model, because, as pointed out in a previous paper [2] , with the K - D model the representation of the viscous and buffer regions turns out to be much more difficult. In fact with the one-equation model was possible to try to model the pressure work term [8], this term play a very important role in the near wall region. Instead using the two-equation model the introduction of the pressure work term brings about a larger dissipation, which gives a negative turbulent energy at a distance of 1 in wall coordinate dimension. The one-equation model, here used, is satisfactory to evaluate a large number of steady flows, chosen as test cases to evaluate the capability of a model to compute complex turbulent flows. In particular numerical results, which will be presented at the 1980-81 AFOSR-HTTM Stanford Conference on Complex Turbulent Flows, show a very good agreement with the experimental ones, measured in the case of steady adverse pressure gradient boundary layers with and without transpiration at the wall. In this paper it can be seen that the extension of this model to the study of unsteady boundary layers is worth.

PHYSICAL MODEL

Let U,V be the mean velocity components in the steramwise dierction x and in the normal direction y, respectively. Let U_e be the free-stream velocity , function both of time and x, U', V' the fluctuating velocity components , ν the kinematic viscosity. The unsteady boundary layer equations , for two-dimensional incompressible flow are

$$\frac{\partial U}{\partial x} + \frac{\partial V}{\partial y} = 0 \tag{1}$$

$$\frac{\partial U}{\partial t} + U\frac{\partial U}{\partial x} + V\frac{\partial U}{\partial y} = \frac{\partial U_e}{\partial t} + U_e\frac{\partial U_e}{\partial x} + \nu\frac{\partial^2 U}{\partial y^2} - \frac{\partial \overline{U'V'}}{\partial y} \tag{2}$$

It is assumed here, that for unsteady flow,the Reynolds stress $\overline{U'V'}$ is modeled as usually done for steady flow , using the Boussinesq hypothesis in which the tubulent analog of the molecular viscous stress is used

$$\overline{U'V'} = -\nu_T\frac{\partial U}{\partial y} \tag{3}$$

ν_T is the eddy viscosity, related to the velocity scale,the turbulent kinetic energy, and lenght scale of the turbulence. The length scale is related to the dimension of the eddies carrying energy in the boundary layer. In the outer region the eddies are of the dimension of the boundary layer thickness, and in the near wall region they have dimensions proportional to the distance from the wall. An analytical expression which take into consideration these physical aspects can be introduced

$$l = C_o\,\delta\left(1 - (1- y/\delta)^{C_1}\right) \tag{4}$$

A damping factor has been introduced, in the tubulent viscosity expression , to model the wall suppression of the turbulent transport.

$$\nu_T = C_2\,Q^{1/2}\,\ell\left(1 - e^{-C_6\frac{Q^{1/2}y}{\nu}}\right) \tag{5}$$

More details to support this point of view are reported in previous papers [1],[8],[9].

The dynamics of the turbulent kinetic energy, $Q = \overline{U'^2} + \overline{V'^2} + \overline{W'^2}$ is given by a transport equation that under the boundary layer semplification is written as

$$\frac{\partial Q}{\partial t} + U\frac{\partial Q}{\partial x} + V\frac{\partial Q}{\partial y} = -2\,\overline{U'V'}\frac{\partial U}{\partial y} + \nu\frac{\partial^2 Q}{\partial y^2} - \frac{\partial \overline{Q'V'}}{\partial y} - 2\frac{\partial \overline{P'V'}}{\partial y} - 2D \tag{6}$$

advection production viscous diff. turb. diff. press. work. dissipation

where D is the "isotropic dissipation", that has to be modelled as function of the turbulent kinetic energy and of the mixing lenght. The rate of energy dissipation, at high turbulent Reynolds number , does not depend on the viscosity , on the contrary near the wall the viscosity is a dissipative agent for the largest eddies, and the dissipation depends on the viscosity. These conditions, by the dimensional analysis,bring to expression [9]

$$D = C_3 \frac{Q^{3/2}}{\ell} \left(1 + C_5 \frac{\nu}{Q^{1/2} \ell} \right) \qquad (7)$$

The turbulent kinetic energy diffusion has been treated by a gradient-diffusion model , in which the pressure work term has not been included , as is usually done in one and two-equation tubulence models. The pressure work term should play a significant role mostly in the near wall region.
In the just referenced paper [8] are explained, by comparison with experimental results and with numerical results yielded by more sophisticated turbulence models [10] (Large Eddy Simulation), the reasons because it is necessary to model explicitly the pressure work term in the near wall region. By dimensional consideration, this term is modelled as

$$- \frac{\partial \overline{P'V'}}{\partial y} = C_3 \frac{Q^{3/2}}{\ell} e^{-\left(\frac{Q^{1/2} \ell}{C_8 \nu} \right)^2} \qquad (8)$$

The values of the constants in the equation (4)(5) and (6) can be evaluated following the procedure described by Norris and Reynolds [9] and by Orlandi and Reynolds [1] . It consist on take into consideration the experimental data, in wall coordinate,of steady flat-plate turbulent boundary layers at $y^+ = 100$ and in the near wall region, then, by theoretical consideration, evaluate the constants. From the experimental results of adverse pressure gradient turbulent boundary layers, it can be seen that the near wall and logarithmic regions are not affected by the external pressure gradients, thus the values of the constants, obtained for the zero pressure gradient case, have been retained . They are

C_0	C_1	C_2	C_3	C_5	C_6	C_8
0.096	4.16	0.37	0.0425	3.93	0.012	4.5

SOLUTION METHOD

A coordinate trasformation in the normal direction with respect to the boundary layer thickness is worthwhile to solve the system of continuity , momentum and turbulent energy conservation equations. Being the boundary layer thickness unknown at the new time step, the normal coordinate is normalized with respect ro the boundary layer thickness at the old time step. The independent variables have been trasformed as

$$x = \xi(x_1) \; ; \qquad y = \delta(x, t-\Delta t)\, \eta(x_2) \qquad (9)$$

If the governing equation have to be integrated down to the wall, some computational points have to be inside the viscous layer, extending from $y^+ = 0$ till $y^+ \simeq 5$. Since the boundary layer thickness in wall coordinates is of the order of thousands, a finite difference scheme with uniform mesh would require a number of grid points too large. To avoid enormous computational time a further coordinate transformation has been introduced

$$\eta = \eta_\infty \left\{ 1 - \frac{tgh(1 - x_2)c}{tgh(c)} \right\} \qquad (10)$$

where x_2 is the "new" coordinate respect to which governing equations will be discretized. Large values of the trasformation parameter c give more computational points near the wall. In order to have the whole boundary layer inside the computational domain, having introduced in the equation (9) $\delta(x, t-\Delta t)$, a value $\eta_\infty = 1.25$ has been assumed. The coordinate transormation in the streamwise direction has been chosen in such a way to have more computational points at the beginning of the unsteady region.

The transformations (9) and (10) introduced in the momentum and turbulent energy transport equations, yield

$$\frac{\partial U}{\partial t} + AU \frac{\partial U}{\partial x_1} + BV \frac{\partial U}{\partial x_2} + C \frac{\partial U}{\partial x_2} + DU \frac{\partial U}{\partial x_2} = \frac{\partial U_e}{\partial t} + U_e \frac{\partial U_e}{\partial x} + + F \frac{\partial}{\partial x_2} \left(E(1+\nu_T) \frac{\partial U}{\partial x_2}\right) \qquad (11)$$

$$\frac{\partial Q}{\partial t} + AU \frac{\partial U}{\partial x_1} + BV \frac{\partial Q}{\partial x_2} + C \frac{\partial Q}{\partial x_2} + DU \frac{\partial Q}{\partial x_2} = 2 F \nu_T E \left(\frac{\partial U}{\partial x_2}\right)^2 + + F \frac{\partial}{\partial x_2} \left(E(1+\nu_T) \frac{\partial Q}{\partial x_2}\right) - 2D + \text{Press. Work} \qquad (12)$$

where A,B,C,D,E and F are the function of the coordinate transormation.

The continuity equation has been differentiated with respect to the y coordinate for numerical reasons [11]. The coordinate transformation (9) and (10) introduced into the differentiated continuity equation gives

$$G \frac{\partial^2 U}{\partial x_1 \partial x_2} + H\left\{\eta \frac{\partial}{\partial x_2} \left(E \frac{\partial U}{\partial x_2} \right) + \frac{\partial U}{\partial x_2}\right\} + L \frac{\partial}{\partial x_2} \left(E \frac{\partial U}{\partial x_2}\right) = 0 \qquad (13)$$

where G,H and L are functions of the coordinate transformation.

The boundary conditions associated to the system of equation (11), (12) and (13), are

$$\begin{aligned} x_2 &= 0 \qquad U = V = Q = 0 \\ x_2 &= 1 \qquad U = U_e(x,t) , \; E \left(\frac{\partial V}{\partial x_2}\right) = - \frac{\partial U_e}{\partial x} , \; \frac{\partial Q}{\partial x_2} = 0 \end{aligned} \qquad (14)$$

The non linear system of unsteady boundary layer equations, have been solved by an implicit procedure that is worthwhile for two reasons: first to avoid stability conditions on the mesh size in the streamwise direction; second and more important, to have a fast coupling between the velocity field and the turbulent quantities. The system of equations (11), (12) and (13) has been linearized along the streamwise direction. All the details of the method, of the finite difference scheme and of the metodology of solution of the algebraic system of equations can be found in [11] and [1].

INITIAL CONDITIONS

To solve the system of governing equations descibed in the previous paragraph initial conditions must be assigned at $x = x_0$ and $t \geq 0$ and at $t=0$ and $x_0 \leq x \leq x_F$. As pointed out in the introduction, the experimental apparatus at Stanford [7], after the convergent, has a flate-plate developing region, $x_u = 2.05$ meters long. In the numerical simulation the horizontal velocity profile, corresponding to a fully turbulent zero pressure gradient boundary layer at $R_{ex_0} = 1.1 \cdot 10^6$ ($U_{\infty 0} = 0.656$ m/sec, $x_0 = 1.97$ m, $\nu = 1.17 \cdot 10^{-6}$ m^2/sec) has been prescribed. The turbulent kinetic energy profile has been calculated, by an iterative procedure, solving the non linear equation (12) with the assumption of zero advection. This assumption gives a satisfactory distribution in the inner region, where the advection term is negligible. The same velocity and turbulent energy profiles have been assigned upstream as initial conditions ($x = x_0$ and $t > 0$).

The boundary layer thickness distribution given by the empirical law $\delta \propto x^{4/5}$ has been assigned and the system of equations (11), (12) and (13) has been solved in the steady case. Then the unsteady equations have been solved, for $U_e(x,t) = U_{\infty 0}$, till the maximum percentual error on $C_f/2$ was less than 10^{-3}. With this procedure at the beginning of the unsteady part, a boundary layer with the following characteristics is obtained

Tab. 1

H	$C_f/2$	Re_δ	Re_{δ_2}	G
1.403	$1.812 \cdot 10^{-3}$	$2.20 \cdot 10^4$	$2.505 \cdot 10^3$	6.73

The calculation from $x = x_0$ to $x = x_u$ is necessary for two reasons: to have at $x = x_u$ a vertical velocity profile, which can not be assigned because usually it is not measured by the experimentalist and, more important, to have a turbulent kinetic energy profile evaluated taking into consideration the effect of the advection term, playing a large role in the outer region ($y/\delta > 0.6$).

RESULTS

The initial conditions, evaluated as previously described, allow to start the numerical simulation of unsteady boundary layers with the following free-stream velocity distrubution

$$\frac{U_e(x,t)}{U_{\infty 0}} = 1 \qquad \text{for } x_0 \le x \le x_1$$

$$\frac{U_e(x,t)}{U_{\infty 0}} = 1 - \alpha \cdot \frac{(x-x_u)}{(x_F - x_u)} \cdot \frac{1-\cos\omega t}{2} = \overline{U}_0 + U_{e1}\cos\omega t \qquad \text{for } x_u \le x \le x_F \qquad (15)$$

For the whole range of α and frequency considered the calculation has been done using both in the x_1 and x_2 coordinates 41 mesh points equally spaced. The magnitude of the time step depends on the frequency considered, it has been chosen in such a way to have 40 time steps in one period. The calculation was carried over for more than one cycle, at low frequencies ($f < 0.5$ Hz) two cycle are sufficient. All the results which will be reported are evaluated considering only the last cycle. Five minutes of CPU time are required for the calculation of two cycles on a UNIVAC 1100/82.

Reynolds and Parik [7] did measurements for unsteady boundary layers with a free-stream velocity distribution almost equal to the distribution given by equation (15). At the inlet of the unsteady section they had a fully developed steady turbulent boundary layer with the following characteristics

Tab. 2

H	$C_f/2$	Re_δ	$Re_{\delta 2}$	G
1.396	1.658	$2.58\cdot 10^4$	$3.075\cdot 10^3$	6.96

The initial conditions of the numerical calculation differ from the experimental ones, thus the comparison will be done dividing the instantaneous quantities by the mean quantities. Each instantaneous quantity can be expressed as

$$A = A_M + A_1 \cos(\omega t + \phi_A) + \text{Higher Armonics} \qquad (16)$$

In the numerical calculation the quantities A_M are calculated as suggested by Cebeci [12] , instead in the experiments they are evaluated by the measured mean velocity profile obtained by a sufficiently long time averaging.

The displacement thickness time history at different frequencies is shown in Fig. 1. The numerical results are in a very good agreement with the experimental ones, they are able to predict the reduction of the amplitude, increasing the frequency and the phase angle variations. At $f = 0$ the displacement thickness is in opposite phase with respect to the free-stream velocity, then increasing the frequency both the numerical and the experimental results show a reduction of the phase angle till it reaches a minimum value and then increases again at higher frequencies. Fig. 2 shows the behavior of $\phi_{\delta 1}$ versus $\omega(x-x_u)/U_{\infty 0}$, calculated by the numerical model at different reduced frequencies $\beta = \frac{\omega x_0}{U_{\infty 0}}$, to be in a very good agreement with the three experimental results available The maximum of $(\Pi - \phi_{\delta 1})$ occours in the range of frequencies of interest in helicopter problems. The phase angle of the shape factor is predicted as well as the phase angle of the displacement thickness, as shown in Fig. 3. The amplitude decrease, increasing the frequency is obtained by the numerical model but, it is not able to give such a large amplitude experimentally meausured at low frequencies.

The quantity more interesting to ana-

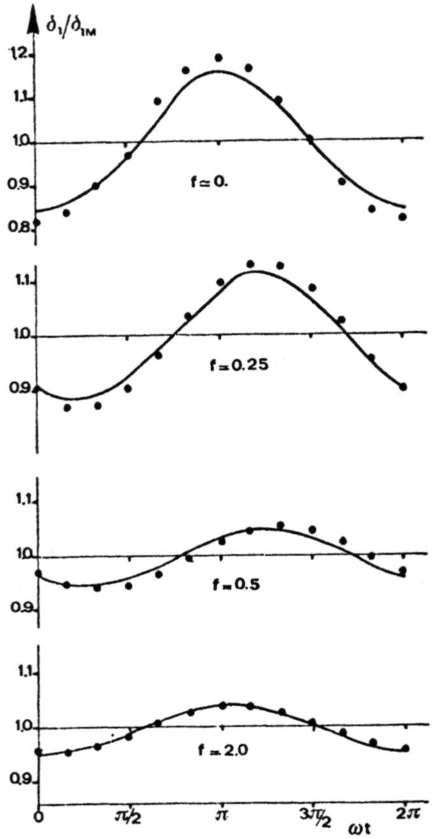

Fig.1 Time history of displacement thickness.

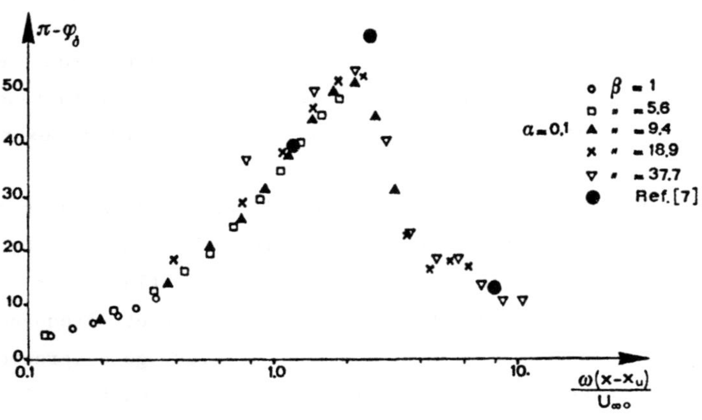

Fig.2 Displacement thickness phase angle.

lyze in the study of turbulent boundary layers is the skin friction $\frac{C_f}{2} = \frac{\tau_W}{\rho U_e^2(x,t)}$. Fig. 4 shows the time history of the skin friction at different frequencies. For the quasi-steady case and at low frequencies a very good agreement is obtained between the experimental and the numerical results. Instead at higher frequencies the numerical results show a behavior completely different from the experimental ones. The first ones show a reduction of the C_f amplitude and then, at very high frequencies, a pronounced increase only lightly emphasized by the experiments. The phase angle behavior is

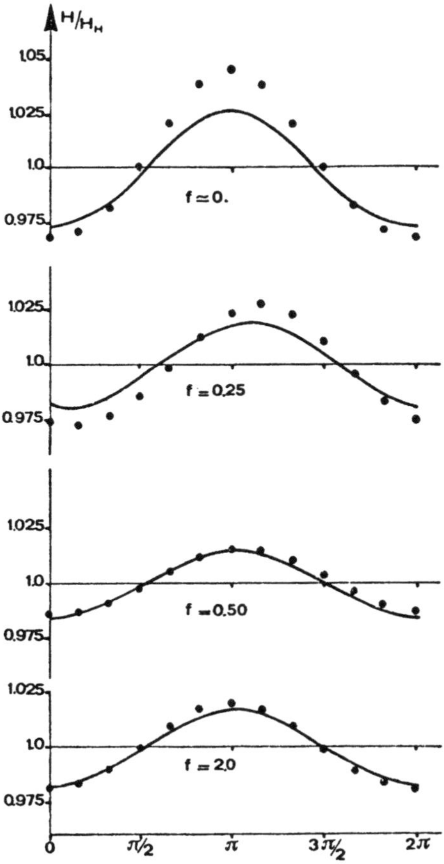

Fig.3 Time history of shape factor.

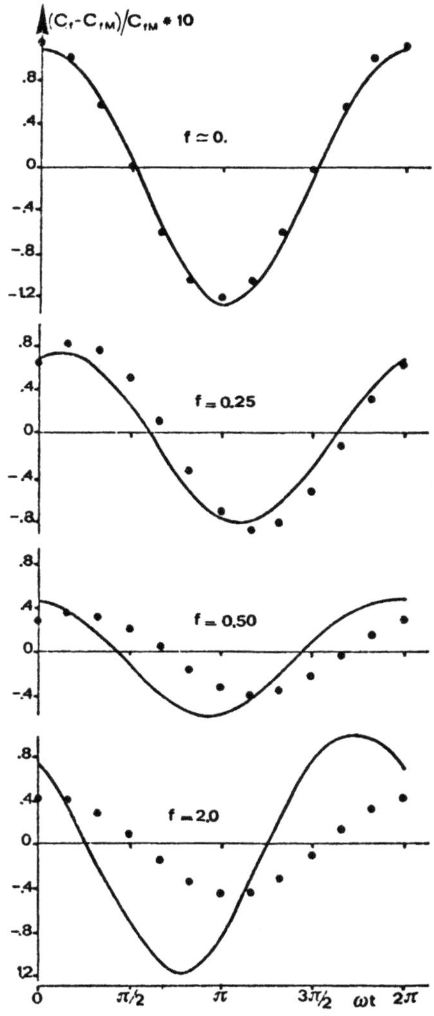

Fig.4 Time history of skin friction.

quite well represented at low frequencies, on the contrary at intermediate and high frequencies the computation gives a C_f that always more lags the free-stream velocity. The experiments, instead show a C_f that leads the $U_e(x,t)$ in a smaller extent but never presenting a phase lag. Fig. 5 shows the phase angle of the fluctuating wall shear as a function of $\omega(x-x_u)/U_{\infty 0}$ calculated at various frequencies. It can be seen that our model, at low and intermediate frequencies, fulfiles the fact that the reduced frequency is a similarity parameter. At high frequencies some discrepancies is obtained, hovewer at different β the same trend, bringing to large phase lags, can be observed.

In the introduction has been affirmed that this model was not able to predict the overshoot effect of the streamwise velocity for the unsteady flat-plate boundary layer with unsteady initial conditions. Fig. 6a shows the profile of U_{11}/U_{e1} (U_{11} is the oscillating part of the velocity $U = U_{1M} + U_{11}\cos(\omega t + \phi_U)$) versus y/δ compared with the experimentally observed one, in the case of unsteady adverse pressure gradient. The calculation gives lower overshoots and located closer to the wall than the experiments do. But the qualitative trend increasing the frequency, is predicted. In fact higher values are obtained at low frequencies, then at f = 0.5 Hz the overshoot disappears complitely to occur again in a lesser extent and closer to the wall at very high frequencies. A better agreement is observed in the near wall region, as shown in Fig. 6b. The calculation has not been done at

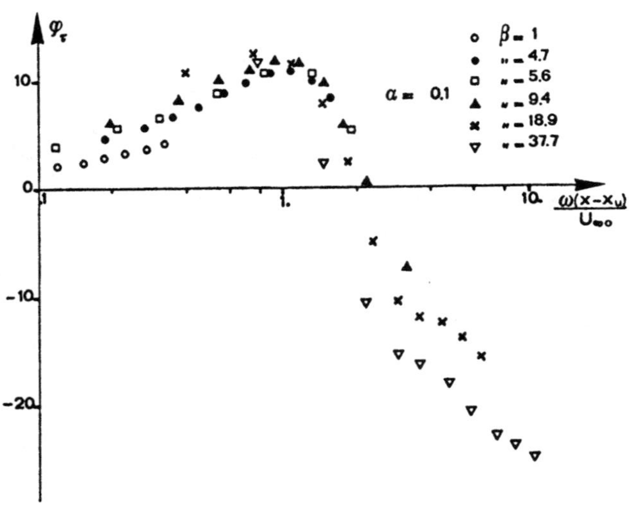

Fig.5 Skin friction phase angle.

f= 0, but at a very low frequency (f=0.025), it means that the largest decreasing of the ratio U_{11}/U_{e1} is obtained giving a small unsteadiness to the free-stream velocity. The effect of the streamwise distance on U_{11}/U_{e1} has not been experimentally investigated. The numerical results show (Fig. 7) that the streamwise distance variations affect the boundary layer behavior as frequency variations do. Fig. 7 shows, also, that further a certain distance the overshoot disappear.

The experiments were done only at α = 0.1, the calculation has been carried out also at α = 0.2 and α = 0.3. Results, not reported in here, show

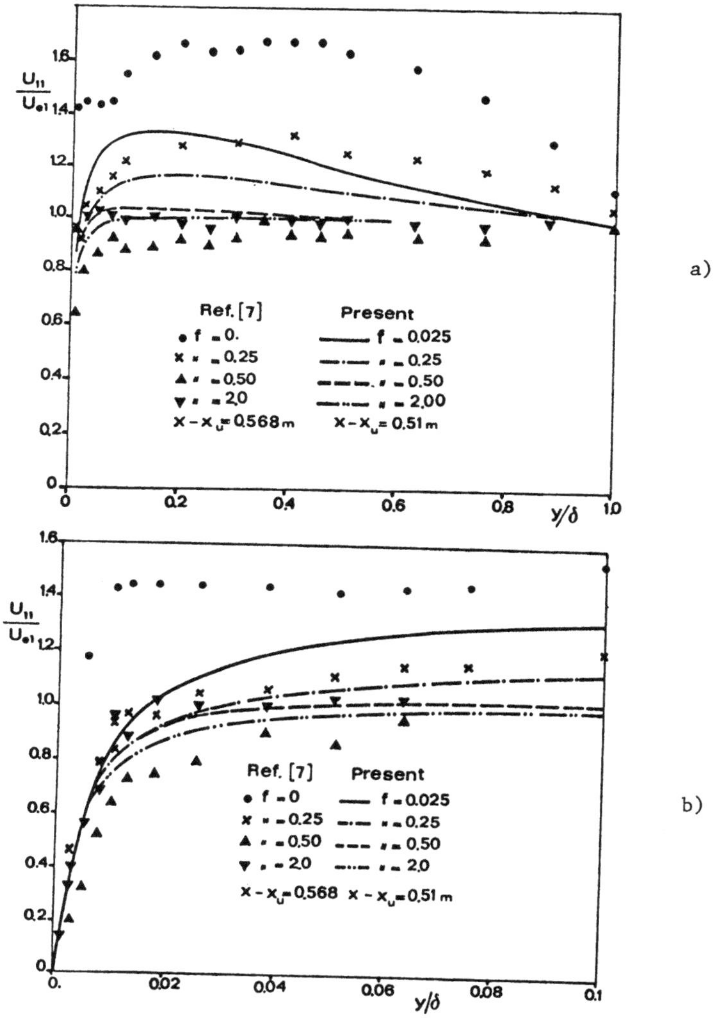

Fig.6 Streamwise velocity amplitude ratio, a) outer region, b) wall region.

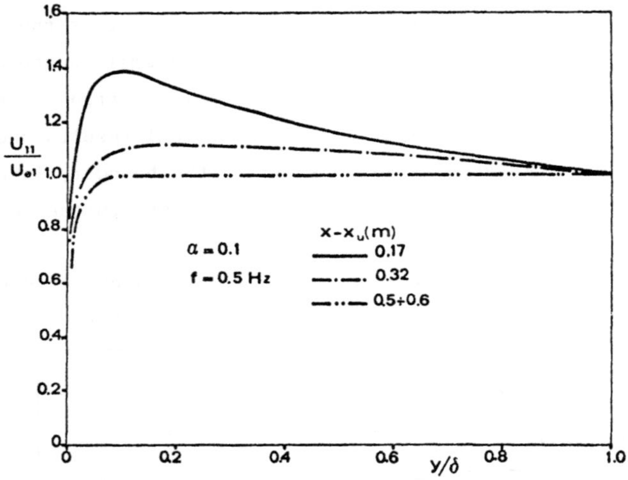

Fig.7 Streamwise velocity amplitude ratio.

the phase angles behaviors and the ratio U_{11}/U_{e_1} are not widely affected by the parameter α. The calculation at α=0.3 and at low frequency f<0.2, showed the occurence of the separation, that disappeared at higher frequencies. These results, if will be gratified by the experimental analysis will be the subject of a next paper.

This work was supported by Consiglio Nazionale delle Ricerche CT80.02709.07

Errata: In Figs. 6 The line - - belongs to f=2.0 and -··- line to f=0.50

REFERENCES

1. Orlandi, P., Reynolds, W.C. : " A Provisional Model for Unsteady Turbulent Boundary Layers", 1979. Submitted to AIAA Journal.
2. Orlandi, P.: " Implicit Non-Iterative Scheme for Turbulent Unsteady Boundary Layers", Proceedings of VII International Conference on Numerical Methods in Fluid Dynamics, June 1980, Stanford, U.S.A. .
3. Karlsson, S.K.F., Journ. Fluid Mech., Vol. 5, n°4,1956 pp. 467-474.
4. McCroskey, W.J., Philippe, J.J., AIAA Journal, Vol. 13, n°1,'75,pp.71-79.
5. Tellionis, D.P., Journal of Fluids Engineering Trans. ASME,1979,pp.29-43.
6. AGARD Confererence Proceedings n° 227 : " Symposium on Unsteady Aerodynamics", Ottawa, Canada, September 1977.
7. Reynolds, W.C., Parikh, P.G.: "Dynamic Behavior of an Unsteady Turbulent Boundary Layer", IUTAM Symposium on Unsteady Turbulent Shear Flows, Toulouse, May 1981.
8. Orlandi, P.: " Model of Low Reynolds Number Wall Turbulence for Equilibrium Layers", Submitted to Trans. of ASME, Journ. of Fluids Engineering.
9. Norris,H.L. and Reynolds , W.C., Report TF-7 Dept. of Mech. Eng.Stanford University , 1975.
10. Moin, P. Reynolds, W.C. and Ferzinger, J.H., Report n° TF-12,Mech. Eng. Dept., Stanford University, 1978.
11. Orlandi, P. and Ferziger,J.H.: " Implicit Non-Iterative Schemes for Unsteady Boundary Layers". To Appear on AIAA Journal.
12. Cebeci, T., Proc. R. Soc. Lond. A., 1977, pp. 225-238.

Influence of the Laminar and Turbulent Boundary Layers in Unsteady Two-Dimensional Viscous-Inviscid Coupled Calculations (*)

A. DESOPPER

Office National d'Etudes et de Recherches Aérospatiales (ONERA)
Centre d'Etudes et de Recherches de Toulouse (CERT)
31055 Toulouse Cedex (France)

Abstract

Within the framework of research on unsteady flows performed at ONERA, a calculation method of two-dimensional unsteady flows in which the effects of the unsteady boundary layers on the profile are taken into account has been developed in incompressible and in transonic regime.
The methods used are briefly described. The results obtained show, by comparison with results of inviscid flow calculations, the influence of viscous effects on unsteady pressure distributions, on total force coefficients and on shock-wave intensity and location.

Nomenclature

ρ density
M_∞ Mach number at infinity upstream
c chord
τ thickness-to-chord ratio
α incidence of the profile
δ_m mean flap deflection
δ_i amplitude of flap oscillations
$f = \omega/2\pi$ oscillation frequency
$k = \omega c/2V_\infty = \pi f c/V_\infty$ reduced frequency
p instantaneous pressure
Δp pressure fluctuation
$K_p = (p - p_\infty)/\frac{1}{2}\rho_\infty V_\infty^2$ instantaneous (or steady) pressure coefficient
$C_p = \Delta p/\frac{1}{2}\rho_\infty V_\infty^2 \delta_i$ unsteady pressure coefficient
C_{p_1} modulus of the first harmonic of the pressure coefficient
K_p^* value of K_p for $M_\ell = 1$
C_{z_δ} unsteady lift coefficient (value of C_{z_1} for $\delta_i = 1$ radian)
C_{m_δ} unsteady pitching moment coefficient
C_{mc_δ} unsteady hinge moment coefficient
φ phase angle
φ_1 phase angle of first harmonic
R_e Reynolds number, relative to chord c
δ boundary layer thickness
U longitudinal component of velocity in the boundary layer
$\theta = \int_0^\delta \rho U(1 - U/U_e)/\rho_e U_e \, dy$ boundary layer momentum thickness
$\delta_1 = \int_0^\delta (1 - \rho U/\rho_e U_e)\, dy$ boundary layer displacement thickness
$\delta_3 = \int_0^\delta \rho U(1 - U^2/U_e^2)/\rho_e U_e \, dy$ kinetic energy thickness ($H_{32} = \delta_3/\theta$)
$\delta_\rho = \int_0^\delta (1 - \rho/\rho_e)\, dy$ boundary layer density thickness
$H = \delta_1/\theta$ boundary layer shape factor
C_f local skin friction coefficient

1. Introduction

At ONERA, within the framework of research aiming at a better understanding and a better prediction of unsteady phenomena, a calculation method of two-dimensional unsteady flows taking into account boundary layer effects has been developed.

This paper will first briefly describe the calculation method which comprises :
– for the inviscid flow calculation, either a singularity method in incompressible regime or a method solving the velocity potential small perturbation equation in transonic regime ;

(*) Work performed with financial support of DRET and STPA.

– an integral method for the determination of the unsteady boundary layer characteristics ;
– a weak coupling technique, of "at the wall" coupling type, not taking into account the separated regions.

The results obtained show, by comparison with results of inviscid flow calculations, the influence of viscous effects on the unsteady pressure distributions, on total force coefficients and on shock-wave intensity and location :
– in incompressible regime, the evolution of viscous effects with reduced frequency is studied for a profile oscillating in pitch and for a profile with an oscillating flap ;
– in transonic regime, the flow around a NACA 64A 010 oscillating in pitch is studied ; the influence of taking into account a laminar region for the boundary layer and that of the thickness effect of the wake developing behind the profile are brought to light.

2. Review of the calculation method

The calculation method and the first results obtained are described in detail in Ref. [1] ; only the most important points will be briefly recalled here.

2.1. Calculation methods in inviscid flow :

– Incompressible flow – To take into account the influence of the mean flow on the unsteady response, we use a calculation program of incompressible unsteady flow developed by the Bertin Company [2], based on the Giesing method [3] (singularity method).
– Transonic flow – We use a program initially developed, for the study of flows over helicopter blades, by F.X. Caradonna [4], and adapted to the case of an oscillating profile or flap by J.J. Thibert. This program solves the unsteady velocity potential (Φ) equation with the assumption of transonic small perturbations and low reduced frequency. In dimensionless form, this equation writes :

$$4 M_\infty^2 \frac{K}{\tau^{2/3}} \Phi_{xT} = \Phi_{xx} \left\{ \frac{1-M_\infty^2}{\tau^{2/3}} - (\gamma+1) M_\infty^2 \Phi_x \right\} + \Phi_{yy}$$

2.2. Boundary layer calculation methods

Since 1974, detailed experimental and theoretical studies are carried out at the Toulouse Research Center on unsteady boundary layers [5 through 8]. Among the various methods developed, integral methods have been retained in order to maintain reasonable computing times.

The equations used in these methods are the total momentum equation, or von Karman equation [1], the total kinetic energy equation [2] and the total continuity, or entrainment equation [3] :

$$\frac{C_f}{2} = \frac{\partial \theta}{\partial x} + \theta \frac{H+2}{U_e} \frac{\partial U_e}{\partial x} + \frac{1}{\rho_e} \frac{\partial \rho_e}{\partial x} + \frac{1}{\rho_e U_e^2} \frac{\partial (\rho_e U_e \delta_1)}{\partial t} - \frac{1}{\rho_e U_e} \frac{\partial (\rho_e \delta_e)}{\partial t} \quad (1)$$

$$\frac{\partial \delta_3}{\partial x} + \frac{1}{U_e} \frac{\partial (\theta + \delta_1)}{\partial t} = 2 C_D - \delta_3 \left(\frac{3}{U_e} \frac{\partial U_e}{\partial x} + \frac{1}{\rho_e} \frac{\partial \rho_e}{\partial x} \right) - 2 \frac{(\delta_1 - \delta_{1i})}{U_e} \frac{\partial U_e}{\partial x} + 2 \frac{(\delta_{1i} - \theta - \delta_1)}{U_e^2} \frac{\partial U_e}{\partial t} - \frac{(\theta + \delta_1)}{\rho_e U_e} \frac{\partial \rho_e}{\partial t} + \frac{1}{\rho_e U_e} \frac{\partial (\rho_e \delta_e)}{\partial t} \quad (2)$$

$$\frac{1}{\rho_e U_e} \frac{\partial [\rho_e U_e (\delta - \delta_1)]}{\partial x} + \frac{1}{U_e} \frac{\partial \delta}{\partial t} = P \frac{C_f}{2} - \frac{\delta}{\rho_e U_e} \frac{\partial \rho_e}{\partial t} - \frac{1}{\rho_e U_e} \frac{\partial (\rho_e \delta_e)}{\partial t} \quad (3)$$

These methods are described in detail in Refs [8] and [9], and we shall only recall their principle :
– Laminar region – The equations solved are the total momentum (1) and kinetic energy (2) equations. The closure relations necessary for the solution of this system of equations are the same as those used in steady flow [10].

– Turbulent region – The equations solved are the momentum equation (1) and the entrainment equation

(3) ; the closure relations are the same as those used in steady flow [11].

— Transition region — Several options are possible :
i) The transition point x_{t_r} is fixed ; we may thus assume, for instance, that the boundary layer is turbulent over the whole profile.
ii) Utilization of the Michel criterion [12] for determining the transition point.
iii) Utilization of the "modified" Michel criterion : at each time $t+dt$ the transition point is determined on the one hand by the Michel criterion and on the other hand, knowing the transition position at time t, by imposing a convection velocity U_c ($\simeq 0.5\, U_e$) of the transition point. The point retained is the more upstream one.

Remarks

— In these calculations, the region around the stagnation point is treated in a quasi steady manner.
— When a laminar separation occurs before transition, the separation point is taken as transition point.
— Wake region — The calculation of the wake taken into account with the transonic small perturbation method is carried out, with the assumption of two symmetrical half wakes, from the values of the boundary layers characteristics at the profile trailing edge. The equations solved are the continuity and momentum equations (in which we assume $C_f=0$) ; the closure relations are those used in steady-flow.

2.3. Coupling technique

The method used is a weak coupling method of "at the wall coupling" type. In incompressible or transonic flow, the method consists in carrying out the inviscid flow calculation with a modification of the boundary conditions on the profile for taking into account the development of the boundary layer. Concretely, this is expressed by a fluid flow through the wall, characterized by a normal velocity V_e in the singularity method and by a modification of the local profile slopes, of the quantity (V_p / U_p), in the transonic small perturbation program ; V_e and (V_p / U_p) are given by :

$$\frac{V_e}{U_e} = \frac{V_p}{U_p} = \frac{1}{\rho_e U_e} \frac{\partial (\rho_e U_e \delta_1)}{\partial x} + \frac{1}{\rho_e U_e} \frac{\partial (\rho_e \delta_e)}{\partial t} \tag{4}$$

Taking into account a wake thickness effect in the transonic small perturbation method is expressed by a jump of the normal velocity on the wake line. This normal velocity discontinuity is given by Eq. (4), in which the thicknesses are the characteristic thicknesses of the wake.

At each time step a few inviscid flow-boundary layer iterations are performed, with possibly utilization of a relaxation method. The shock wave-boundary layer interactions, and the separations which might occur at the trailing edge for some configurations, are treated in a simplified manner, and it is quite obvious that this calculation method can only be used in the case of weak shock wave-boundary layer interaction and of small separated zones.

3. Results obtained. Influence of viscous effects

3.1. Incompressible flow

In all the results presented here, the boundary layer is assumed turbulent over the whole profile.

The method developed has first been used to calculate the flow around a supercritical profile equipped with an oscillating flap at the trailing edge. Experimental results, as well as detailed calculation results, can be found in Refs. [1, 2 and 13].

For instance figure 1, taken from Ref. [1], shows that the non linear character of dependency on the mean flap deflection, which affects the unsteady performance of the flap at low speeds, is essentially of viscous origin.

With a view to better characterize the importance of viscous effects on unsteady performance at low speeds, we considered the case of a NACA 0012 profile oscillating in pitch, or that of this same profile equipped with an oscillating flap. In both cases, the oscillation amplitude is 1° and the range of reduced frequency varies from zero to $K = 2$.

a) NACA 0012 profile oscillating in pitch

Two configurations were investigated : an oscillation around a zero mean incidence, and one around a 5° mean angle.

– Influence of viscosity on unsteady pressure distributions

On figure 2, the unsteady pressures (modulus and phase of C_{P_1}) provided by a linear theory [14] are compared with results of the singularity method, account being taken or not of the boundary layer. The inviscid flow calculation predicts unsteady pressure moduli higher than the linear theory over most of the profile. On the other hand, the result of taking the boundary layer into account is to reduce the unsteady pressure moduli ; phases are practically unchanged, except around the trailing edge where viscous effects are most important.

– Influence of viscosity on total force coefficients

The results obtained by the singularity method with and without coupling are shown in Table I. The evolution of the unsteady coefficients (lift C_{z_α} and pitching moment at the leading edge $C_{m_{BA_\alpha}}$)as a function of reduced frequency, for a 5° mean incidence is presented on figure 3. Up to a reduced frequency of the order of 1, the result of taking into account the thickness effect in the singularity method is to increase the amplitude of these coefficients. The influence of viscosity is to decrease these amplitudes ; phases are very little modified, either by viscosity or by mean incidence (see Table I and figure 3).

On figure 4 the evolution of the ratio : amplitude of the coefficient obtained by inviscid flow calculation relative to that given by a coupled calculation, as a function of reduced frequency, makes it possible to quantify the importance of viscous effects. Viscosity reduces the amplitude of the lift coefficient by 10 to 15%, an that of the pitching moment at the leading edge by 20 to 25%. For a mean incidence $\alpha_m = 5°$, viscous effects are more important than for $\alpha_m = 0$; this is particularly obvious at low frequencies, but the difference decreases relatively fast with frequency. In both cases, the influence of viscous effects decreases with reduced frequency.

Figure 5 shows the evolution with reduced frequency of the ratio : amplitude of the coefficient at a reduced frequency K relative to its value for $K = 0$ (quasi steady value). The evolution obtained by the singularity method is compared with that provided by the linear theory. Up to a reduced frequency of the order of 0.5, the various calculation methods give very similar results ; for higher reduced frequencies, we observe that the results obtained with viscosity taken into account are located between inviscid flow results (singularity method) and those provided by the linear theory.

b) NACA 0012 profile with an oscillating flap

The flap has a 25% chord, and two configurations have been investigated : a flap oscillation around zero mean deflection, and one around a 5° mean deflection.

– Influence of viscosity on unsteady pressure distributions

Figure 6 shows that, as in the case of oscillating profile the result of taking account of the boundary layer is to decrease the unsteady pressure moduli (particularly C_{p_1}, on the flap) ; phases are practically unchanged on the fixed part of the profile, but differ on the flap, where viscous effects are more important ; this difference may be slightly enhanced by the presence of a small separation at the trailing edge during part of the oscillation.

It can again be noted that thickness effects and viscous effects are in opposite senses.

– Influence of viscosity on total force coefficients

The unsteady coefficients considered are the lift (C_{z_δ}), pitching moment (C_{m_δ}) and hinge moment (C_{mc_δ}) coefficients. The results are presented in the same form as for the oscillating profile, and shown in Table II. The values of this Table show that the results obtained by the inviscid flow calculation (singularity method) are practically insensitive to the mean deflection, and that the evolution of unsteady performance with mean deflection is essentially of viscous origin.

As in the case of oscillating profile, the result of taking into account the boundary layer is to decrease the amplitude of the total force coefficients (fig. 7) ; except for the hinge moment, phases are little modified by viscous effects. The reduction of the lift coefficient amplitude is 15 to 20%, that of the pitching moment 15 to 25%, and that of the hinge moment may reach 50% at low frequencies, for a mean flap deflection $\delta_m = 5°$ (fig. 8). This should be attributed to the fact that viscous effects increase with deflection and that they are particularly important on the flap. We may also observe that the influence of viscous effects decreases with reduced frequency.

We plotted on figure 9 the amplitude of a coefficient at reduced frequency K relative to its quasi steady value ($K \simeq 0$). The various calculation methods give similar results for the lift coefficient ; for the pitching moment coefficient, the results obtained with viscosity taken into account are close to those of the linear theory (for $0° < \delta_m < 5°$) ; on the other hand, for the hinge moment for which viscous effects are very important and vary rapidly with reduced frequency (see figure 8), the results obtained with boundary layer effects accounted for are very different from those provided by the linear theory.

3.2. Transonic flow

Within the assumption of a turbulent boundary layer over the whole profile, the method has already been used for calculating the flow around a section of helicopter blade and for determining the aerodynamic characteristics of a NACA 64A006 profile with oscillating trailing edge flap. Detailed comparisons between calculation and experiment may be found in Ref. [1, 13 and 15].

The calculations presented here concern a NACA 64A010 profile oscillating in pitch ; the profile used is not the theoretical one, but that used in experiments performed at the NASA Ames Research Center [16].

– Influence of viscosity on steady and unsteady pressure distributions (within the assumption of fully turbulent boundary layer). The unsteady program of transonic small perturbation makes it possible to calculate transient configurations, and by convergence a steady configuration. The mean pressure distribution on the profile upper surface (fig. 10) obtained this way makes it possible to characterize the influence of viscous effects in steady flow : reduction of the shock intensity, and travel of this shock toward the leading edge.

Results obtained in unsteady conditions are presented on figures 11 and 12. Pressure distributions at azimuths $\varphi = 0$ ($\alpha = -0.21°$), $\varphi = 90°$ ($\alpha = 0.79°$), $\varphi = 180°$ ($\alpha = -0.21°$) and $\varphi = 270°$ ($\alpha = -1.21°$) clearly emphasize the unsteady character of the flow ; in particular we observe that on the lower surface and for the same incidence ($\alpha = \alpha_m = -0.21°$), the shock is stronger when the incidence is increasing ($\varphi = 0$) than when it is decreasing ($\varphi = 180°$).

The upper surface boundary layer thicknesses resulting from the calculation presented in section 2.2, plotted on figure 11, reveal a rapid variation of these parameters at the shock (when it exists) and a very fast thickening of the boundary layer at the trailing edge, over the last 10% of the chord.

The distributions of instantaneous (fig. 11) and unsteady (fig. 12) pressure bring to light the fact that the boundary layer effect is to reduce the shock intensity and to locate it further upstream. The boundary layer, through reducing the size of the supersonic zone, source of delay, slightly reduces the phase lag in front of the shock ; this reduction of shock intensity is accompanied by an increase of the moduli upstream of the shock.

On figures 10 and 12 the calculation results are compared with experimental ones [16]. In the experiment we observe a shock of higher intensity than in the calculation, which is characterized, in unsteady conditions, by a peak of higher level on the modulus curve, and by a more important phase lag in the supersonic region.

– Influence of taking into account a laminar region in the boundary layer.

On figure 13, the unsteady pressure distribution obtained by assuming the boundary layer turbulent over the whole profile is compared with that obtained if the laminar region is taken into account. We observe a reduction of the boundary layer effect, in particular a less important increase of the phase shift toward the trailing edge, certainly due to a thinner boundary layer.

– Influence of taking the wake into account

As already mentioned, only a effect of wake thickness is considered. The mean pressure distribution around the profile at a $-1.21°$ incidence is presented in figure 14.

The result of taking into account the wake in the coupled calculation is to decrease the viscous effects : the shock is stronger and its position is further back as compared with the calculation without wake. This is certainly due to the evolution of the wake thickness, which decreases rapidly downstream of the trailing edge (fig. 14).

The reduction of viscous effects while the wake is accounted for is also found in unsteady flow, as can be seen on figure 15.

Remarks :

– In the calculations with a laminar region in the boundary layer, the motion of the transition point due to the unsteady character of the flow makes it necessary to reduce the time step, and this considerably increases computing times.

– For low frequencies ($K < 0.1$), convergence problems arise in unsteady calculations. These difficulties may be due to reflexions of the perturbations on the external boundaries of the calculated domain.

– A method of unsteady transonic calculation involving viscous effects by a strong coupling technique is being developed at ONERA [17]. This method should make it possible to treat in a more rigorous manner the shock-boundary layer interaction and to account for separated regions.

4. Conclusions

In incompressible flow, the calculations presented emphasize the following points :

– The influence on unsteady pressures of thickness and curvature effects on the one hand, and of viscous effects on the other hand, act in contrary senses, so that it is necessary to take into account the boundary layer effects when a nonlinear method is used for the inviscid flow calculation.

– The result of taking into account boundary layer effects is to decrease the amplitude of total force coefficients. For instance, for a NACA 0012 profile with oscillating flap, the reduction of the hinge moment coefficient amplitude may reach 50% at low reduced frequencies.

– For a configuration of profile with oscillating flap, the nonlinear behavior of the total force coefficients with mean deflection is essentially of viscous origin.

– The influence of viscous effects decreases with reduced frequency.

In unsteady transonic flow there appears, as for steady flows, that the shock wave position and intensity are very sensitive to viscous effects. The influence of viscosity is a shock of lower intensity and located further upstream. For the configuration investigated, taking the wake into account reduces the effect of the boundary layers developing over the profile.

In view of the experimental results obtained, the effects brought to light by calculation may be considered as realistic, and this makes it possible to envisage new developments of the methods presented in this paper.

References

1. Desopper, A., Grenon, R., Couplage fluide parfait - fluide visqueux en écoulement instationnaire bidimensionnel incompressible et transsonique, in Computation of Viscous-Inviscid Interactions, AGARD CP 291 (1980).
2. Baudu, N., LE, Th., Etude du comportement aérodynamique d'un profil avec gouverne oscillante. R.S.F. Bertin Note 75 CC06 (1975).
3. Giesing, J.P., Non linear two dimensional unsteady potential flow with lift. J. of Aricraft, vol. 5 No 2, (1968).
4. Caradonna, F.X., Philippe, J.J., The flow over a helicopter blade tip in the transonic regime. Vertica, vol. 2 No 1, 1978.
5. Desopper, A., Etude expérimentale des couches limites turbulentes et transitionnelles en écoulement pulsé. Essais de prévision théorique. Thesis presented at ENSAE, Toulouse, June 1977.
6. Houdeville, R., Cousteix, J., Premiers résultats d'une étude sur les couches limites turbulentes en écoulement pulsé avec gradient de pression moyen défavorable. 15e Colloque d'Aérodynamique Appliquée, Marseille, 7-9 November 1978.
7. Cousteix, J., Houdeville, R., Turbulent boundary layer calculations in unsteady flow. Presented at "Numerical Methods in Applied Fluid Dynamics", Univ. of Reading, 4-6 January 1978.
8. Cousteix, J., Houdeville, R., Javelle, J., Méthodes de calcul des couches limites instationnaires, in Boundary Layer Effects on Unsteady Airloads, AGARD CP 296 (1980).
9. Houdeville, R., Cousteix, J., Programme de calcul des couches limites instationnaires laminaires et turbulentes par une méthode ingégrale, ONERA internal report (1979).
10. Eppler, R., Praktishe Berechnung laminarer und turbulenter Absauge Orenzschicten. Ing. Arch. 32 (1963).
11. Michel, R., Quémard, C., Cousteix, J., Méthode pratique de prévision des couches limites turbulentes bi et tridimensionnelles, La Recherche Aérospatiale No 1972-1.

12. Grenon, R., Thers, J., Etude d'un profil supercritique avec gouverne oscillante en écoulement subsonique et transsonique, in Unsteady Aerodynamics AGARD CP 227 (1977).
13. Grenon, R., Desopper, A., Sidès, J., Effets instationnaires d'une gouverne en écoulement bidimensionnel subsonique et transsonique, in Aerodynamic characteristics of controls AGARD CP 262 (1979).
14. Albano, E., Rodden, W.P., A doublet lattice method for calculating lift distributions on oscillating surfaces in subsonic flow. AIAA Journal vol. 7 No 2 (1969).
15. Philippe, J.J., Lafon, P., Bohl, J.C., Etudes expérimentales d'aérodynamique instationnaire sur des maquettes de rotors d'hélicoptère en soufflerie. 15e Colloque d'Aérodynamique Appliquée AAAF, Marseille, 7-9 November 1978. NT AAAF 79-21.
16. Davis, S.S., Malcolm, G.N., Experiments in unsteady transonic flow, AIAA.ASME/ASCE/AHS 20th Structures, Structural Dynamics and Materials Conf., April 4-6, 1979, St Louis (Missouri), AIAA Paper 79-0769.
17. Couston, M., Angélini, J.J., Le Balleur, J.C., Girodroux-Lavigne, P., Prise en compte d'effets de couche limite instationnaire dans un calcul bidimensionnel transsonique, in Boundary Layer Effects on Unsteady Airloads, AGARD CP 296 (1980).

a

	Inviscid flow calculation				Inviscid flow + boundary layer calculation			
K	$\|C_{z\alpha}\|$	$\varphi_{C_{z\alpha}}$	$\|C_{m_{BA\alpha}}\|$	$\varphi_{C_{m_{BA\alpha}}}$	$\|C_{z\alpha}\|$	$\varphi_{C_{z\alpha}}$	$\|C_{m_{BA\alpha}}\|$	$\varphi_{C_{m_{BA\alpha}}}$
0	7	0	1.86	0	6.28	0	1.49	0
0.025	6.72	− 1.3	1.79	1	5.95	− 1.4	1.44	0.9
0.05	6.445	− 1.8	1.73	2.5	5.73	− 1.9	1.39	2.3
0.1	5.98	− 2.9	1.62	6.8	5.34	− 1.2	1.31	6.5
0.2	5.34	4.6	1.51	18.5	4.79	4	1.22	18.3
0.4	4.89	21.2	1.62	43.2	4.37	20.2	1.31	43.9
0.7	5.32	44.9	2.19	68.5	4.72	44.2	1.81	70.3
0.9	6.11	56.7	2.72	78.6	5.29	57.2	2.26	81.7
1.1	6.84	67.3	3.27	87.4	6.04	67.8	2.76	90.4
1.3	7.82	75.6	3.88	94	6.92	76.7	3.32	97.9
1.6	9.47	85.7	4.89	102.1	8.46	87.5	4.25	106.1
2	12	96.2	4.41	110.8	10.89	98.6	5.71	115

b

	Inviscid flow calculation				Inviscid flow + boundary layer calculation			
K	$\|C_{z\alpha}\|$	$\varphi_{C_{z\alpha}}$	$\|C_{m_{BA\alpha}}\|$	$\varphi_{C_{m_{BA\alpha}}}$	$\|C_{z\alpha}\|$	$\varphi_{C_{z\alpha}}$	$\|C_{m_{BA\alpha}}\|$	$\varphi_{C_{m_{BA\alpha}}}$
0	6.97	0	1.85	0	5.96	0	1.41	0
0.025	6.69	− 1.3	1.78	1	5.73	− 1.3	1.35	1.2
0.05	6.42	− 1.8	1.72	2.5	5.55	− 1.7	1.32	2.8
0.1	5.95	− 1	1.61	6.8	5.18	− 0.9	1.25	7.2
0.2	5.31	4.6	1.51	18.6	4.66	4.5	1.18	19.6
0.4	4.85	21.1	1.61	43.3	4.28	20.7	1.29	45.3
0.6	5.04	37.7	1.96	61.7	4.42	37.2	1.60	64.1
0.8	5.59	51.5	2.43	74.4	4.87	51.5	2	77.2
1	6.35	62.7	2.97	83.7	5.52	63.4	2.47	87.2
1.3	7.76	75.7	3.86	94.1	6.80	77.5	3.28	98.3
1.6	9.40	85.8	4.87	102.2	8.33	88.4	4.23	107
2	11.92	96.4	6.39	110.9	10.77	99.4	5.69	115.6

Table 1 : NACA 0012 profile oscillating in pitch.
a) $\alpha = 1^\circ \cos\omega t$, b) $\alpha = 5^\circ + 1^\circ \cos\omega t$.

a

	Inviscid flow calculation						Inviscid flow + boundary layer calculation					
K	$\|C_{z\delta}\|$	$\varphi_{C_{z\delta}}$	$\|C_{m\delta}\|$	$\varphi_{C_{m\delta}}$	$\|C_{mc\delta}\| \times 10^2$	$\varphi_{C_{mc\delta}}$	$\|C_{z\delta}\|$	$\varphi_{C_{z\delta}}$	$\|C_{m\delta}\|$	$\varphi_{C_{m\delta}}$	$\|C_{mc\delta}\| \times 10^2$	$\varphi_{C_{mc\delta}}$
0	4.20	0	0.69	0	4.76	0	3.55	0	0.552	0	2.91	0
0.025	4.03	− 2.89	0.691	2.23	4.73	3	3.42	− 2.89	0.555	2.33	2.91	3.76
0.05	3.86	− 4.89	0.694	3.72	4.72	5.56	3.29	− 4.94	0.56	3.82	2.93	6.74
0.1	3.57	− 7.22	0.697	5.82	4.73	9.77	3.05	− 7.41	0.566	5.82	2.39	11.39
0.2	3.13	− 8.11	0.70	9	4.79	16.47	2.69	− 8.62	0.571	8.73	3.07	18.5
0.4	2.64	− 3.93	0.699	15.14	4.97	27.82	2.26	− 4.67	0.571	14.87	3.25	31.6
0.6	2.40	2.66	0.704	21.67	5.27	38.2	2.05	2.11	0.573	21.9	3.52	44.33
0.8	2.29	9.74	0.719	28.28	5.71	47.28	1.94	9.13	0.585	29.32	3.95	55.57
1	2.24	16.75	0.743	34.73	6.26	55.28	1.9	17.32	0.607	36.64	4.52	64.81
1.3	2.26	26.7	0.794	43.77	7.26	65.21	1.92	28.2	0.658	46.77	5.56	75.2
1.6	2.34	35.8	0.86	51.9	8.4	73.19	2.01	37.98	0.725	55.53	6.74	82.69
2	2.51	46.57	0.966	61.34	10.1	81.71	2.19	49.23	0.833	65.24	8.42	90.18

b

	Inviscid flow calculation						Inviscid flow + boundary layer calculation					
K	$\|C_{z\delta}\|$	$\varphi_{C_{z\delta}}$	$\|C_{m\delta}\| \times 10$	$\varphi_{C_{m\delta}}$	$\|C_{mc\delta}\| \times 10^2$	$\varphi_{C_{mc\delta}}$	$\|C_{z\delta}\|$	$\varphi_{C_{z\delta}}$	$\|C_{m\delta}\| \times 10$	$\varphi_{C_{m\delta}}$	$\|C_{mc\delta}\| \times 10^2$	$\varphi_{C_{mc\delta}}$
0	4.18	0	6.84	0	4.7	0	3.31	0	5.00	0	2.18	0
0.025	4.01	− 2.88	6.86	2.25	4.67	3.07	3.20	− 2.86	5.04	2.47	2.2	4.78
0.05	3.84	− 4.89	6.89	3.75	4.67	5.67	3.08	− 4.83	5.10	4.17	2.24	8.89
0.1	3.55	− 7.22	6.93	5.84	4.69	9.94	2.86	− 7.14	5.17	6.53	2.33	15.38
0.2	3.12	− 8.12	6.95	9	4.75	16.71	2.53	− 8.26	5.24	9.8	2.47	24.15
0.4	2.63	− 3.95	6.94	15.18	4.93	28.11	2.13	− 4.29	5.26	16.5	2.74	39.27
0.6	2.39	2.67	6.99	21.74	5.24	38.42	1.94	2.53	5.33	24.1	3.14	52.79
0.8	2.27	9.77	7.13	28.39	5.68	47.61	1.85	10.25	5.51	32.05	3.72	63.67
1	2.23	16.81	7.37	34.88	6.24	55.59	1.81	17.79	5.78	39.41	4.37	71.59
1.3	2.24	26.80	7.88	43.96	7.24	65.51	1.84	28.61	6.35	49.42	5.49	80.27
1.6	2.32	35.93	8.54	52.11	8.38	73.47	1.93	38.23	7.08	57.72	6.71	86.28
2	2.43	46.73	3.6	61.56	10	81.96	2.11	40.45	8.17	67.0	8.38	92.77

Table 2 : NACA 0012 profile with an oscillating flap.
a) $\delta = 1^\circ \cos\omega t$, b) $\delta = 5^\circ + 1^\circ \cos\omega t$.

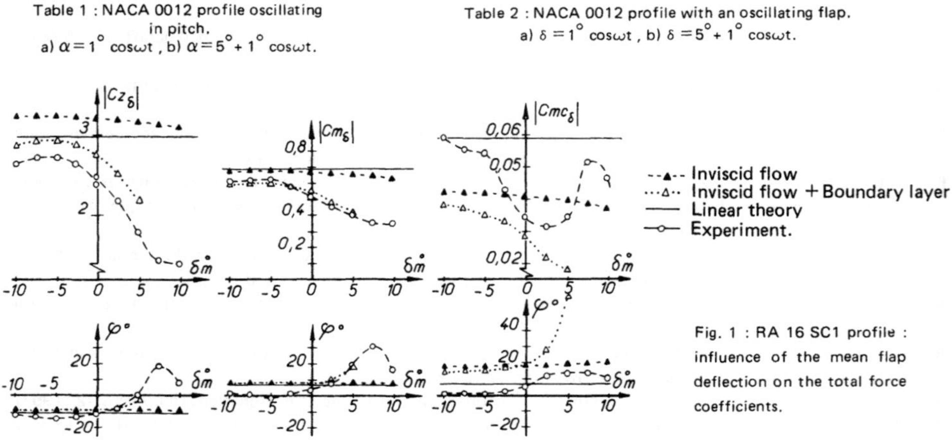

Fig. 1 : RA 16 SC1 profile : influence of the mean flap deflection on the total force coefficients.

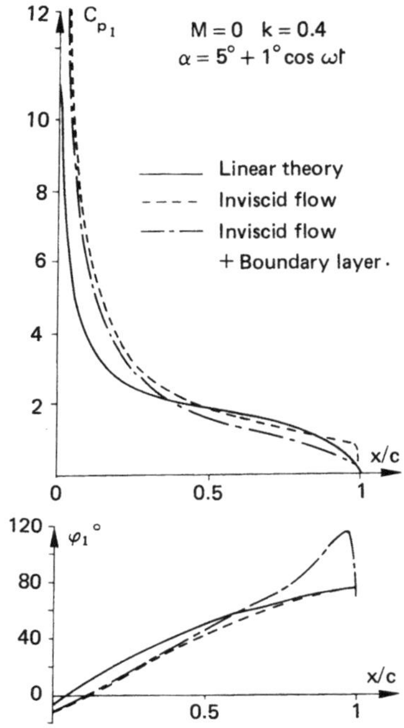

Fig. 2 : NACA 0012 profile oscillating in pitch.

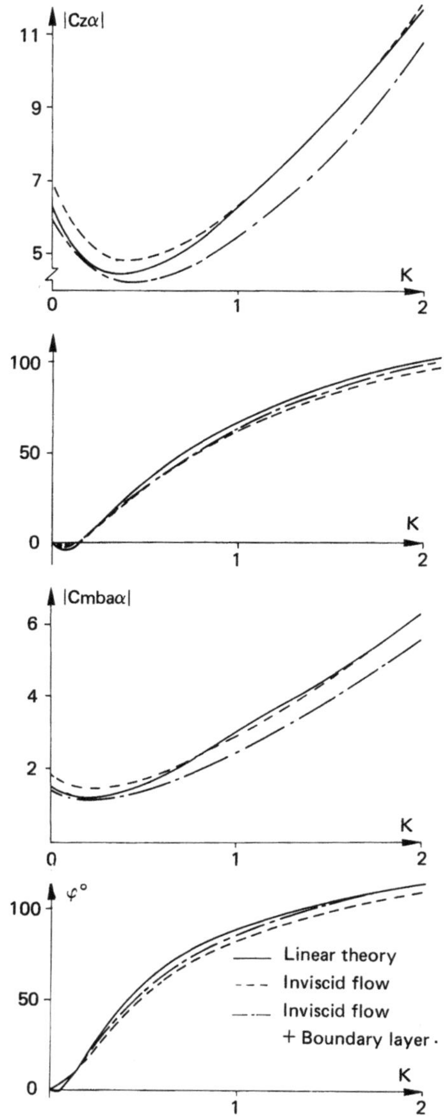

Fig. 3 : Influence of the reduced frequency on the total force coefficients.

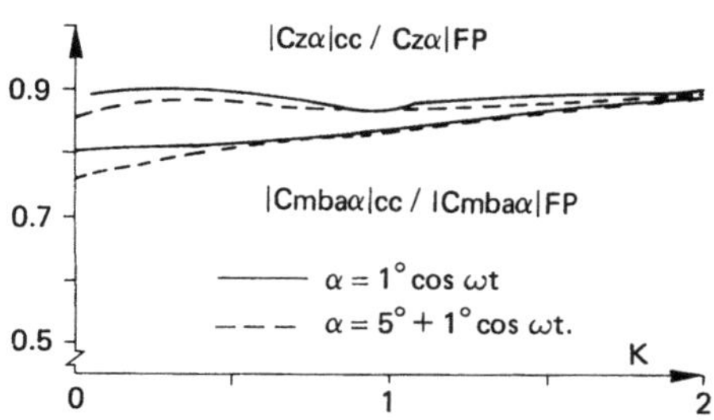

Fig. 4 : Influence of the viscous effects on the total force coefficients.

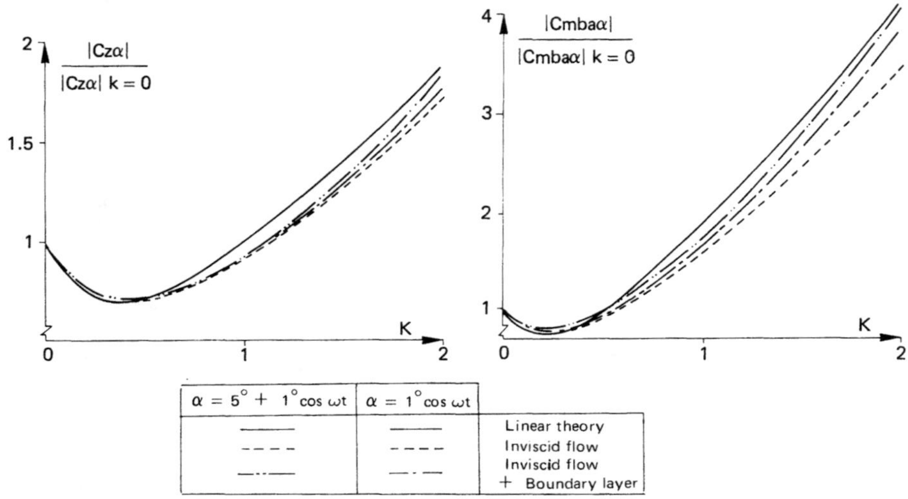

Fig. 5 : Influence of the reduced frequency on the total force coefficients.

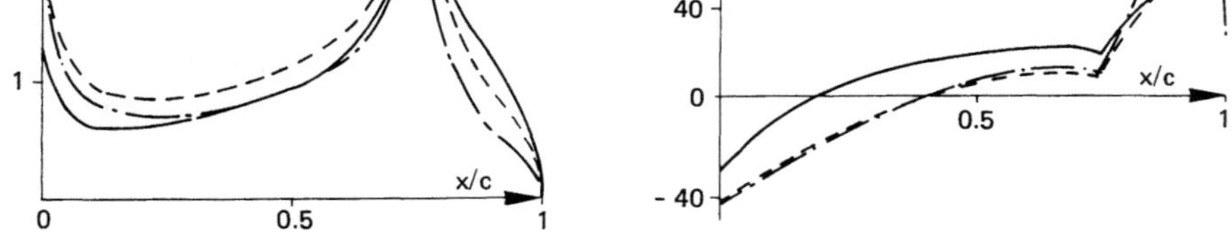

Fig. 6 : NACA 0012 profile with an oscillating flap.

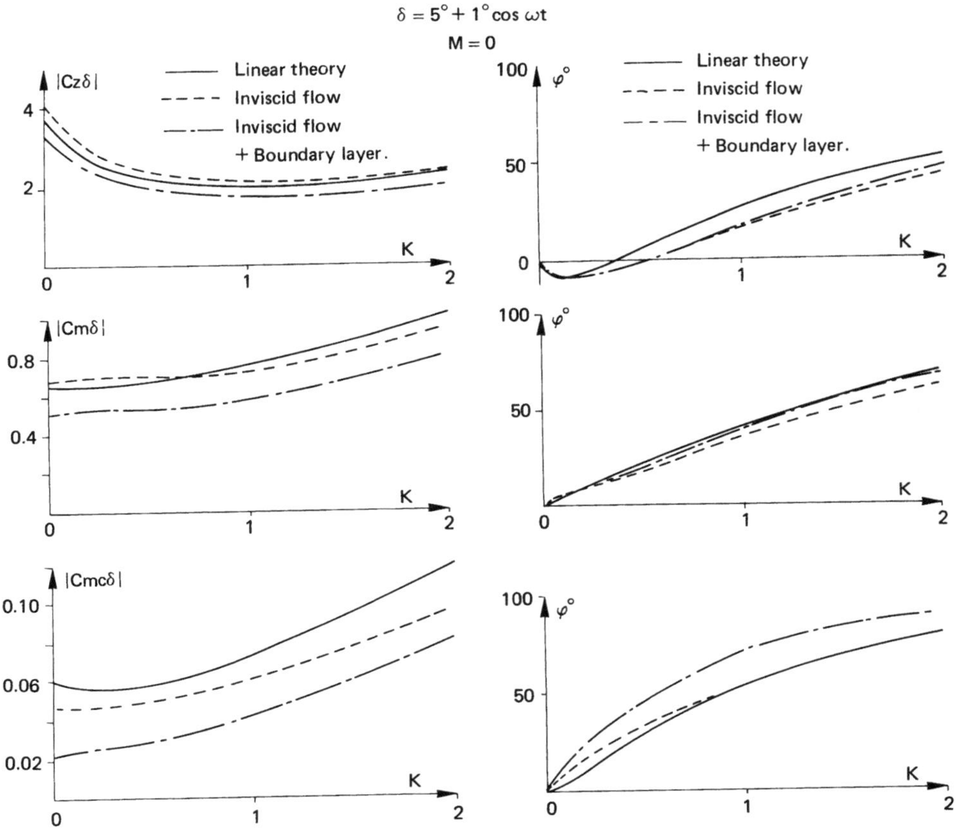

Fig. 7 : Influence of the reduced frequency on the total force coefficients.

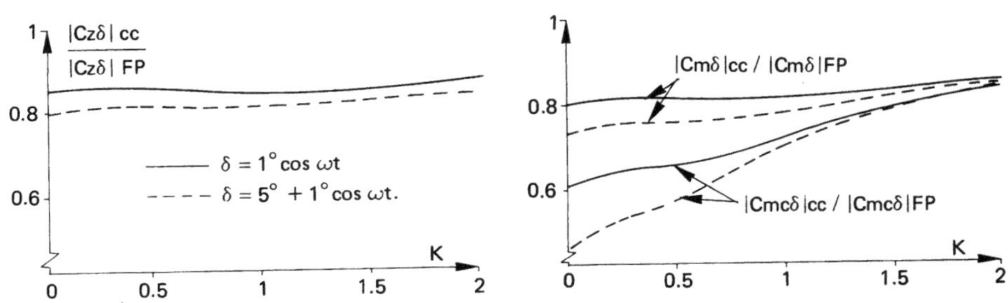

Fig. 8 : Influence of the viscous effects on the total force coefficients.

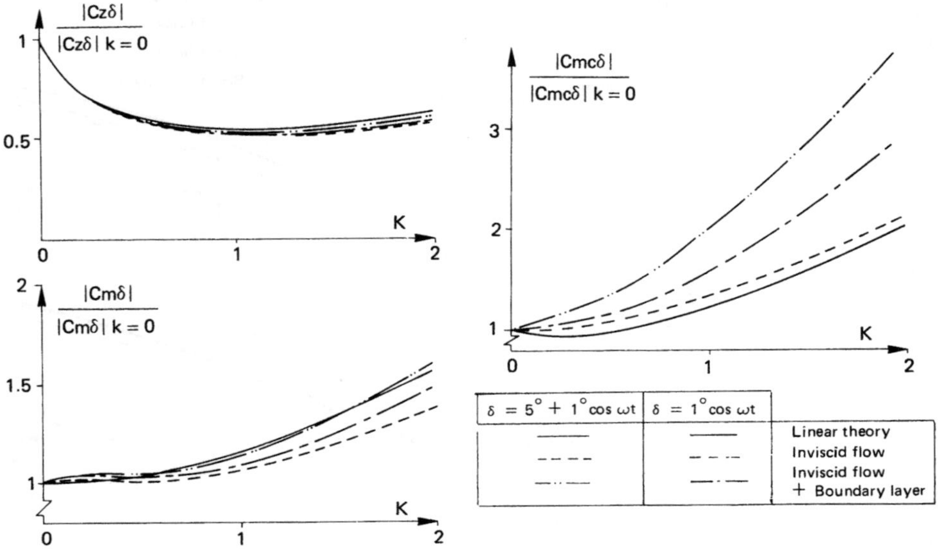

Fig. 9 : Influence of the reduced frequency on the total force coefficients.

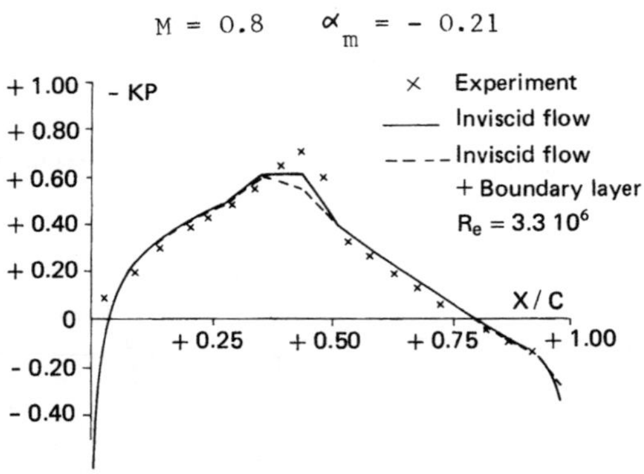

Fig. 10 : NACA 64 A 010 profile. Steady pressure on the upper side.

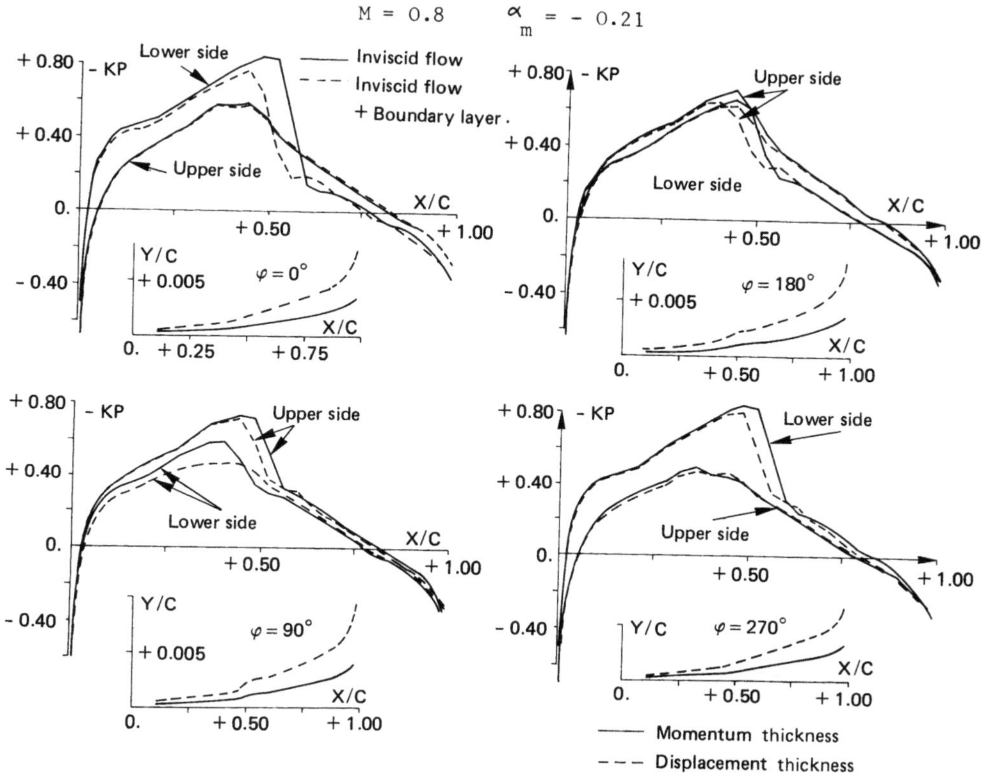

Fig. 11 : NACA 64 A 010 profile oscilatting in pitch.

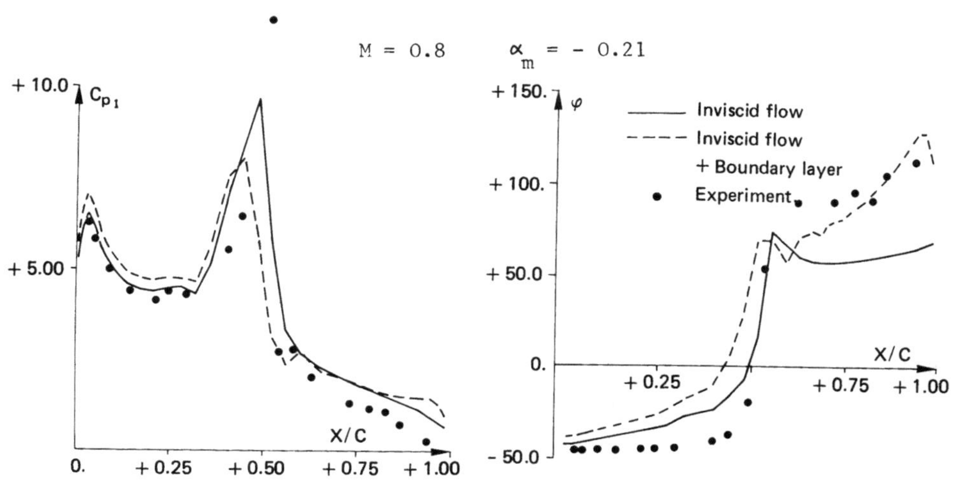

Fig. 12 : NACA 64 A 010 profile. Unsteady pressure.

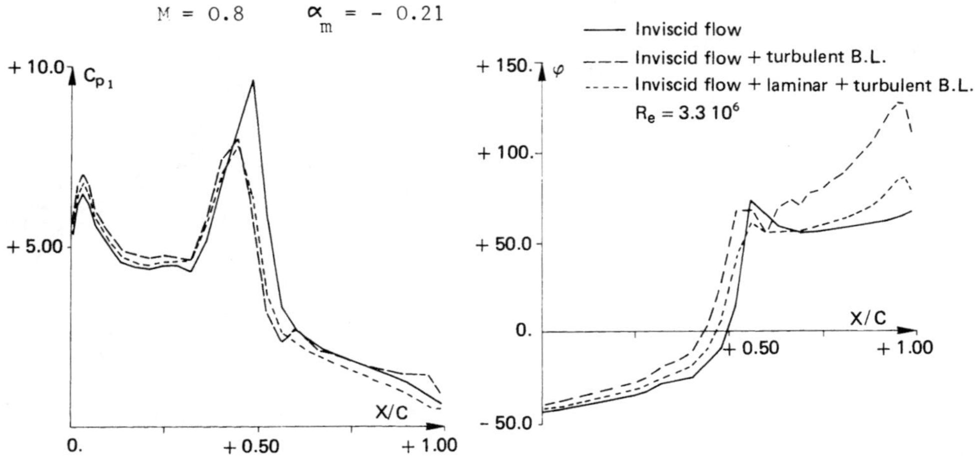

Fig. 13 : NACA 64 A 010 profile oscillating in pitch.

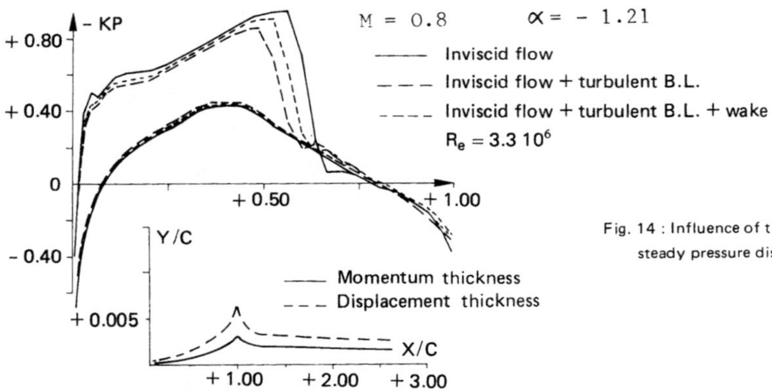

Fig. 14 : Influence of the wake on the steady pressure distribution.

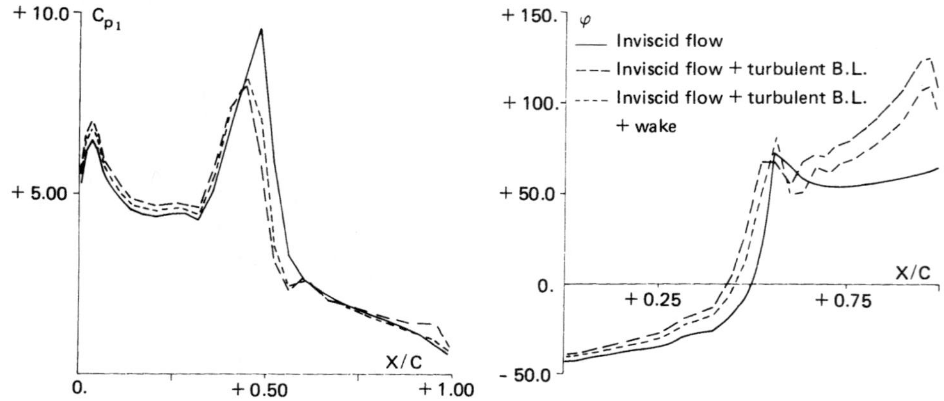

Fig. 15 : Influence of the wake on the unsteady pressure.

A Turbulent Flow Navier-Stokes Analysis for an Airfoil Oscillating in Pitch

S. J. Shamroth

Scientific Research Associates, Inc., Glastonbury, CT 06033 USA

Abstract

A time-dependent compressible turbulent Navier-Stokes analysis is applied to the oscillating airfoil flow field problem. The turbulence model is based upon a turbulence energy equation. Results obtained for an airfoil oscillating in pitch are compared to data.

Introduction

The present paper addresses a fluid mechanics problem containing important turbulent and unsteady effects, the high Reynolds number flow about a two-dimensional airfoil section oscillating in pitch. This effort has been motivated primarily by the flow field surrounding a helicopter rotor. As the helicopter blade travels through the rotor disc in forward flight, the blade experiences a continuous change in incidence angle and, therefore, prediction of airfoil characteristics including lift and moment coefficient, airfoil stall, etc. requires an analysis which includes both unsteady and turbulent effects. The focus of the present paper concerns the application of a compressible, turbulent, time-dependent Navier-Stokes calculation procedure to the oscillating airfoil problem.

Relatively early applications of the Navier-Stokes equations to incompressible, laminar airfoil flow fields were developed by Mehta and Lavan (Ref. 1) and Lugt and Haussling (Ref. 2), both of whom investigated impulsively started airfoil flow fields; some more recent incompressible laminar analyses are cited in Ref. 3. Although arguments can be made in favor of one of these procedures versus the other, it is clear that as a group they have demonstrated the practicality of using Navier-Stokes procedures to predict these complex flow fields. However, these efforts were limited by their assumptions of laminar incompressible flow.

More recent analyses including explicit, hybrid and implicit methods (see Ref. 3) have aimed at the compressible problem. Implicit schemes, although more complicated to code than explicit schemes, do not present the formidable coding problems associated with the hybrid scheme and are not subject to the severe stability limitations of explicit methods. Therefore, at present implicit schemes appear to be the most promising approach for the airfoil flow field Navier-Stokes analysis. An implicit solution of the full laminar, compressible Navier-Stokes equations has been obtained by Gibeling, Shamroth and Eiseman (Ref. 4) who applied the Briley-McDonald split linearized block implicit procedure (Ref. 5) to flow about a Joukowski airfoil. A similar procedure has been used by Sankar and Tassa (Ref. 6) to study an oscillating airfoil in a low Reynolds number compressible fluid and in Ref. 7 Steger applied the thin shear layer equations (a simplified form of the full Navier-Stokes equations) to the airfoil flow field problem.

Although Refs. 4 and 6 focused upon the compressible problem, they were confined primarily to laminar flow (Ref. 7 represented turbulence via a two layer eddy viscosity model) whereas most flow fields of practical interest are turbulent. In principle, a laminar procedure can be extended to turbulent flow in a straightforward manner if eddy viscosity and forced transition concepts are assumed viable. However, in the general airfoil flow field the eddy viscosity assumption which relates eddy viscosity to the mean flow via an algebraic equation is expected to be inadequate both in regions of strong pressure gradients and in separated flow. In addition, an important component of the flow field development may be the transition process since early transition may inhibit separation. In this regard a forced transition model, where the transition location is uniquely related to some mean flow parameter such as a boundary layer integral thickness, may lead to serious errors in the predicted results. Thus, a more general turbulence model is sought.

With these considerations in mind, a model aimed at predicting the flow in the laminar, transitional and turbulent regimes has been applied to the airfoil flow field by Shamroth and Gibeling (Ref. 8). The model combines a turbulence energy equation and an algebraic length scale equation and makes the specified turbulence structural coefficients a function of the local turbulence Reynolds number. The analysis was used to predict high Reynolds number turbulent flow about unstalled NACA 0012 airfoils at zero and six degrees incidence. As discussed in Ref. 8, converged airfoil solutions were obtained in a relatively few number of time steps (≈ 150) and the predicted results including the transition location showed reasonable agreement with experimental data.

This same procedure was applied by Shamroth and Gibeling (Ref. 3) to high Reynolds number flows about airfoils both in ramp motion and in stall.

As shown in Ref. 3, comparisons between predicted flow behavior of a stalled NACA 0012 airfoil at 19 degree incidence was in reasonable agreement with the data of Young, Meyers and Hoad (Ref. 9) for this same case. More recently this same procedure has been used to compute the transonic flow around a loaded cascade of cambered airfoils (Ref. 10). The present paper focuses upon applying this turbulent Navier-Stokes calculation procedure to an airfoil oscillating in pitch. The results presented focus upon an airfoil oscillating below stall conditions; however, the eventual objective is to consider the stalled regime.

Turbulence Model

The turbulence model utilized in the present effort is the turbulence energy-algebraic length scale model previously used in Refs. 3 and 8. The approach assumes an isotropic turbulent viscosity, μ_T, relating the Reynolds stress tensor to mean flow gradients. The turbulent viscosity is related to the turbulence energy, k, and the turbulence energy dissipation rate, ε, via the Prandtl-Kolmogorov constitutive equation

$$\mu_T = \rho C_\mu k^2/\epsilon \; f(y/\delta) \tag{1}$$

where C_μ is a turbulence structural coefficient and $f(y/\delta)$ is a factor used to ensure small turbulent viscosities at locations far from the airfoil. The function $f(y/\delta)$ is taken as

$$f(y/\delta) = 1.0 \qquad y \le \delta \qquad f(y/\delta) = e^{-b(y/\delta - 1.0)} \qquad y > \delta \tag{2}$$

where b is a constant, δ is the local boundary layer thickness and y represents normal distance from the airfoil surface.

The turbulence energy, k, is obtained from the turbulence energy equation

$$\frac{\partial \rho k}{\partial t} + \frac{\partial \rho u k}{\partial x} + \frac{\partial \rho u k}{\partial y} = \frac{\partial}{\partial x_k}\left[\left(\mu + \frac{\mu_T}{\sigma_k}\right)\frac{\partial k}{\partial x_k}\right] + \mu_T\left(\frac{\partial u_i}{\partial x_k} + \frac{\partial u_k}{\partial x_l}\right)\frac{\partial u_i}{\partial x_k} - \rho\epsilon - 2\mu\frac{\partial k^{1/2}}{\partial x_j}\frac{\partial k^{1/2}}{\partial x_j} \tag{3}$$

The turbulence energy dissipation rate ε, is related to a length scale ℓ, the turbulence energy, k, and the structural coefficient C_μ via the equation

$$\epsilon = C_\mu^{3/4}\frac{k^{3/2}}{\ell} \tag{4}$$

The length scale is taken as a minimum value of two lengths; a wall length and a wake length. The wall length is assumed to be given by a conventional wall damped Prandtl's mixing length, via

$$\ell = \kappa y_s\left(1 - e^{-y_s^+/27}\right) \tag{5}$$

with a maximum value of 0.09 δ. In Eq. (5) κ is the von Karman constant taken as 0.43, y^+ is the dimensionless distance from the airfoil surface and δ is the boundary layer thickness. The wake length scale was taken as $\ell = .05\ L_w$ where L_w is the wake thickness. In regions of separated flow the length scale is modified so that $\ell \geq \ell min$ where

$$\ell_{MIN} = 0.1h\left(1 - e^{-y^+/27}\right) \tag{6}$$

where h is the local height of the separated region. Finally, the coefficient C_μ is evaluated following the procedure of McDonald and his coworkers (e.g. Ref. 11) and is taken as a function of the local turbulence Reynolds number as discussed in Refs. 3 and 8.

It should be noted that with the current turbulence model, the turbulence equations are solved in conjunction with the mean flow equations throughout the flow field. The analysis predicts some regions having a turbulent viscosity much larger than the laminar viscosity (turbulent regions), other regions having a turbulent viscosity on the order of the laminar viscosity (transitional regions), and finally, some regions having turbulent viscosity less than the laminar viscosity (laminar regions). No transition location per se is input into the analysis.

Mean Flow Equations

The present approach utilizes the ensemble-averaged, compressible, time-dependent Navier-Stokes equations to predict the airfoil flow field. With the assumed turbulence model, the 'turbulent shear stress' is incorporated within the eddy viscosity, μ_T. The equations solved are obtained by transforming the continuity and two momenta equations into general spatial coordinates (ξ, η) where

$$\xi = \xi(x,y,t) \qquad \eta = \eta(x,y,t) \qquad \tau = t \tag{7}$$

and where x and y are the Cartesian spatial coordinates and t is time. The equations are expressed in the form

$$\frac{\partial W}{\partial \tau} + \xi_t \frac{\partial W}{\partial \xi} + \xi_x \frac{\partial F}{\partial \xi} + \xi_y \frac{\partial G}{\partial \xi} + \eta_t \frac{\partial W}{\partial \eta} + \eta_x \frac{\partial F}{\partial \eta} + \eta_y \frac{\partial G}{\partial \eta} = \frac{1}{Re}\left[\xi_x \frac{\partial F_1}{\partial \xi} + \eta_x \frac{\partial F_1}{\partial \eta} + \xi_y \frac{\partial G_1}{\partial \xi} + \eta_y \frac{\partial G_1}{\partial \eta} \right] \tag{8}$$

where

$$W = \begin{pmatrix} \rho \\ \rho u \\ \rho v \end{pmatrix}, \quad F = \begin{pmatrix} \rho u \\ \rho u^2 + p \\ \rho uv \end{pmatrix}, \quad G = \begin{pmatrix} \rho v \\ \rho uv \\ \rho v^2 + p \end{pmatrix} \quad F_1 = \begin{pmatrix} 0 \\ \tau_{xx} \\ \tau_{xy} \end{pmatrix} \quad G_1 = \begin{pmatrix} 0 \\ \tau_{xy} \\ \tau_{yy} \end{pmatrix} \tag{9}$$

The set of equations represented by Eq. (8) in conjunction with the turbulence model and the assumption of constant total temperature governs the flow field development.

The Numerical Procedure

The numerical procedure used to solve the governing equations is a consistently split linearized block implicit scheme originally developed by Briley and McDonald (Ref. 5). A recent comprehensive description of the method is given by Briley and McDonald in Ref. 12. The method can be outlined as follows: the governing equations are replaced by an implicit time difference approximation. Terms involving nonlinearities at the implicit time level are linearized by Taylor expansion about the solution at the known time level, and spatial difference approximations are introduced. The result is a system of multidimensional coupled (but linear) difference equations for the dependent variables at the unknown or implicit time level which are solved by the Douglas-Gunn procedure for generating alternating-direction implicit (ADI) schemes. Details of the procedure are given in Refs. 4, 5 and 8.

Boundary Conditions

An important component of the airfoil analysis concerns specification of boundary conditions. The present analysis utilizes a constructive coordinate system and requires boundary conditions to be set along the lines $\xi = \xi_{min}$, $\xi = \xi_{max}$, $\eta = \eta_{min}$ and $\eta = \eta_{max}$. With the coordinate system sketched in Fig. 1, $\xi = \xi_{min}$ (line EH) and $\xi = \xi_{max}$ (line DF) are downstream boundaries. The present formulation follows the suggestion of Briley and McDonald (Ref. 13), and specifies total pressure, incidence and density derivative along arc JNK, and static pressure and velocity derivative along the remainder of the outer boundary. No slip velocity conditions as well as a transverse momentum equation are satisfied on the airfoil surface. It should be noted that when using the full ensemble averaged Navier-Stokes equations there is no need to apply a Kutta condition at the trailing edge.

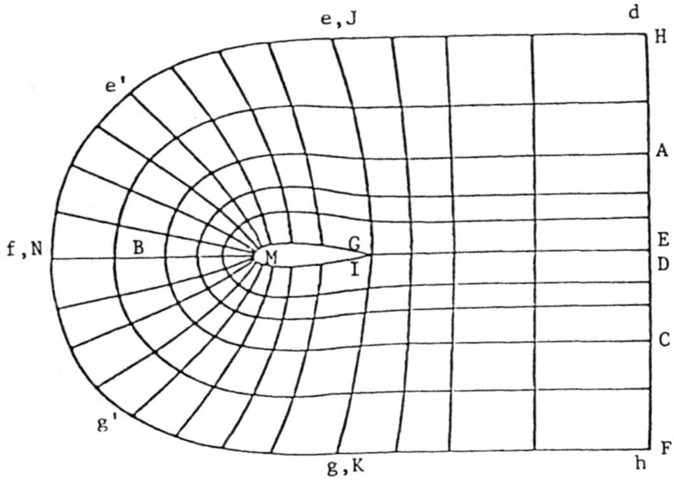

Figure 1. - Sketch of coordinate system.

Grid Spacing and Artificial Viscosity

The solution of the Navier-Stokes equations for an isolated airfoil at high Reynolds number presents a formidable grid resolution problem. If the regions having rapid changes in dependent variables are to be adequately resolved and if the outer boundary is to be placed in a region only modestly perturbed by the airfoil flow field, then considerable grid stretching must be used. The present calculation utilized a grid having 81 points in the pseudo-azimuthal (ξ) direction and 39 grid points in the pseudo-radial (η) direction (see Fig. 1). The radial grid was very highly resolved in the vicinity of the airfoil where the first grid point was placed 0.00002 chords from the airfoil surface. In contrast, the grid spacing in the outer region of the flow was of the order of 0.6 chords. Similarly 'streamwise' grid points were concentrated in the airfoil leading edge region.

A second problem which arises in high Reynolds number flow is the spurious oscillations associated with the so-called "cell Reynolds number problem". In the present approach these oscillations were damped by adding a normal diffusion term to the equations in which the artificial or numerical viscosity μ_a, was set by the criterion that the cell Reynolds number be less than or equal to 2. More recent calculations (Ref. 10) have indicated that it should be possible to reduce the artificial viscosity by an order of magnitude and still suppress the spatial oscillations. It should be noted that unless the flow gradients are large and poorly resolved, the artificial viscosity should not introduce significant artificial diffusion.

Results

The analysis described in the preceding section has been applied to a NACA 0012 airfoil oscillating in pitch. In the case considered the mean incidence is 5.25 degrees, the amplitude of oscillation is 5.25 degrees, the reduced frequency is 0.253, the approach Mach number is 0.20 and the free stream Reynolds number based upon airfoil chord is 0.26×10^7. The case was initiated from a turbulent zero degree airfoil calculation with the incidence angle assumed to follow the equation

$$\alpha = 5.25° + 5.25° \cos \omega (t - t_o + \pi)$$

The calculation was run for one and one-quarter cycles and required approximately 650 time steps.

Predicted values of pressure coefficient are compared to the steady data of Gregory and O'Reilly (Ref. 14) in Fig. 2 for three instantaneous incidence angles. At each incidence angle the pressure distribution is shown for two points on the cycle; one cycle point corresponds to $\dot{\alpha}>0$ and the second corresponds to $\dot{\alpha}<0$. At the first incidence presented, $\alpha = 2°$, the predicted suction peaks are both less than that measured; this discrepancy is discussed subsequently. The predicted suction surface pressure distribution for $\dot{\alpha}>0$ and $\dot{\alpha}<0$ differ primarily in the leading edge region where the suction peak for $\dot{\alpha}<0$ is somewhat more pronounced than that for $\dot{\alpha}>0$. This difference in results for $\dot{\alpha}>0$ and $\dot{\alpha}<0$ is consistent with the hysteresis effects. When $\dot{\alpha}>0$, the peak pressure is increasing with time and when $\dot{\alpha}<0$, the peak pressure is decreasing with time. Therefore, the different peak pressures at the same α, but different $\dot{\alpha}$ reflect a time lag in the development of the suction peak.

The next set of results are at 6° and these are compared to data for both pressure and suction surfaces. This is the lowest incidence at which Gregory and O'Reilly present pressure surface pressure coefficient data. The suction surface results are similar to those previously presented. On the pressure surface the $\dot{\alpha}>0$ prediction gives higher pressures than does the $\dot{\alpha}<0$ prediction. Although not shown, this trend also held for 2°, 4° and 8°, and it is this different pressure differential on the rear half of the airfoil which gives the major contribtuion to the lift hysteresis curve presented

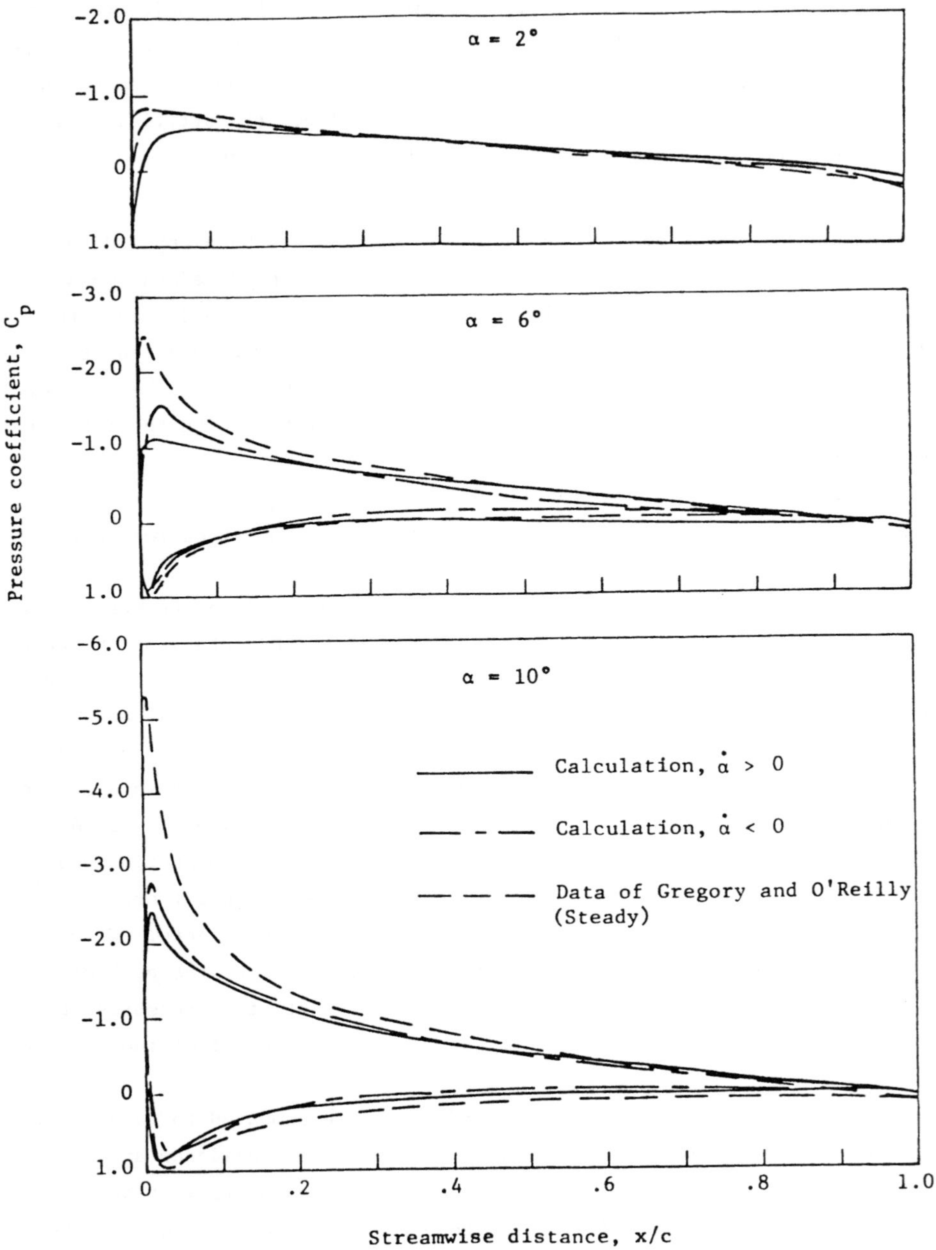

Fig. 2 - Surface pressure distribution for NACA 0012 airfoil in sinusoidal pitching motion, $\bar{\alpha} = 5.25°$, $\Delta\alpha = 5.25°$, $k = 0.253$.

subsequently. The final set of data on this figure shows surface pressure distribtuions at $\alpha = 10°$. As can be seen, the major qualitative discrepancy between the predicted time-dependent analysis and the steady data occurs in the vicinity of the airfoil leading edge. As discussed in Refs. 3 and 8, this is at least partially due to the lack of resolution of the leading edge suction peak region. The strong, favorable pressure gradient leading to the suction peak occurs over a very small distance between $x/c \approx 0$ and $x/c \approx .01$, and although the computational grid was packed in this region to gain resolution only four pseudo-radial lines were placed in this small region. Obviously, the strong pressure peak at high incidence angles could not be resolved.

The predicted lift versus incidence curve is presented along with the steady and unsteady data of Grey and Liiva (Ref. 15) in Fig. 3. Considering first the experimental data, the unsteady curve shows a hysteresis loop. Furthermore, the general slope of the curve is less than that of the steady data and the unsteady lift at zero incidence is higher than that of the steady data (which is zero). The prediction shows the same general characteristics. The calculation was initiated at zero degrees incidence from a steady calculation and followed the theoretical quasi-steady lift-incidence curve until $\alpha \approx 4°$. After reaching 4°, the lift predicted is less than the inviscid value and this is primarily a result of the under-prediction of the suction peak (see Fig. 2). Upon reaching the maximum incidence, $\alpha = 10.5°$, the curve forms a hysteresis loop as incidence decreases. This loop is somewhat more pronounced than that measured. After reaching the minimum value of $\alpha = 10°$, the lift increases with incidence and at the last time calculated the loop is closing. Although the thickness of the predicted hysteresis loop is somewhat greater than that of the measured loop, the average slopes agree. In addition, both prediction and data show significant lift at zero incidence; this is in contrast to the quasi-steady calculation.

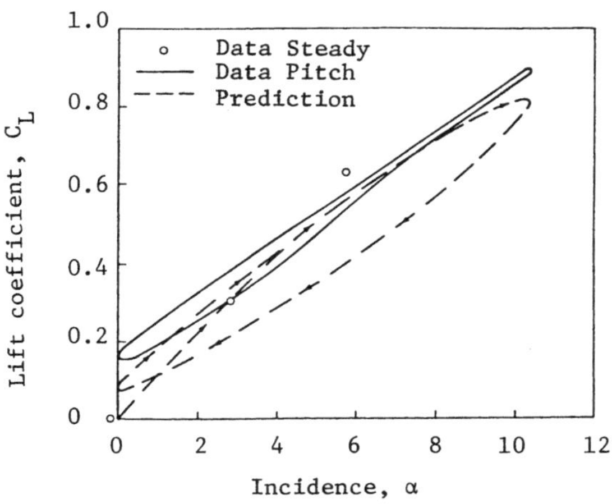

Fig. 3 - Lift vs. incidence curve for NACA 0012 airfoil in pitch, k - 0.25.

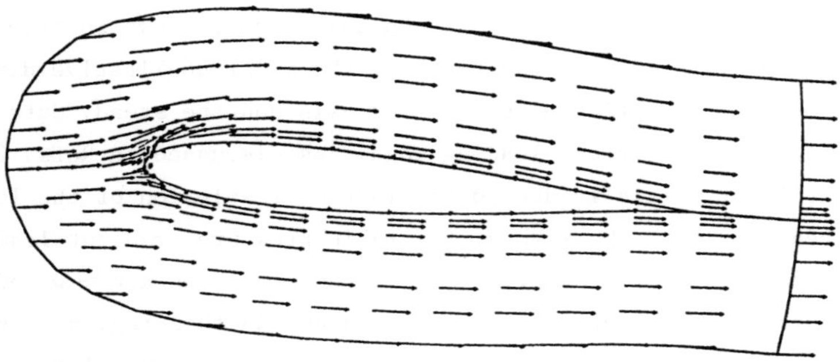

Velocity field, $\alpha = 5°$, $\dot{\alpha} > 0$.

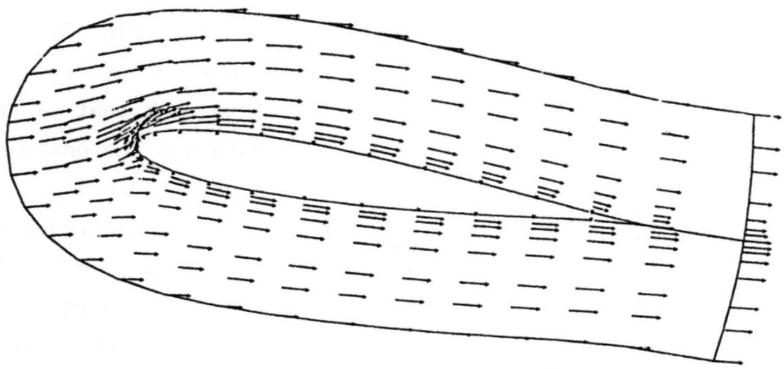

Velocity field, $\alpha = 10°$, $\dot{\alpha} > 0$.

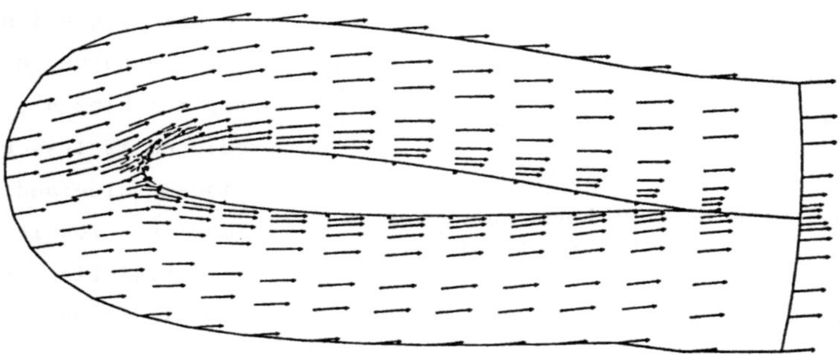

Velocity field, $\alpha = 5°$, $\dot{\alpha} > 0$.

Fig. - 4

A detailed examination of the flow field prediction shows the major contribution to the lift loop results from the suction surface boundary layer thickness for $\dot{\alpha}<0$ being greater than that for $\dot{\alpha}>0$ at the same value of α. This result represents a lag in the boundary layer reaction to the pressure gradientwhich modifies the mid chord and trailing pressure distribution. The mid chord and trailing edge effect is somewhat modified by differences in the leading edge where the suction peak for $\dot{\alpha}<0$ is more pronounced than that for $\dot{\alpha}>0$. It should be noted that the loop calculation is a very sensitive one and its formation results from relatively small pressure changes on both the pressure and suction surfaces. Velocity vector plots are given in Fig. 4. These figures clearly show the general flow pattern which includes the approach to the leading edge stagnation point, acceleration around the leading edge and the boundary layer and wake development. A comparison of the vector plots shows that during the upstroke ($\dot{\alpha}>0$) the flow along the aft portion of the airfoil tends to align with the suction (upper) surface whereas on the downstroke it tends to align with the pressure (lower) surface. Furthermore, the differences in the suction surface boundary layer thickness and wake position are clearly shown.

Concluding Remarks

The present paper applies the full, compressible, ensemble-averaged, time-dependent Navier-Stokes equations to the problem of an airfoil oscillating in pitch between zero and ten degrees. The analysis includes a turbulence model based upon the turbulence energy equation and an algebraic length scale equation. Although further studies must be made with a refined grid, the results obtained show the qualitative features of the oscillating airfoil flow field and the predicted lift curve shows qualitative agreement with experimental data.

Acknowledgement

This work was supported under a joint NASA/Army program between the Structures Laboratory, U.S. Army Research and Technology Laboratory (AVRADCOM) and NASA Langley Research Center under Contract NAS1-15214.

REFERENCES

1. Mehta, U.B. and Lavan, Z.: Starting Vortex, Separation Bubble and Stall: A Numerical Study of Laminar Unsteady Flow about an Airfoil. J. Fluid Mech., Vol. 67, 1975, pp. 227-256.
2. Lugt, H.J. and Haussling, H.J.: Laminar Flow about an Abruptly Accelerated Elliptic Cylinder at 45° Incidence. J. Fluid Mech., Vol. 65, 1974, pp. 611-734.
3. Shamroth, S.J. and Gibeling, H.J.: Analysis of Turbulent Flow about an Isolated Airfoil Using a Time-Dependent Navier-Stokes Proceudre. Paper presented at AGARD Specialists Meeting on Boundary Layer Effects on Unsteady Airloads, Aix-en-Provence, September 1980.
4. Gibeling, H.J., Shamroth, S.J., and Eiseman, P.R.: Analysis of Strong-Interaction Dynamic Stall for Laminar Flow on Airfoils. NASA CR-2969, April 1978.
5. Briley, W.R. and McDonald, H.: Solution of the Multidimensional Compressible Navier-Stokes Equations by a Generalized Implicit Method. J. Comp. Physics, Vol. 24, No. 4, August 1966, p. 372.
6. Sankar, N.L. and Tassa, Y.: Reynolds Number and Stability Effects on Dynamic Stall of an NACA0012 Airfoil. AIAA Paper No. 89-0010, 1980.
7. Steger, J.L.: Implicit Finite Difference Simulation of Flow About Arbitrary Two-Dimensional Geometries. AIAA Journal, Vol. 16, 1978, pp. 679-686.
8. Shamroth, S.J. and Gibeling, H.J.: A Compressible Solution of the Navier-Stokes Equations for Turbulent Flow About an Airfoil. NASA CR-3183, 1979. (See also AIAA Paper 79-1543).
9. Young, W.H., Jr., Meyers, J.F. and Hoad, D.R.: A Laser Velocimeter Flow Survey Above a Stalled Wing, NASA Technical Paper 1266, AVRADCOM Technical Report 78-50, 1978.
10. Shamroth, S.J., McDonald, H. and Briley, W.R.: A Navier-Stokes Solution for Transonic Flow Through a Cascade. Report in preparation.
11. McDonald, H. and Fish, R.W.: Practical Calculation of Transitional Boundary Layers. Int. J. Heat and Mass Transfer, Vol. 16, No. 9, 1973, pp. 1629-1744.
12. Briley, W.R. and McDonald, H.: On the Structure and Use of Linearized Block Implicit Schemes. J. of Comp. Phys., Vol. 34, 1980, pp. 54-73.
13. Briley, W.R. and McDonald, H.: Computation of Three-Dimensional Horseshoe Vortex Flow Using the Navier-Stokes Equations. Seventh International Conference on Numerical Methods in Fluid Dynamics, 1980.
14. Gregory, N. and O'Reilly, C.L.: Low Speed Aerodynamic Characteristics of NACA0012 Airfoil Section, Including the Effects of Upper Surface Roughness Simulating Hoarfrost. Aero Report 1308. National Physics Laboratory, 1970.
15. Grey, L. and Liiva, J.: Two-Dimensional Tests of Airfoil Oscillating Near Stall. Volume II: Data Report. USAAVLABS Report 68-13B, 1968.

Unsteady Kutta Condition of a Plunging Airfoil

HO, CHIH-MING and CHEN, SHIN-HSING

Department of Aerospace Engineering
University of Southern California
Los Angeles, California 90007

Abstract

Unsteady Kutta condition is an important criterion for theoretical analyses in unsteady aerodynamics and in aerodynamic noise generations. The experimental studies on this subject are very limited. In the present investigation, the stagnation streamline at the trailing edge of a NACA 0012 airfoil in plunging motion is measured from phase averaged streamwise and transverse velocity components. The Kutta condition is examined for reduced frequency up to 1.0 and at different angles of attack.

Introduction

In many engineering applications, the airfoils either are in unsteady motion itself, e.g. helicopter blades and rotors in turbines, or encounter unsteady incoming flows, e.g. stator blades. The unsteady flows around the airfoil are rather complicated (McCroskey [1] as well as McAlister and Carr [2]). The lift of an unsteady airfoil varies during one cycle of a revolution. Unsteady pressure fluctuations on the airfoil produce vibrations and radiates noise. The Kutta condition is applied in order to determine the pressure distribution on the airfoil. For an airfoil with a sharp trailing edge in the flow, the Kutta condition requires the rear stagnation point to be on the trailing edge, so that the singularity can be removed. The surface pressure distribution and the lift then can be calculated. In a steady flow, the Kutta condition is well established. When the flow is unsteady, the validity of the Kutta condition is still an unsettled problem.

Only a few experiments on the unsteady Kutta condition are available. These existing experiments are performed in different types of flow configurations. The results are not conclusive as far as the applicable range of the reduced frequency is concerned. This could be a physical fact, because the validity of the Kutta condition might not be a function of the reduced frequency only and could vary with flow configurations. Kovasznay and

Fujita [3] examined a flat plate placed in the potential core of a jet perturbed by a rotating rod. The streamline pattern near the trailing edge was constructed. The evolution of the stagnation streamline indicated that the Kutta condition is valid up to $K = (\omega c)/(2U_o) = 4.9$. Archibald [4] studied the trailing edge loading of a flat plate and an airfoil under self-excited acoustic perturbations. The unsteady Kutta condition does not hold for $K>7$. Osdiek [5] found that the trailing edge pressure distribution of blades in cascade did not follow theoretical prediction even at $K = 0.08$. Commerford and Carta [6] investigated the pressure fluctuations at 90% of a circular arc airfoil and concluded the Kutta condition is satisfied at $K = 3.9$, but the agreement with phase is very poor. Satyanarayana and Davis [7] concluded that the Kutta condition is applicable for a pitching airfoil up to $K = 0.6$. Bechert and Pfizenmaier [8] studied the trajectory of particles leaving a jet nozzle under excitation. They found the Kutta condition was not valid in this case. Fleeter [9] studied the isolated airfoil or airfoils in cascade. The Kutta condition held for flat airfoils up to $K = 10$, but not for cambered airfoils in cascade.

Apparently, the flow configuration plays a dominant role in deciding the applicability of the Kutta condition. For helicopters in forward flight, the blades experience three different unsteady modes, the plunging mode, the pitching mode and the translational mode. The validity of Kutta condition should be examined in individual modes. In the present study, the near wake of a plunging mode is studied. The reduced frequency has more then an order of magnitude variation and ranges from 0.05 to 1.0. Three angles of attack, $\theta = 5°$, $7.5°$ and $10°$, are used in the experiment.

Experimental Facilities

The airfoil and the driving mechanism are located in a low turbulence ($u'/U_o<0.02\%$) wind tunnel. The speed of the tunnel can be varied from 2.5 m/sec to 33 m/sec. The airfoil has a NACA 0012 profile and is molded from high strength epoxy filled with aluminum powder (DEVCON F2). No detectable deformation is found, even the airfoil is under the most severe test conditions. i.e. $f = 20$Hz and $U_o = 33$ m/sec. The chord of the airfoil is $c = 10$cm, the span is 53cm.

The driving mechanism of the unsteady airfoil has a simple and versatile design. A motor and gear combination can provide a wide variation of oscillation frequencies. The amplitude of oscillation is 0.32cm, which can be changed by using wheels with different eccentricities. The angle of attack can be adjusted by set screws at the connections, between airfoil and the driving mechanism. A photo-electric phase reference device is attached to a gear of the driving mechanism. The triggering signal provides a phase reference to ensemble average the measured signals.

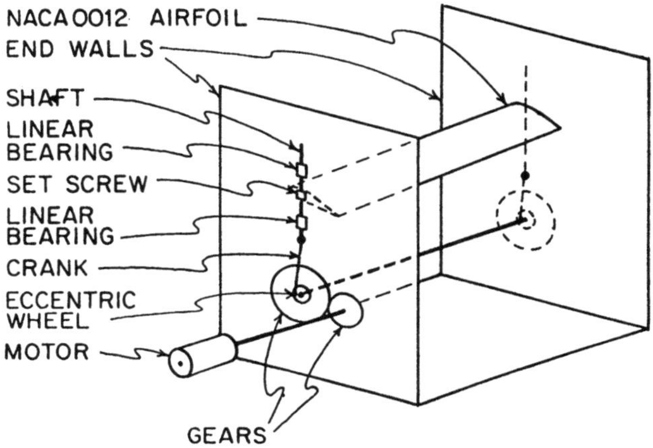

Fig. 1 Experimental Set-up.

A miniature hot-wire rake was made to survey the flow. There are ten hot-wires (five x-wires) on the rake. The cross section of each x-wire is 1.0mm x 1.4mm. The distance between the x-wire is 3mm. The hot-wires are calibrated against a capacitive pressure gauge which is accurate to 5×10^{-5}mm Hg. Consequently, both the streamwise and transverse velocity components can be accurately measured. The hot-wire rake is mounted on a traverse mechanism which is driven by stepping motors. The spatial resolution can be as fine as 2×10^{-4}cm.

All of the ten hot-wire outputs together with the phase reference signal are directly connected to the analog - digital converter of a PDP 11/55 minicomputer. The phase averaging of the velocity is processed digitally in the computer. About one hundred and fifty ensembles are used in each phase averaged velocity profile.

Experimental Results

The Unsteady Wake: the flow properties of a wake after a plunging airfoil, $\theta = 5°$, were investigated in detail by Ho and Chen [10]. Many interesting features were observed. The mean velocity has an asymmetric bell shaped profile. The instantaneous velocity traces measured simultaneously by the five hot-wire rake revealed that the fluctuating velocities are also asymmetric. In the lower half of the wake, the turbulent structure has a low frequency intermittent pattern. In the upper half of the wake, high frequency and low intermittancy fluctuations are the general features. It is very difficult to measure the mean transverse velocity component because it is only a few per cent of the streamwise velocity. We are able to obtain the accurate mean transverse velocity component through a careful calibration procedure. Both the mean streamwise and mean transverse velocity profiles indicate that the wake consists of two parts; a viscous wake with a large velocity defect and narrow width as well as an inviscid wake with a small velocity defect and about a two chord length in width. Except for the mean streamwise velocity components, the other flow properties, e.g. the turbulence levels and the Reynolds stress, do not reach self-similar within a chord length from the trailing edge.

<u>The Phase Averaged Velocity Profiles</u>: both the phase averaged streamwise, $\tilde{U}$, and transverse, $\tilde{V}$ velocity profiles were measured in the near wake ($x/c<1$) for all the test cases. Examples of the profiles are shown in Fig. 2. The abscissa is the vertical position. The origin of the y axis is chosen at the lowest position of the trailing edge. Both $\tilde{U}$ and $\tilde{V}$ are normalized to the free stream velocity U_o and the scales are shown at the right of the figures. The profiles at each phase angle are displaced in the ordinate as being shown on the left of the figures, so that variations can be easily observed.

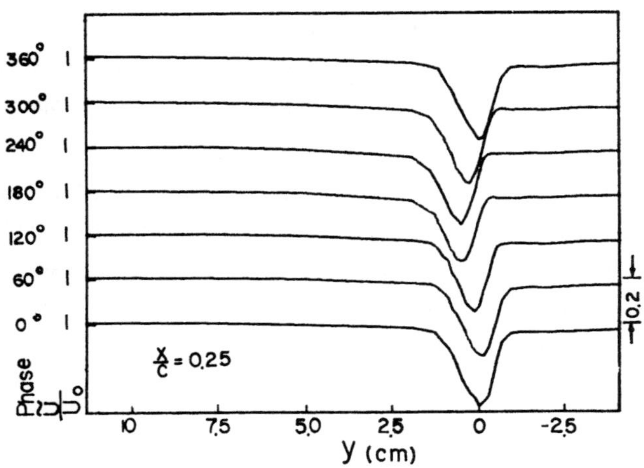

Fig. 2a Phase averaged streamwise velocity.

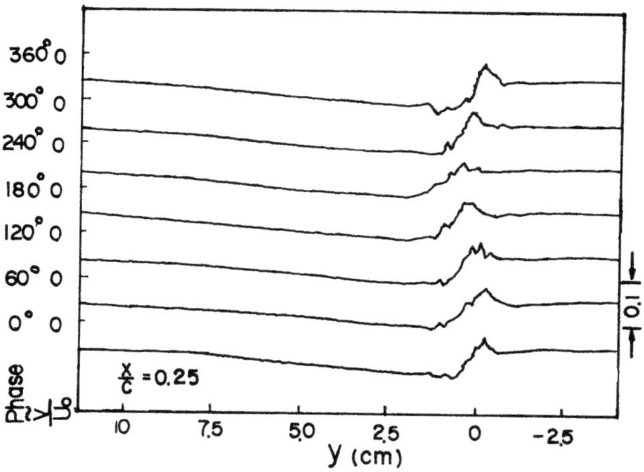

Fig. 2b Phase averaged transverse velocity

$\alpha_o = 5°$

$Re = 2.71 \times 10^4$

$K = 0.256$

The phase averaged streamwise velocity distributions have asymmetric bell shaped profiles which are similar to the time averaged streamwise velocity distribution. A slight variation at each phase angle is observed. The location of the minimum velocity follows the motion of the plunging airfoil. The transverse velocity is always less than zero across the wake because of the mean circulation. Inside the viscous region, the $\tilde{V}$ at the lower side of the wake is less negative than that at the upper side of the wake, since the boundary layer from the lower side has a counter clockwise vorticity, while the boundary layer from the upper side has a clockwise vorticity. These features mentioned above are common in other angles of attack, and at higher reduced frequencies up to K = 1 which is the highest reduced frequency measured in the experiment.

<u>The Phase Averaged Rear Stagnation Streamlines</u>: while the rear stagnation streamline was studied, the phase averaged velocity components were measured at about one hundred fifty stations distributed within a quarter chord distance from the trailing edge. The location of the stagnation streamline at each phase angle was obtained by taking the following steps:

(1) Let the stream function at the trailing edge, (x_o, y_o) equal zero $\psi(x_o, y_o) = 0$

(2) Integrate the velocity components along path I (Fig. 3) until the stream function equals zero again at (x_I, y_{PI}).

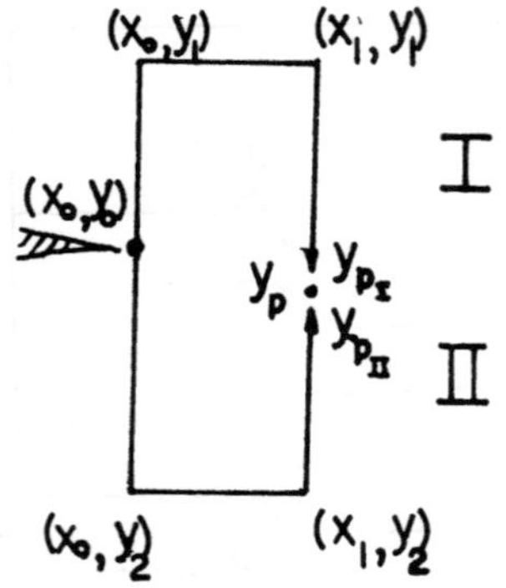

Fig. 3 The integration paths.

$$\text{I}: \int_{y_0}^{y_1} \tilde{U}(x_0, y')dy' - \int_{x_0}^{x_1} \tilde{V}(x', y_1)dx' = \psi(x_1, y_1)$$

$$\psi(x_1, y_1) + \int_{y_1}^{y_{p_I}} \tilde{U}(x_1, y')\, dy' = \psi(x_1, y_{p_I}) = 0$$

The point, (x_1, y_{PI}), indicates the location of the phase averaged stagnation streamline at $x = x_1$.

(3) Second integration path is used to locate the stagnation streamline at the same downstream location, $x = x_1$. The value of the stream function becomes zero at (x_1, y_{PII}) along path II.

$$\text{II}: \int_{y_0}^{y_2} \tilde{U}(x_0, y')dy' - \int_{x_0}^{x_1} \tilde{V}(x', y_1)dx' = \psi(x_1, y_2)$$

$$\psi(x_1, y_2) + \int_{y_2}^{y_{p_{II}}} \tilde{U}(x_1, y')dy' = \psi(x_1, y_{p_{II}}) = 0$$

In principle, the point (x_1, y_{PI}) should coincide with (x_1, y_{PII}). However, the two points usually are not the same in practice. The difficulty in obtaining the extremely accurate transverse velocity component is the main reason for the discrepancy.

(4) The averaged value, y_P, of y_{PI} and y_{pII} is taken as the position

of the stagnation streamline, $y_p = (y_{pI} + y_{pII})/2$.

The same procedure is carried out at several downstream locations (Fig. 4) so that the stagnation streamline at a certain phase angle can be determined. The error bars in the diagram indicate the limit of y_{PI} and y_{PII}. In the present experiment, the typical error is less than one mm.

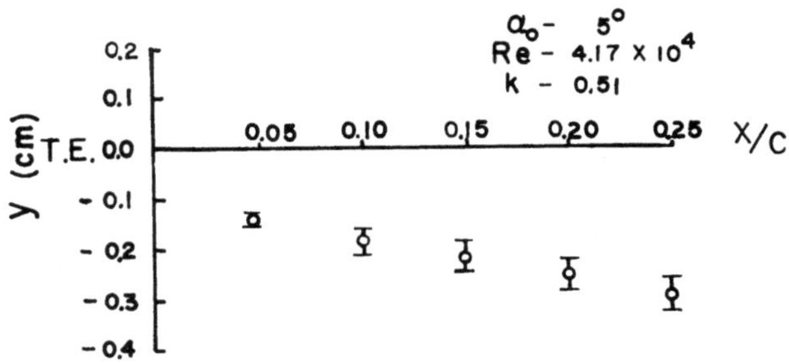

Fig. 4 Phase averaged rear stagnation streamline.

The Unsteady Kutta Condition: After the locations of the phase averaged stagnation streamlines are determined, the inclination angle between the stagnation streamline and the mean flow direction can be calculated from the vertical displacement of the streamline and the streamwise distance from the trailing edge. In Fig. 5, the difference between the inclination angle and the mean angle of attack are plotted against the phase angles. The solid line represents the difference between the instantaneous angle of attack of the airfoil and the mean angle of attack. Figure 5 reveals that the rear stagnation streamlines follow the instantaneous angle of attack at all phase angles. In other words, the phase averaged stagnation streamlines leave the trailing edge tangentially to the chord line of the plunging airfoil. The same characteristics are observed for other reduced frequencies lower than 0.51 and at angles of attack of 5° and 7.5°. Therefore, the unsteady Kutta condition is validated by the experimental evidence for $K<0.51$ and $\alpha_o<7.5°$. At $\alpha_o = 10°$ and $K = 0.51$, the scattering of data is very much increased, the validity of unsteady Kutta condition is in doubt. While the reduced frequency is increased to the order of one, the inclination of the stagnation streamlines do not follow the instantaneous angles of attack at all (Fig. 6).

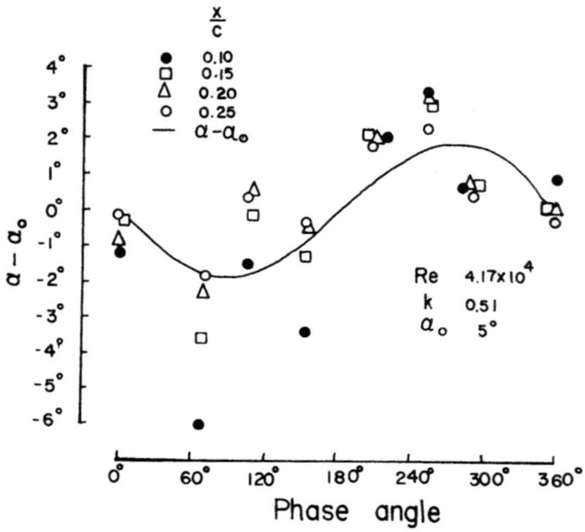

Fig. 5 The inclination angle of stagnation streamline at K = 0.51.

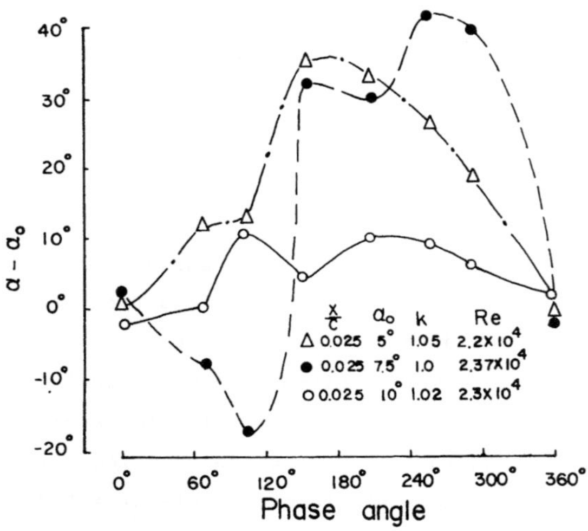

Fig. 6 The inclination angle of stagnation streamline at K≃1.0.

Large displacements of the stagnation streamlines are detected at various phase angles. As much as 40° of deviation is observed in Fig. 6. Hence, the unsteady Kutta condition is violated at K = 1.0.

Conclusions

The velocity field in the wake of a NACA 0012 airfoil is investigated in detail. Both the streamwise and transverse velocity components are measured at about one hundred and fifty stations downstream from the trailing edge. Three angles of attack, 5°, 7.5° and 10°, are used and the reduced frequency is varied from 0 to 1.0. The time averaged and phase averaged flow properties are investigated. The wake is asymmetric because the turbulence structures are different in the upper and lower portions of the wake. Except for the time mean streamwise velocity profiles, other flow profiles do not reach self-similar within one chord length downstream from the trailing edge [10]. Observable variations in the phase average velocity profiles are detected during a cycle of oscillation, but no dramatic change appears even at the highest angle of attack, $\alpha_o = 10°$, and the highest reduced frequency, $K = 1.0$. The validity of the unsteady Kutta condition is investigated from the phase averaged rear stagnation streamline. The unsteady Kutta condition holds for $K<0.51$ and $\alpha_o<7.5°$. Clear violation of the unsteady Kutta condition is observed for $K = 1.0$ at all the angles of attack studied in the present experiment.

Acknowledgment

This work is supported by Army Research Office under Contract No. DAAG29-78-G-0023.

References

1. McCroskey, W.J.: Recent Developments in Dynamic Stall. Proc. of Unsteady Aerodynamics, vol. 1, p. 1, 1975.

2. McAlister, K.W. and Carr, L.W.: Water Tunnel Visualization of Dynamic Stall. Nonsteady Fluid Dynamics (ed. by Crow, P.E. and Miller, J.A.) p. 103, 1978.

3. Kovasznay, L.S.G. and Fujita, H.: Unsteady Boundary Layer and Wake Near the Trailing Edge of a Flat Plate. Proc. IUTAM Symp. at Laval, Canada p. 805, 1972.

4. Archibald, F.S.: Unsteady Kutta Condition at High Values of the Reduced Frequency Parameter. J. of Aircraft, Vol. 12, p. 545, 1975.

5. Osdiek, F.R.: A Cascade in Unsteady Flow. AGARD CP 177 1975.

6. Commerford, G.L. and Carta, F.O.: Unsteady Aerodynamic Response of a Two Dimensional Airfoil at High Reduced Frequency. AIAA J., Vol. 12, p. 43, 1974.

7. Satyanarayana, B. and Davis, S.: Experimental Studies of Unsteady Trailing Edge Conditions. J. AIAA, Vol. 16, No. 2, p. 125, 1978.

8. Bechert, D. and Pfizenmaier, E.: Optical Compensation Measurements on the Unsteady Exit Condition at a Nozzle Discharge Edge. JFM, Vol. 71,

p. 123, 1975.

9. Fleeter, S.: Trailing Edge Conditions for Unsteady Flows at High Reduced Frequency. AIAA Paper No. 79-0152, 1979.

10. Ho, C.M. and Chen, S.H.: Unsteady Wake of a Plunging Airfoil. AIAA Paper No. 80-1446, 1980.

Dynamic Stall of an Oscillating Airfoil in Turbulent Flow Using Time Dependent Navier-Stokes Solver

Y. TASSA AND N. L. SANKAR

Lockheed-Georgia Company, Marietta, Georgia

Summary

The unsteady compressible Reynolds time averaged Navier-Stokes equations which include an algebraic turbulence model have been applied to an oscillating airfoil in turbulent flow. The governing equations are written in conservation form in a body fitted coordinate system and solved using an Alternating Direction Implicit (ADI) procedure. Results are presented for turbulent flow about NACA 0012 and the ONERA-CAMBRÉ airfoils whose incidence oscillate from 0 degree to 20 degrees. The effects of reduced frequency and leading edge camber on the normal force and pitching moment coefficients are analyzed and qualitatively good agreement has been obtained with experimental data.

Introduction

The problem of dynamic stall of an airfoil has been a topic of great interest and active research in recent years, experimentally as well as theoretically. The problem presents a unique combination of unsteady effects, non-linearity and strong viscous inviscid interaction. It has been observed in wind tunnel experiments [1] that the basic feature of dynamic stall of an airfoil is the mechanism of an abrupt turbulent leading edge separation forming a strong vortex that is convected downstream over the upper surface of the airfoil, distorting the pressure distribution, thus causing transient forces and moments that are basically different from the static stall condition. Related analytical research mainly in the area of unsteady thin airfoil theory [2] and in the area of unsteady boundary layer theory coupled with unsteady full potential flow theory [3,4,5] have added insight into the effects of reduced frequency, Reynolds number and the external unsteady potential flow over the onset of dynamic stall. These approaches are limited to mild viscous inviscid interaction where the boundary layer remains thin. In flow cases such as deep stall which are characterized by the shedding of a strong vortex from the leading edge region and convected downstream, the viscous layer is of the order of the airfoil chord during the vortex shedding process, consequently, this class of flow problems may be properly simulated only by the use of the full Navier-Stokes equations. An early attempt of the Navier-Stokes analysis to simulate the dynamic stall problem was done by Mehta [6] who solved the incompressible laminar stream function-vorticity formulation at low Reynolds numbers. The computed streamlines contours resemble the flow visualization by Werlé [7]. Sankar and Wu [8] applied the integro-differential formulation to the oscillating airfoil. These methods are however restricted to incompressible and laminar flows, whereas most flowfield of practical interest are compressible and turbulent. Sankar and Tassa [9] studied the oscillating airfoil in a compressible low Reynolds number fluid and the first attempt to study the flow about an airfoil in a ramping motion in high Reynolds number turbulent flow was done by Shamroth and Giebling [10]. The main objectives of this paper are first to analyze the effects of the reduced frequency on the dynamic stall characteristics in the regime of deep stall under compressible high Reynolds number turbulent flow conditions, and secondly to study the leading edge camber effects on the leading edge vortex generation, shedding and its corresponding forces and moments coefficients loops. In the following sections we first outline the mathematical formulation, then the numerical procedure and finally the numerical results.

Mathematical Formulation

The two dimensional unsteady compressible Navier-Stokes equations may be written in a strong conservation form in a general non-orthogonal curvilinear coordinate system as (cf. Peyret and Viviand [11], Steger [12] and Vinokur [13])

$$\frac{\partial}{\partial \tau}\{\hat{q}\} + \frac{\partial}{\partial \xi}\{\hat{F}\} + \frac{\partial}{\partial \eta}\{\hat{G}\} = 0 \tag{1}$$

here ξ, η, and τ are the independent variables subject to the general transformation:

$$\begin{aligned} \xi &= \xi(x,y,t) \\ \eta &= \eta(x,y,t) \\ \tau &= t \end{aligned} \tag{2}$$

and:

$$\begin{aligned} \{\hat{q}\} &= \frac{1}{J}\{\vec{q}\} \\ \{\hat{F}\} &= \frac{1}{J}\{\xi_t\vec{q} + \xi_x\vec{F} + \xi_y\vec{G}\} \\ \{\hat{G}\} &= \frac{1}{J}\{\eta_t\vec{q} + \eta_x\vec{F} + \eta_y\vec{G}\} \end{aligned} \tag{3}$$

where:

$$\vec{q} = \begin{Bmatrix} \rho \\ \rho u \\ \rho v \\ \rho(e + \frac{u^2+v^2}{2}) \end{Bmatrix} \qquad \vec{F} = \begin{Bmatrix} \rho u \\ \rho u^2 + p - \tau_{xx} \\ \rho uv - \tau_{xy} \\ \rho u(\gamma e + \frac{u^2+v^2}{2}) - \beta_x \end{Bmatrix}$$

$$\vec{G} = \begin{Bmatrix} \rho v \\ \rho uv - \tau_{xy} \\ \rho v^2 + p - \tau_{yy} \\ \rho v(\gamma e + \frac{u^2+v^2}{2}) - \beta_y \end{Bmatrix} \tag{4}$$

$$J = \xi_x\eta_y - \xi_y\eta_x \tag{5}$$

$$\xi_t = -x_\tau\xi_x - y_\tau\xi_y \tag{6}$$

$$\eta_t = -x_\tau\eta_x - y_\tau\eta_y \tag{7}$$

$$\tau_{xx} = \frac{\lambda}{R_e}\mu_T(u_x + v_y) + 2\frac{\mu_T}{R_e}u_x \tag{8}$$

$$\tau_{xy} = \frac{\mu_T}{R_e}(u_y + v_x) \tag{9}$$

$$\tau_{yy} = \frac{2}{R_e}\mu_T v_y + \frac{\lambda\mu_T}{R_e}(u_x + v_y) \tag{10}$$

$$\beta_x = \frac{\gamma K}{R_e P_r}e_x + u\tau_{xx} + v\tau_{xy} \tag{11}$$

$$\beta_y = \frac{\gamma K_T}{R_e P_r}e_y + u\tau_{xy} + v\tau_{yy} \tag{12}$$

$$\mu_T = \mu + \mu_t \tag{13}$$

where p is the pressure, ρ is the density, e the specific internal energy and λ is taken as $- 2/3\ \mu_T$ according to the Stoke's hypothesis. Here x_τ and y_τ are the cartesian velocities of the grid point (x,y). For oscillating airfoil problems as the body fitted coordinate system rotates with the airfoil one can show that x_τ and y_τ are the x and y components of $\vec{r} \times \vec{\Omega}$ where $\vec{\Omega}$ is the angular velocity vector of the airfoil, and x and y are the coordinates of the grid as observed in the inertial frame. The origin of the inertial system coincides with the axis of rotation. In the above equations all distances are normalized with respect to the airfoil chord C, the velocities are normalized with respect to free stream velocity V_∞, the density is normalized with respect to the free stream density ρ_∞ and the specific internal energy is normalized with respect to V_∞^2. R_e and P_r are the Reynolds number and Prandtle number respectively and μ_T is the total viscosity.

Grid Generation

The main advantage of a generalized curvilinear coordinate system is that boundary surfaces in the physical plane are mapped onto rectangular surfaces in the transformed plane and boundary conditions may be treated more accurately. Also grid points may be clustered in regions where rapid changes in the flow field gradients occur. The grid generation used in the present work is based on Thompson et al [14] method that solves two poisson equations. An '0' type grid such as shown in Fig. 1 has been used in the present work since it gives the best airfoil resolution for the same number of grid points. In the present calculation the first grid point is placed 0.0001 chords from the airfoil surface, and 31 grid points are placed in the η direction. The airfoil surface is represented by 49 grid points and the outer boundary is placed at 16 chord lengths from the airfoil surface.

Numerical Formulation

The numerical procedure used to solve the governing equations is a modified form of the Briley-McDonald ADI procedure described in Reference [15]. It is also closely related to the Warming-Beam algorithm [16]. The method can be outlined as follows: The governing equations are parabolic with respect to time. Assuming the flow field is known at a time level t_n, the ADI procedure is used to advance the solution to a new time level t_{n+1} using a fairly large time step. The metric terms ξ_x, ξ_y

etc. are evaluated numerically at an intermediate time level $t_{n+\frac{1}{2}}$. The mixed derivatives that arise from terms such as $(\xi_x u \tau_{xx})_\xi$ etc. are lagged one time step. The flow quantities ρ, u, v and e at the new time level are written in terms of their values at the known time level and incremental quantities. For example;

$$\rho^{n+1} = \rho^n + \Delta\rho \tag{14}$$

Terms involving non-linearities at the time level t_{n+1} are linearized by Taylor expansion about the solution at known time level t_n. The time derivative $\frac{\partial \hat{q}}{\partial t}$ is written as two point backward difference formula at the new time level. Performing these operations and taking all the quantities at the known time level to right hand side one obtains a linear matrix equation for the incremental quantities at each grid point in the computational plane, except at grid points on the boundaries. The matrix equation may be written as:

$$[A]\{\Delta q\} + \frac{\partial}{\partial \xi}[B]\{\Delta q\} + \frac{\partial}{\partial \eta}[C]\{\Delta q\} = \{R\}^n \tag{15}$$

The Douglas-Gunn [17] procedure for generating alternating-direction implicit scheme is introduced now to solve the above system of equations by approximately splitting equation [16] into two equations where each involves only one dimensional operator.

$$[A]\{\Delta q\}^* + \frac{\partial}{\partial \xi}[B]\{\Delta q\}^* = \{R^n\} \tag{16}$$

$$[A]\{\Delta q\} + \frac{\partial}{\partial \eta}[C]\{\Delta q\} = [A]\{\Delta q\}^* \tag{17}$$

note that

$$\{\Delta q\} = \{\Delta\rho, \Delta u, \Delta v, \Delta e\}^T \tag{18}$$

Equations (17) and (18) are discretized using second order accurate central difference formulas for the spatial derivatives. This technique leads to a system of block tridiagonal matrix structure which may be solved efficiently by a standard block elimination procedure. One needs to provide boundary conditions for the unknown $\{\Delta q\}$ as well as for $\{\Delta q\}^*$ at the boundaries. Once $\{\Delta q\}$ is obtained the flow field variables at the new time level is explicitly known. In the present application fourth order artificial dissipation terms have been added explicitly to the right hand side in the manner suggested by Steger [12] to suppress the high frequency components associated with numerical instability that appear in high Reynolds number problems.

Boundary Condition

The present procedure requires boundary conditions to be set on the solid boundary $\eta = \eta_{min}$, the far-field boundary $\eta = \eta_{max}$ and at the fictitious cut $\xi = \xi_{min}$ and $\xi = \xi_{max}$ (see Fig. 1). At the solid boundary the condition of no slip requires the fluid velocity to be the same as that of the solid; the solid motion is known. Also, adaibatic flow condition $\frac{\partial e}{\partial \eta} = 0$ has been applied on the solid surface. The density at the surface may be evaluated in various ways; in the present calculation we used a two point extrapolation of the form:

$$\rho_{i,1} = \frac{4}{3}\rho_{i,2} - \frac{1}{3}\rho_{i,3} \tag{19}$$

Algebraic Turbulence Model

In the present work the effects of turbulence are simulated using an algebraic eddy viscosity model. Recently, progress has been made in using multi-equation turbulence models in conjunction with Navier-Stokes equations to simulate the behavior of separated turbulent flows. Yet numerical results using multi-equation turbulence models for separated flows are not completely satisfactory. For this reason the simpler, algebraic model was chosen. The algebraic turbulence model used in the present work is that of Baldwin-Lomax [15] who modified Cebeci turbulence model and is more suitable for use in Navier-Stokes solvers.

Numerical Results

The two basic geometries analyzed are NACA 0012 and the ONERA-CAMBRE' airfoils. The effects of reduced frequency and leading edge camber on the dynamic stall characteristics are calculated. Before detailed discussion of the numerical results, it is useful to study experimental data obtained at High Reynolds number turbulent flows, and observe the effects of reduced frequency and leading edge camber on the dynamic stall. It has been observed that the NACA 0012 first experiences trailing edge separation which appear as flow reversal and progressing upstream as the incidence increases, up to about 40 percent chord and at this point abrupt flow reversal up to the leading edge occurs.

The effects of reduced frequency for the range of 0.02 to 0.25 for NACA 0012 have been studied experimentally [1], indicating a delay in stall-onset as the reduced frequency increases. Another important parameter studied [1] is the effects of leading edge camber on dynamic stall characteristics and was shown that the stall-onset was delayed relative to the basic NACA 0012 by approximately 1 degree in incidence.

NACA 0012 Airfoil

The effects of reduced frequency on the dynamic stall of NACA 0012 airfoil was studied numerically. In this case the Reynolds number was taken as 2.5 X 10^6, free stream Mach number as 0.3, the angle of attack was varied from 0 degrees to 20 degrees according to:

$$\alpha = 10^\circ \ (1 - \cos(\omega t)) \tag{20}$$

The numerical results are shown for two values of reduced frequency $\frac{\omega c}{2V_\infty}$ = 0.15 and 0.25. The normal force and moment coefficients variation with incidence during one cycle of the motion is shown in Fig. 2 and Fig. 3 respectively. The generation and shedding of the leading edge vortex produces C_N and C_M hysteresis loops that are characteristics to deep stall type. A point of interest is the qualitative resemblance between the hysteresis for both frequencies obtained numerically and those obtained experimentally as shown in Fig. 34 of Ref. 16. As shown in both cases the moments stall begins followed by lift stall. The loss in lift is steeper for the lower reduced frequency as observed in wind tunnel data [16]. Also a second vortex like disturbance causing a secondary peak in the C_N and C_M loops for the lower frequency case during the downstroke motion.

Fig. 4 and Fig. 5 show the time history of the pressure coefficient distribution for both reduced frequencies. In each case the stall onset is indicated by a small kink in the pressure distribution on the upper surface, associated with the birth of the leading edge vortex. The stall-onset for the reduced frequency of 0.15 occurs approximately at $\alpha \simeq 16.6$ degree during the upstroke motion and for the reduced frequency of 0.25 it occurs approximately at $\alpha \simeq 17.8$ degrees, hence a delay of about 1.2 degrees. The effect of reduced frequency on stall-onset has been observed experimentally [16]. Time history of velocity profiles on the airfoil is shown in Fig. 6 for the reduced frequency of $\omega c/2V_\infty = 0.15$. As shown separation or flow reversal starts at the trailing edge during the upstroke motion. The flow reversal propagates upstream up to about 50 percent chord and at this point abrupt leading edge separation occurs that generate a strong clockwise leading edge vortex that is shed downstream. This phenomenon also has been observed in the wind tunnel experiments [1]. Vorticity contours at corresponding time levels are shown in Fig. 7. Similar behavior has been obtained for the case of reduced frequency $\omega c/2V_\infty = 0.25$ but with delay of about 1.2 degrees.

ONERA CAMBRE' AIRFOIL

The ONERA CAMBRE' airfoil was developed in France for possible Helicopter applications. Basically, it is NACA 0012 airfoil with extension of the leading edge by 0.025 chord and cambering the leading edge section. In this case the effect of leading edge camber was studied. The Reynolds number was taken as 2.5 X 10^6, free stream Mach number as 0.3 and reduced frequency $\omega c/2V_\infty = 0.25$. The variation of angle of attack with time was kept the same as for NACA 0012 case. The normal force and moment coefficients variation during one cycle of motion is shown in Fig. 8 and Fig. 9 respectively. Comparison with the numerical results obtained for NACA 0012 the normal force hysteresis loop seems almost identical during most of the cycle and differ slightly near the maximum incidence. The moment hysteresis loop indicate more positive damping than the NACA 0012 airfoil and a slight decrease in minimum value of C_M. In the experiment it was observed a larger decrease of the minimum value for C_M. The time history of pressure coefficient distribution is shown in Fig. 10. In this case the small kink in the pressure associated with the birth of the leading edge vortex appears approximately at $\alpha \simeq 19.4$ degrees which is a delay of 1.6 degrees compared to NACA 0012 airfoil under the same flow conditions. McCroskey et al [1] observed in the wind tunnel experiments, a delay of approximately 1 degree in incidence. Velocity profiles and vorticity contours resemble the results shown for NACA 0012 but delayed in time.

Conclusion

The present paper describes the application of the unsteady compressible time averaged Navier-Stokes equations numerical procedure to the dynamic stall of an oscillating airfoil in turbulent flow. The effects of reduced frequency and leading edge camber on the dynamic stall characteristics have been studied. Numerical results obtained showed qualitatively good agreement with experimental data. Increasing the reduced frequency appears to decrease the intensity of the dynamic stall vortex shedding and delay the formation and growth of the leading edge vortex. Cambering the leading edge tends to delay stall-onset by approximately 1.6 degrees, whereas experiments observed a delay of about 1 degree. Finally, the hysteresis loops of the normal force and moment coefficients resemble those obtained in experiments and showed clearly the deep stall characteristics for these flow conditions.

References

1. McCroskey, W. J., Carr, L. W., and McAlister, K. W.: Dynamic stall experiments on oscillating airfoils. AIAA J., Vol. 14, No. 1, January 1976.

2. McCroskey, W. J.: Inviscid flow field of an oscillating Airfoil. AIAA Journal, Vol. II, No. 8, Aug. 1973.

3. Scruggs, R. M., Nash, J. F., and Singleton, R. E.: Analysis of dynamic stall using unsteady boundary layer theory. NASA CR 2462, Oct. 1974.

4. Crimi, P.: Investigation of non-linear inviscid and viscous flow effects in the analysis of dynamic stall. NASA cr 2335, Feb. 1974.

5. McCroskey, W. J., and Phillippe, J. J.: Unsteady viscous flow on oscillating airfoils. AIAA J., Vol. 13, No. 1, January 1975.

6. Mehta, U. B.: Dynamic Stall of an oscillating airfoil. AGARD CP227, AGARD Meeting on Unsteady Aerodynamics, Paper No. 23, Ottawa, Sept. 1977.

7. Werlé, H.: Visualization of hydrodynamic d'Ecoulements Instationnaires. Proc. of IUTAM Symposium on Recent Research in Unsteady Boundary Layers, Quebec, Canada, May 24-28, 1971.

8. Sankar, N. L., and Wu, J. C.: Viscous flow around oscillating airfoil - A numerical study. AIAA paper 78-1225, July 1978.

9. Sankar, N. L., and Tassa, Y.: Reynolds number and compressibility effects on dynamic stall of NACA 0012 airfoil. AIAA paper 80-0010, January 1980.

10. Shamroth, S. J. and Gibeling, H. J.: Analysis of turbulent flow about an isolated airfoil using a time-dependent Navier-Stokes procedure. AGARD Specialist Meeting on Boundary Layer Effects on Unsteady Airloads, Sept. 1980.

11. Peyret, R. and Viviand, H.: Computation of viscous compressible flows based on Navier-Stokes equations. AGARD-ag-212, 1975.

12. Steger,J. L.: Implicit Finite difference simulation of flow about arbitrary geometrics with application to airfoils. AIAA 77-665, June 1977.

13. Vinokur, M.: Conservation equations of gas dynamics in curvilinear coordinate system. J. of Comp. Physics, Vol. 14, Feb. 1974, pp 105-125.

14. Thompson, J. G., Thames, F. C., and Mastin, C. M.: Automatic numerical generation of body fitted curvilinear coordinate system for field containing any number of arbitrary two-dimensional bodies. J. of Comp. Physics, Vol. 15, 1974, pp 299-319.

15. Baldwin, B. S., and Lomax, H.: Thin layer approximation and algebraic model for separated turbulent flows. AIAA 78-257, Jan. 1978.

16. Carr, L. W., McAlister, K. W., and McCroskey, W. J.: Analysis of the development of dynamic stall based on oscillating airfoil experiments. NASA TN D-8382, Jan. 1977.

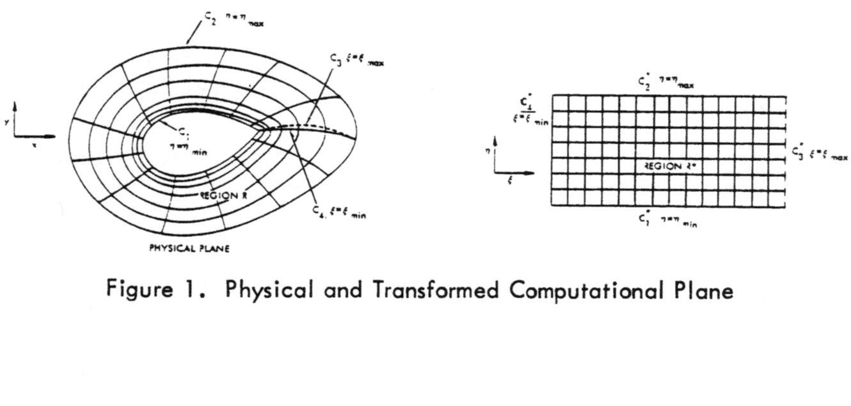

Figure 1. Physical and Transformed Computational Plane

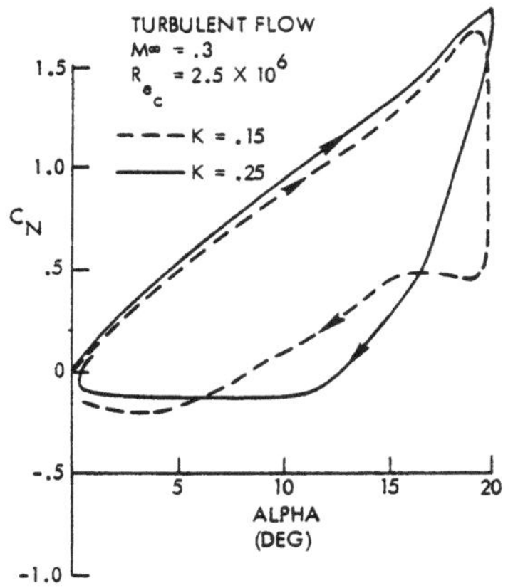

Figure 2. Hysteresis Loop of Normal Force Coefficient for NACA 0012 Airfoil

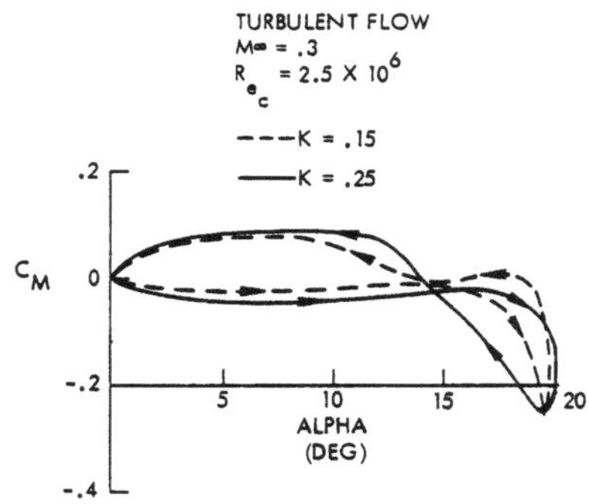

Figure 3. Hysteresis Loop of Moment Coefficient for NACA 0012 Airfoil

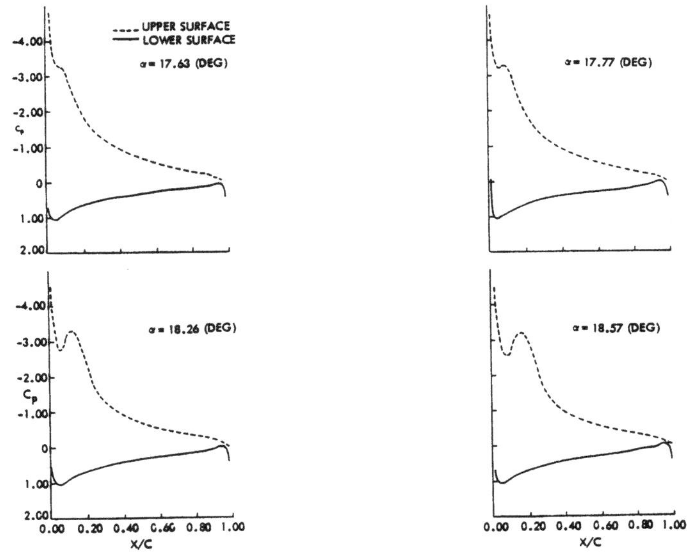

Figure 4. Surface Pressure Distribution During Upstroke Motion for NACA 0012 Airfoil. Mach Number 0.3, Reynolds Number 2.5 X 10^6 and Reduced Frequency 0.25

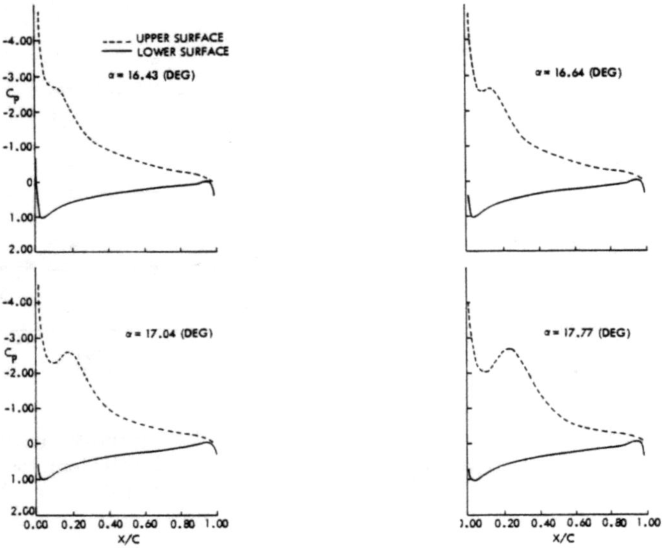

Figure 5. Surface Pressure Distribution During Upstroke Motion for NACA 0012 Airfoil. Mach Number 0.3, Reynolds Number 2.5×10^6 and Reduced Frequency 0.15

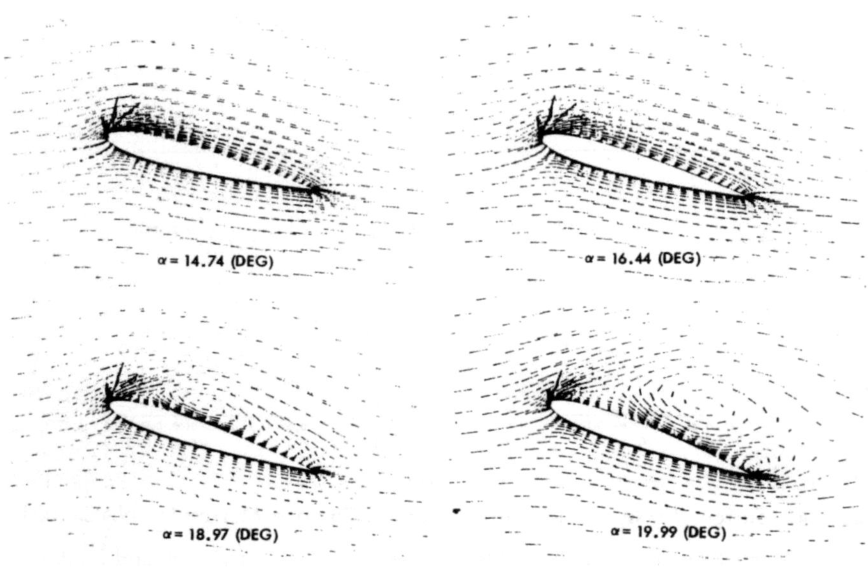

Figure 6. Time History of Velocity Profiles (NACA 0012, $M_\infty = .3$, $Re = 2.5 \times 10^6$, $K = .15$

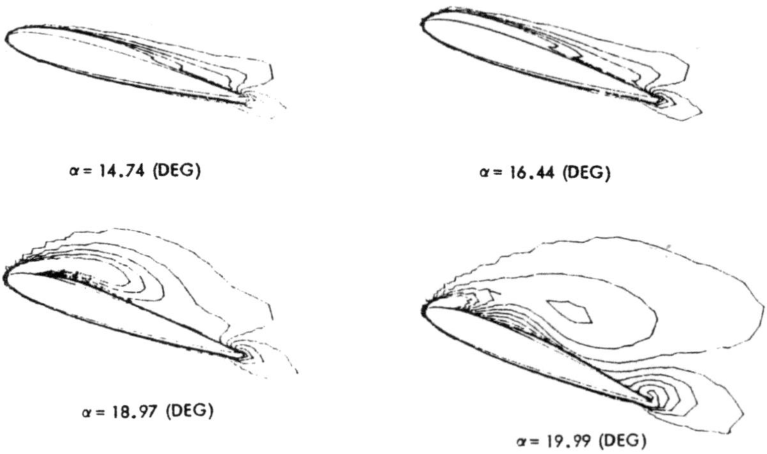

Figure 7. Time History of Vorticity Contour (NACA 0012, $M_\infty = .3$ Re $= 2.5 \times 10^6$, K = .15

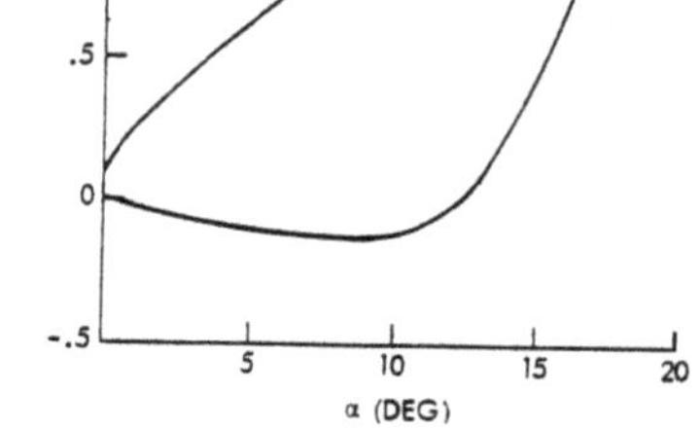

Figure 8. Hysteresis Loop of Normal Force Coefficient for ONERA CAMBRE' Airfoil

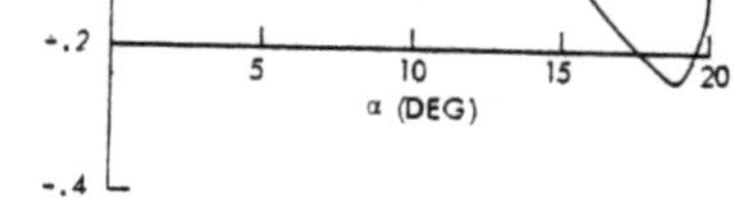

Figure 9. Hysteresis Loop of Moment Coefficient for ONERA CAMBRE' Airfoil .

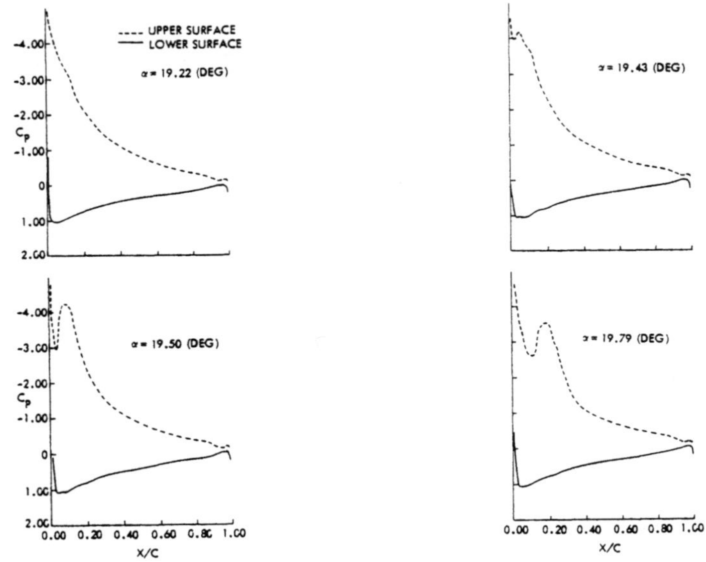

Figure 10. Surface Pressure Distribution During Upstroke Motion for ONERA CAMBRE' Airfoil at Mach Number 0.3, Reynolds Number 2.5 X 10^6 and Reduced Frequency 0.25

Numerical Experiments on Transition Triggering off in a Two-Dimensional Shear Flow*

M. BOURIOT, and L.F. TSEN

Université de Poitiers
Centre d'Etudes Aérodynamiques et Thermiques
86000 Poitiers.

Summary

Numerical experiments on finite difference solutions of time dependent two-dimensional Navier-Stokes equations are used to study the transition triggering off in a mixing layer of initial tanh(y) profile. The inflow is excited by sinusoïdal waves resulting from the linear theory of hydrodynamic instability. Numerical realizations are compared, through streamwise growth of momentum thickness and vorticity plots, with Winant and Browand experiments.

Introduction

The transition control and triggering off are studied through the finite difference solution of the two-dimensionnal time dependant Navier-Stokes equations. In a shear flow, the transition results from hydrodynamic amplification of unstable disturbances stimulated by perturbations of diverse origin. In a unidirectional flow, the most unstable disturbances are two-dimensional transverse waves {2, 3} and experiments have shown a two-dimensional structure of the general motion even far from the area where small disturbances grow exponentially {5, 8, 9, 10}. Thus, a two-dimensional numerical simulation is well suited to the study of transition triggering off.

Numerical simulation description

The simulation field is rectangular Fig. 1. We consider a mixing layer, with an initial mean velocity profile :

$$\bar{u} = U + \frac{\delta U}{2} \tanh(y/d) ,$$

where δU is the velocity difference between the two layers and chosen as velocity scale, U the general translation velocity, and d a characteristic length. In the dimensionless problem, with $\delta U = d = 1$, the momentum thickness is $\theta_o = 0.5$. At upstream boundary, $x = 0$ a second condition $\frac{\partial \bar{v}}{\partial x} = 0$ is imposed on mean velocity. On this limit, some unsteady disturbances could be possibly superimposed. For the two lateral boundaries, $y = \pm h/2$ reflection conditions $\frac{\partial \bar{u}}{\partial y} = 0$, $\bar{v} = 0$ are used in order to simulate free shear boundaries. At $x = \ell$, the outflow conditions are simply $v = 0$, $\frac{\partial v}{\partial x} = 0$ for instantaneous velocity.

The Navier-Stokes equations are formulated in terms of the stream function ψ and vorticity $\omega = -\nabla^2\psi$ and discretized over a grid of square meshes $\delta x = \delta y$. The vorticity advection is approached by an implicit second order scheme {7} which conserves both vorticity and square vorticity. In association with classical five points scheme for the Laplacian operator of Poisson's equation for ψ, this scheme also conserves kinetic energy $(\nabla\psi)^2/2$. In order to keep these important conservation properties for ω in present numerical simulations, the viscous diffusion is approached by a scheme of Crank-Nikolson type. The numerical code being designed in this way, the parabolic growth of a laminar mixing layer for the zero perturbation case is accurately described. After this check the code is used for the simulation of transition. A mixing layer of constant thickness is always unstable for sufficiently large transversal waves, whatever be the value of the Reynolds number $R_\theta = \delta U\theta_o/\nu = 1/2\nu$, {1}. This result transposed in the present spatial growing case, suggests that any unsteady disturbances on the upstream boundary having energy in the low frequency range shall trigger off the transition into the simulation field. In order to reduce the computation field size and simultaneously the simulation cost, the inflow is excited with the most unstable wave for an hyperbolic tangent mean velocity profile, which is represented in dimensionless form :

$$\psi(O,y,t) = \bar{\psi}(O,y) + \varepsilon\left(\psi_R(y)\cos\beta t + \psi_I(y)\sin\beta t\right) .$$

ψ_I, ψ_R (Fig. 1), are imaginary and real parts of normalized eigenfunction, β is the angular frequency of this wave depending on the similitude parameters U and R_θ. Eigen functions for growing spatial case has been numericaly calculated by Michalke {4} for the case $U = 0.5$, $R_\theta = \infty$. In this case the features of the unstable disturbance are, a wave-number $\alpha_r = 0.403129$ a phase velocity $C = 0.5127$, an angular frequency $\beta = \alpha_r C$ and a growth rate $-\alpha_i = 0.228425$. These ψ_R, ψ_I, α_r values can be used in finite difference computations for other values of U and R_θ on condition that new angular frequency β are fixed with a U corrected phase velocity C, on the assumption that eigen-modes, for a moving observator at U velocity, are practically independent of the parameters U and R_θ.

Results

Numerical experiments has been done for a general translation velocity $U = 1.0496$ according to the experimental parameters of Winant and Browand {8}. For this case, the corrected frequency is $\beta = 0.4282$. A grid of 118x32 square meshes, $\delta x = 1$, has been used. The thickness $\delta_{0.99}$ of the hyperbolic tangent profile is discretized by five points and the perturbation wave length $\lambda = 2\pi/\alpha_r$ by approximatively 15 points. The computed field is about 8 wavelength λ long. There is about 23 time steps fixed at $\delta t = \delta x/(0.5+U) = 0.667$, for one perturbation period.

The first realisation R_5 is calculated for $R_\theta = 12.5$, the amplitude of imposed perturbation $\varepsilon = 0.03$ is infered from experimental values of root-mean square velocity {8}. The realization reaches a steady state at a dimensionless time of about $t = 200$. The configuration of the resulting flow is basicaly laminar. The growth of momentum thickness is parabolic according to experiments and theory, Fig. 2.

One other realization R_6 has been done for $R_\theta = 40$ and $\varepsilon = 0,141$. This Reynolds value is related to the onset of the transition in experiments above which the momentum thickness $\theta(x)$ growth becomes linear. Contrary to experiments the streamwise amplification disturbance is quickly bounded by the non-linear interactions. A steady stream of vortex cores passed through the computed field without vortex merging Fig. 4a. The growth of $\theta(x)$ remains parabolic in an over-saturated laminar configuration

Fig. 3, this result seems to confirm the existence of a finite amplitude steady state {3, 6}.

The whole picture changes when there is a noisy disturbance in the flow. The R_8 realization duplicates R_6 with a Gaussian noise of .05 standard deviation superimposed on inflow excitation function. Then the structure of the computed flow Fig.4b is totally different. We observe an unstable vorticity sheet which gives rise to a row of rolling-up vortices. This configuration change, is also observed on the streamwise growth of momentum thickness θ, Fig. 3, which deviates from the laminar regions to catch up with the domain of linear experimental variation.

REFERENCES

1. R. Betchov, A. Szewczyk, "Stability of a shear layer between parallel streams", Phys.of Fluid,V.6, N.10,pp.1391-1396,(1963)

2. A. Michalke, "On the inviscid unstability of the hyperbolic tangent velocity profile", J. of Fluid Mech., V.19, pp.543-556, (1964)

3. H. Shade, "Contribution to the non linear stability theory of inviscid shear layers", Phys. of Fluid, V.7, N.5, pp.623-628, (1964)

4. A. Michalke, "On spatially growing disturbance in an inviscid shear layer", J. of Fluid Mech., V.23, N.3, pp.521-544, (1965)

5. P. Freymuth, "On transition in a separated laminar boundary layer", J. of Fluid Mech., V.25, N.4, pp.683-670, (1966)

6. J.J. Stuart, "On finite amplitude oscillations in laminar mixing layers", J. of Fluid Mech., V.29, N.1, pp.417-440,(1967)

7. A. Arakawa, "Numerical solution of large scale atmospheric motions", SIAM AMS Proc., V.2, pp.24-40, (1970)

8. C.D. Winant, F.K. Browand, "Vortex pairing : the mechanism of turbulent mixing layer, growth at moderate Reynolds number" J. of Fluid Mech., V.63, N.2, pp.237-255, (1974)

9. A. Roshko, "Structure of turbulent shear flow, a new look", AIAA J. V.14, N.10, p.1349, (1976)

10. I. Wygnanski, D. Oster, H. Fielder, B. Dziomba, "On the perseverence of quasi dimensional eddy structure at a turbulent mixing layer", J.of Fluid Mech., V.93, N.2, pp.325-335, (1979)

* - Work under D.R.E.T. contract 81/601.

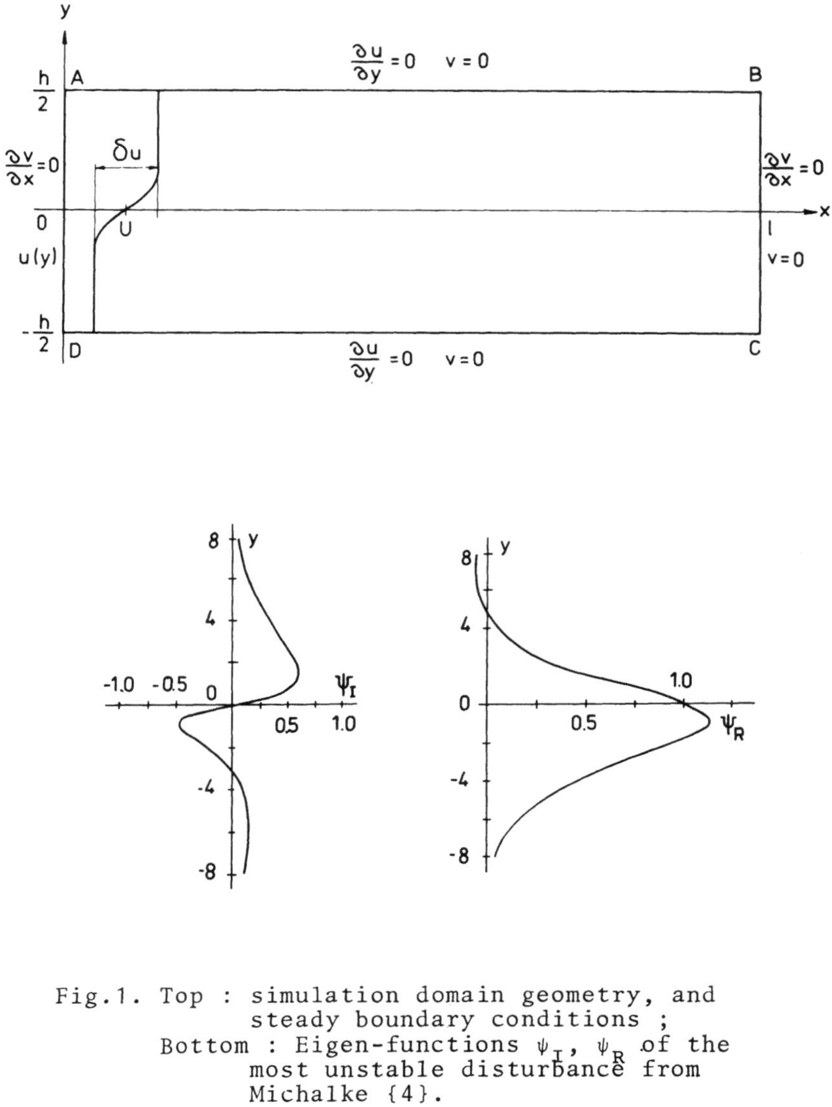

Fig.1. Top : simulation domain geometry, and steady boundary conditions ;
Bottom : Eigen-functions ψ_I, ψ_R of the most unstable disturbance from Michalke {4}.

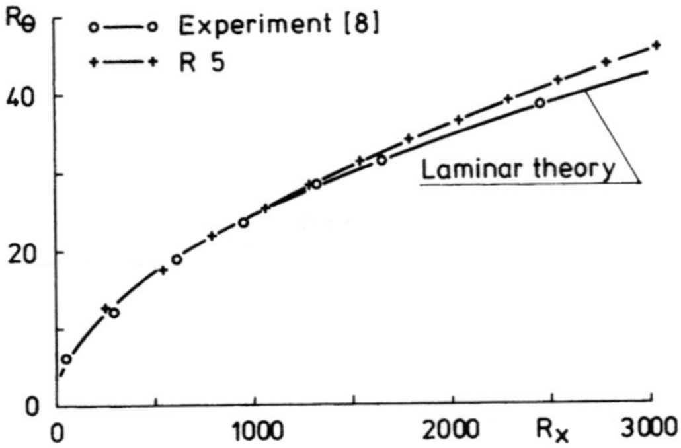

Fig.2. Stream wise growth of momentum thickness θ(x)
$R_\theta = \frac{\delta U.\theta}{\nu}$, $R_x = \frac{\delta U.x}{\nu}$; laminar Reynolds range

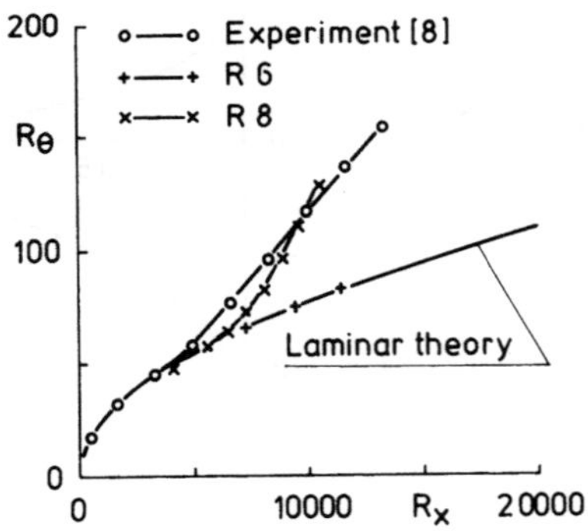

Fig.3. Stream wise growth of momentum thickness θ(x) ; transitional Reynolds range.

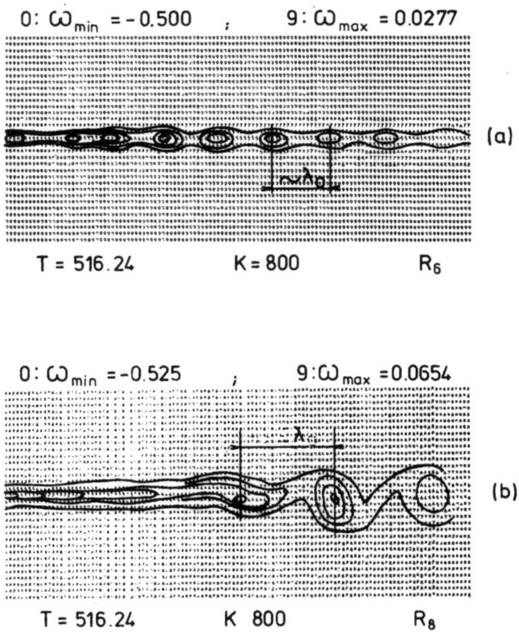

Fig.4. Printer iso-vorticity plot :
a - without noise R_6
b - with Gaussian noise on excitation R_8

Natural and Forced Vortex Shedding

G. S. JONES*, C. BARBI† and D. P. TELIONIS*

*Virginia Polytechnic Institute & State University, Blacksburg, VA 24061
†On leave from Institut de Mecanique des Fluides de Marseille

Abstract

Natural and forced vortex shedding has been investigated experimentally over circular cylinders in the VPI water tunnel. Flow visualizations and Laser-Doppler Anemometry have been employed. Results indicate that in the lock-on region there are three distinct modes of vortex shedding. Most interesting is the second mode, namely simultaneous shedding of wake vortices and the third mode, shedding at the subharmonic of the driving frequency.

Introduction

Vibrations induced on structures due to natural shedding of vortices are of great engineering importance in the design and installation of heat exchangers, power cables, underwater oil pipes and piles, etc. The problem becomes more complex and the constraints it imposes on the design more severe, if the wake "locks on" to the frequency of an external disturbance. Aeroelastic resonance then may drive the structure to unwanted amplitudes.

Strouhal[1] recorded the remarkable fact that the frequency of shedding vortices over a bluff body is proportional to the velocity. For the flow around a circular cylinder at subcritical Reynolds numbers, $Re < 10^5$, the reduced frequency $f_s D/U_\infty$ is a constant equal to 0.21, where f_s is the shedding frequency, D is the cylinder diameter and U_∞ is the free-stream velocity. This frequency is often referred to in literature as the **Strouhal frequency.** The combination of the parameters in the form of a reduced frequency is sometimes termed the **Strouhal Number.** To avoid confusion we call this dimensionless number the **reduced shedding frequency.** Natural shedding experiments have been extensive and were reviewed comprehensively by Morkovin[2]; Berger and Wille[3], Savkar[4]; and McCroskey[5].

Lock ons have been studied only for disturbances introduced by oscillating the body. These studies have emphasized the oscillation of a cylinder perpendicular to the direction of the oncoming flow [see, for example, the review articles of Savkar[4]; McCroskey[5]; and Sarpkaya[6]]. Oscillation of a cylinder in the direction of the oncoming stream has been studied experimentally by Griffin and Ramberg[7], and Tanida, Okajima and Watanabe[8]. In the present investigation we study the phenomenon of **lock on** for fixed models and oscillations of the velocity of the oncoming flow.

The response of aerodynamic surfaces to a fluctuating oncoming stream has been investigated with emphasis on the fluctuations of the lift and drag of an airfoil [Maresca, Favier and Rebont[9]; Maresca, Favier, Rebont, Jones and Telionis[10]]. However, it is believed that in these studies the amplitude of the fluctuations and the range of the reduced frequencies forced the vortex shedding to be always **locked on** to the frequency of the external disturbance.

An important role in the development of an unsteady wake may be played by the instabilities of the free shear layers which emanate from the point of separation. Morkovin[11] demonstrated that the shedding of large-scale vortices is influenced by the interaction of the two shear layers that detach from the body at the point of separation. Many articles have been published on the stability of free shear layers as reviewed by Michalke[12], but all the work is concentrated on shear layers that emanate from a sharp corner or from the trailing edge of a flat plate. The properties of a free shear layer that lifts smoothly from a continuous solid surface have not yet been investigated. This problem in itself may be the topic of extensive work. Here, its possible connection to the vortex-shedding process is briefly discussed.

In this paper we report on preliminary results obtained with three circular cylinders that were tested in a water tunnel. Velocity measurements and flow visualizations provide a comprehensive picture of the lock-on phenomenon. Qualitative agreement with earlier experimental results is quite satisfactory.

1. Facilities and Experimental Rig

Experiments were conducted in the VPI water tunnel which has been described adequately in earlier publications[13-16]. This tunnel has a contraction ratio of 6:1 leading to a 25cmx30cm test section and achieves speeds up to 5m/sec (Fig. 1). The turbulence level ranges from 0.3 to 0.8% depending on the experimental rigging and the speed of the tunnel.

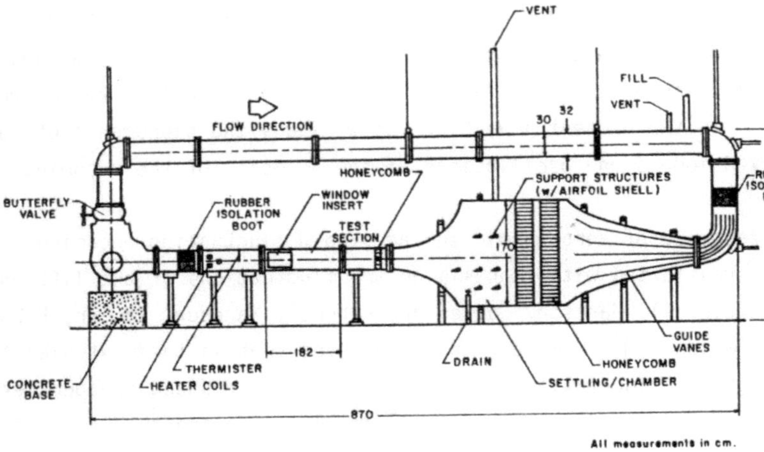

Fig. 1 The VPI & SU low speed water tunnel.

The flow was oscillated by means of flaps downstream of the test section. In this way frequencies ranging from 0.1 to 2 Hz for stream amplitudes of the order of 10% to 25% were achieved. Two sequences of experiments were conducted; one with an extra attachment which increases the contraction ratio to 8:1 and a flexible flap and a second with a free test section and a solid flap. The first set of experiments were conducted in the summer of 1980 and the second in the spring of 1981. The flaps were driven by a variable speed motor through a four-bar-linkage. An LED was mounted on the flywheel to provide a signal for triggering.

To generate a wider range of Strouhal frequencies, three circular cylinders with diameters 1cm, 2cm and 3.5cm were tested. The largest was instrumented with dye ports and a fiber-optic port as shown in Fig. 2. The dye ports allow the emission of two different colors of dyes in the upstream and the downstream region of the cylinder and aligned as close as possible to the direction of the flow, namely, tangent to the cylinder skin. Experience indicated that in order to keep the speed of the ejected fluid as small as possible, the port diameter should be no smaller than 2mm and the supply pressure should be very carefully controlled. The fiber-optic port allowed the marking of the 90° position on the cylinder (see Fig. 2), but proved to be rather ineffective.

The flow was investigated by a DISA 55L series Laser-Doppler Velocimeter (LDV). Since measurements were obtained in the wake where the flow is reversing, this method is superior to any other obtrusive method, as for example the hot wire. The description and operation characteristics

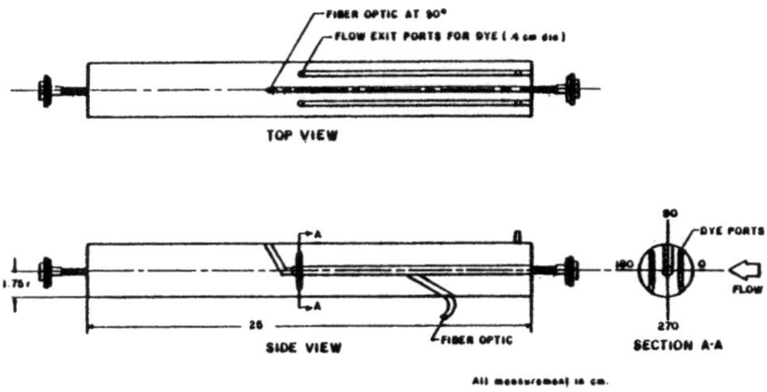

Fig. 2 Schematic of the instrumented circular cylinder showing the dye ducts and reference system.

of the LDV are included in many earlier publications[13-16]. The tracker of the system in operation at VPI & SU has been recently modified with a drop-out circuitry. If tracking is interrupted for any reason, the system is automatically returned to the previous levels of measurement and tracking is recovered. This is necessary if the data are to be averaged over a large number of samples.

The signal from the DISA processor was fed via anti-aliasing filters into an FFT system (ZONIC's DMS 5003). This is a two channel Data Memory System that operates digitally in both the time and the frequency domain. The system controller of the DMS 5003 is a microcomputer based on the Intel 8080 Central Processing Unit (CPU). A TEKTRONIX 4051 intelligent graphics display terminal is interfaced by a hand-shake linkage with this system, adding to the ZONIC's 16K programmable "read-only" memory (PROM) and 4K random access memory (RAM) a 6K block of ROM. A block diagram of the data acquisition system is shown in Fig. 3.

2. Natural Shedding

Measurements were obtained at different positions upstream but mostly downstream of the circular cylinders. Figures 4, 5 and 6 show the 1cm, 2cm and 3.5cm cylinders, respectively, indicating the relative location of the LDV measuring volume.

In the first step of the investigation, we reexamined the phenomenon of natural shedding. At a Reynolds number of 3000, the boundary layer on the cylinder is assumed laminar. Downstream of separation the boundary layer becomes a free shear layer, or a free vortex sheet. This layer may

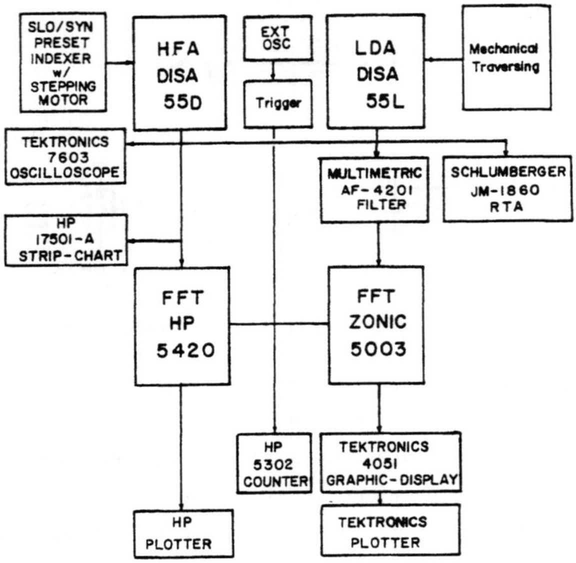

Fig. 3 Block diagram of the VPI water tunnel data acquisition system.

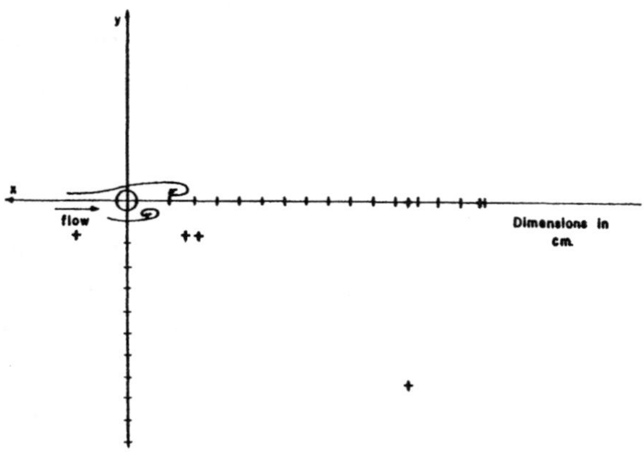

Fig. 4 Measuring points for the 1 cm cylinder

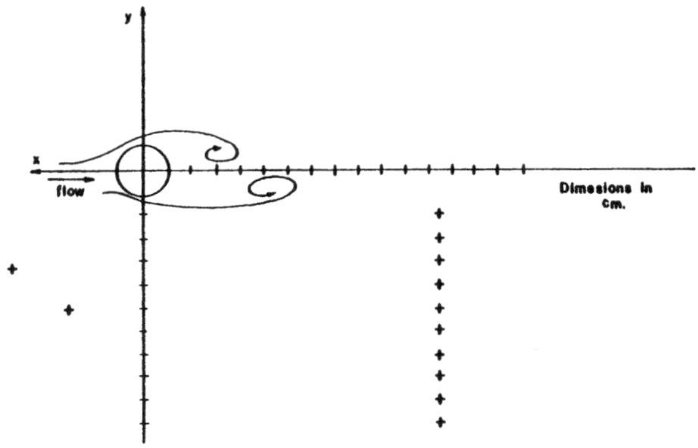

Fig. 5 Measuring points for the 2 cm cylinder

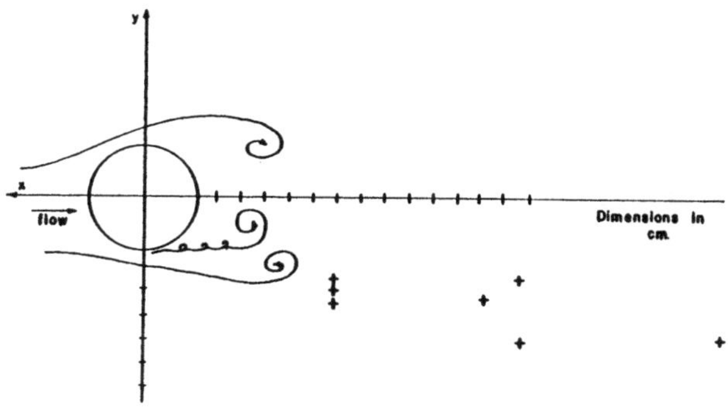

Fig. 6 Measuring points for the 3.5 cm cylinder.

develop instabilities, which appear at first as a rolling-up of the shear layer, and which eventually generate a series of small vortices. The free shear layer is influenced by the opposite layer as described by Morkovin[11]. The communication of two free shear layers leads to the large-scale alternating vortex structures. These large-scale vortex structures are formed and shed periodically downstream of the cylinder.

To gain confidence in the experimental rig some classical tests were repeated. It was shown that the reduced shedding frequency measured at Reynolds numbers between 1350 and 2.1×10^4 remain nearly constant. This can be seen in Fig. 7. In this range of Reynolds numbers we measured a shedding frequency that corresponds to a reduced frequency of 0.21, which is in agreement with all classical and well accepted results.

The frequency of the free-shear layer instabilities was difficult to measure with LDV because of the periodic wandering of the shear layer

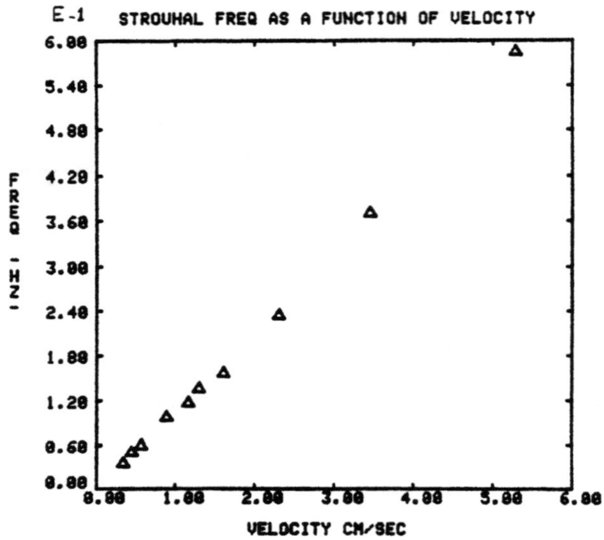

Fig. 7 The dimensional shedding frequency versus the free stream velocity for the 2 cm cylinder.

itself. Since the shear layer was thin, the measuring volume would periodically find itself outside the area of interest. This makes the location of the control volume critical when investigating the free shear layer.

To ensure that the frequency data obtained by LDV was related to the phenomenon under consideration, a comparison was made with flow visualizations. A slow motion movie and a television video system were used for this purpose. This optical information revealed that the free shear layer does not have a constant reduced shedding frequency, as does the Strouhal shedding. It can be argued therefore that there is no direct relationship between the two phenomena. It was observed from flow visualization that the free-shear layer eventually rolls up into bundles comprised of five to seven small vortex structures, that form one large vortex. There was a randomness associated with the number of small vortices engulfed into a large one. This, however, did not seem to influence the frequency of the large scale vortex shedding.

The relative position of the small scale vortex structure of the free shear layer and the large scale vortex structure can be seen in the flow visualization sequence of Fig. 8. Dyes are injected from the upsream ports of the cylinder as shown in Fig. 1. The dye then follows the attached boundary layer into the free shear layer. A second color was often used from the downstream ports to help identify separation.

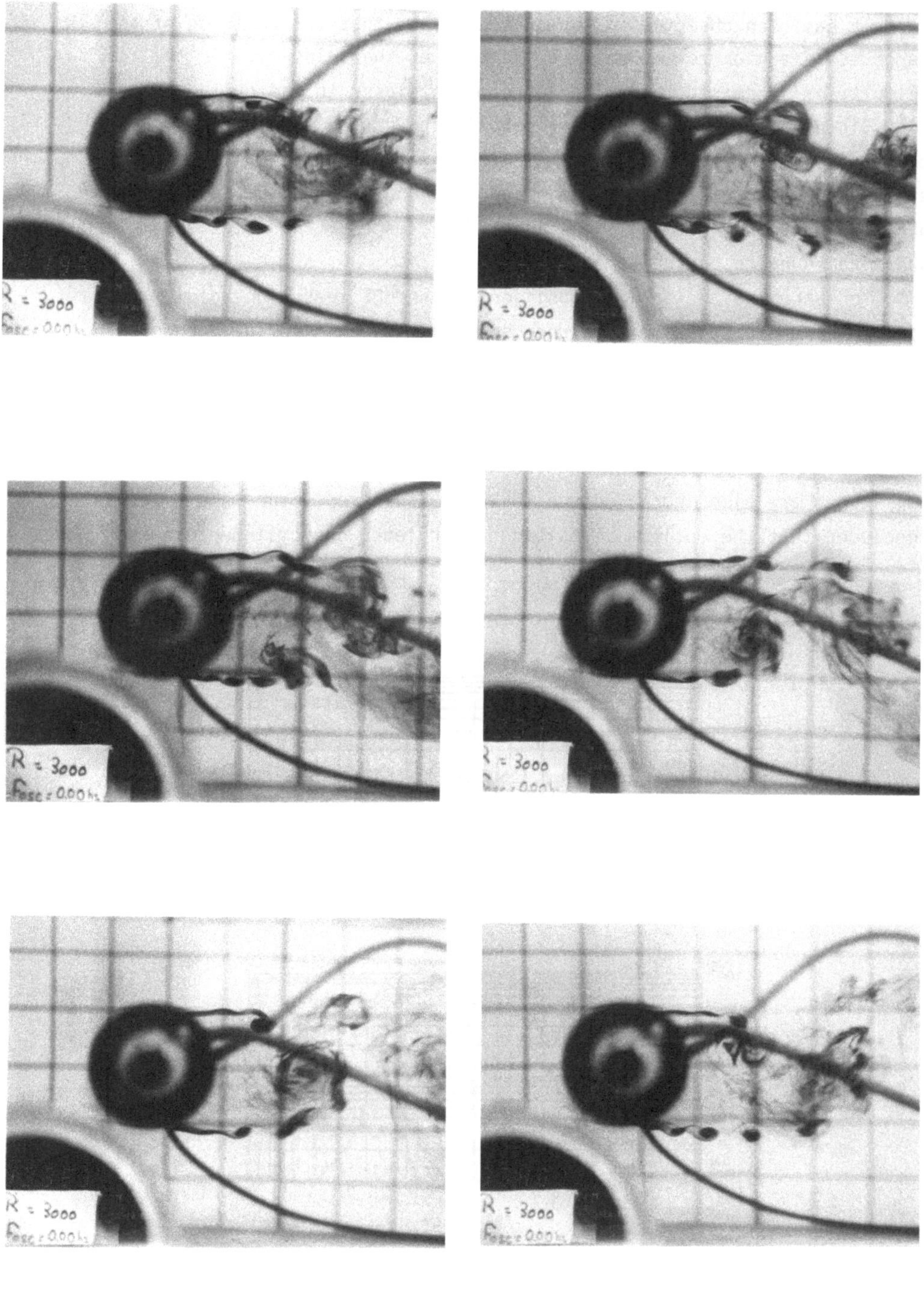

Fig. 8 Instantaneous flow patterns of natural shedding.

It has been observed in the present study that the position of the large scale vortex formation is a function of the velocity. As the velocity is increased the vortex core moves closer to the cylinder body.

Since the domain of interest contains reversing flow regions, Bragg cells were installed to obtain the correct sign of the velocity. An example shown in Fig. 9 of two autospectra, one with Bragg cells and the other without, revealed no significant differences. The relative amplitude was also found to be the same.

To determine the optimum position of the LDV measuring volume, certain horizontal and vertical planes were scanned as shown in Fig. 4 through 6. In a log amplitude-frequency spectrum it appears that the amplitude associated with the Strouhal shedding frequency increases as the LDV control volume approaches the centerline of the cylinder along the y axis. However, this is misleading, because the surrounding frequency component amplitudes are also increased due to the turbulence of the wake. It was concluded that the optimum y-position is at least one-cylinder diameter

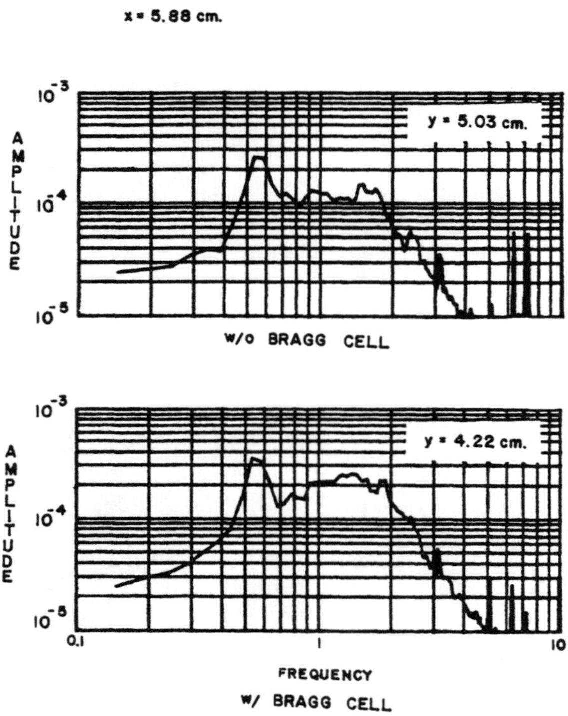

Fig. 9 Comparison of LDV signal with and without Bragg cells.

away from the centerline of the cylinder. This allows the unambiguous detection of the natural shedding frequency at the expense of losing other information concerning the wake. The variation of the signal in the x-direction was also found to vary significantly. The optimum y-position was not known when scanning was performed in the x-direction. However, it was found that the signal at a 3/4 diameter distance downstream is not as clean as at a distance of 1.5 diameters. It was concluded that one cylinder diameter downstream is optimum for obtaining clean frequency spectra.

3. Forced Shedding

The next series of experiments was aimed at locking the natural shedding of the vortex structure to a forced oscillation of the mean flow. The phenomenon of **locking on** has been studied in the past by oscillating the body, either traversely or perpendicular to the flow as referenced in the introduction. Stansby[17] describes the minimum requirements in the amplitude of the oscillation for **lock on**. Even though the same criteria cannot be applied here, it is recognized that the amplitude of the oscillation is a factor that should be considered. When changes in the frequency are made, the amplitude of the disturbance should be altered to maintain a constant reduced amplitude, $\omega A/U_\infty$, where ω is the driving frequency, A is the amplitude of the disturbance, and U_∞ is the averaged mean velocity. It was discovered that at an amplitude of 13%, the **lock on** phenomenon could not be observed on the 1cm or 2cm cylinder. When the amplitude was increased to 25% of the mean flow, **lock on** was obtained for the 2cm and the 3cm cylinder. The 1cm cylinder was not tested at the 25% amplitude, while the 3.5cm cylinder was not checked at the 13% amplitude. Since it was difficult to vary the amplitude of oscillation while simultaneously varying the frequency, little attention was given to amplitude once **lock on** was achieved. An example of these amplitude variations showning the deviation from the desired sine-wave oscillation is given in Fig. 10. These measurements were taken at least ten diameters upstream of the cylinder where the influence of the cylinder on the free stream is undetected.

It appears that a very interesting sequence of phenomena occurs when the flow is oscillated. Three characteristic modes of the **lock on** wake structure can be identified. Schematically these are shown in Fig. 11. Mode I is qualitatively very similar to natural shedding. In Mode II it was discovered that the vortices on the two sides of the cylinder are shed

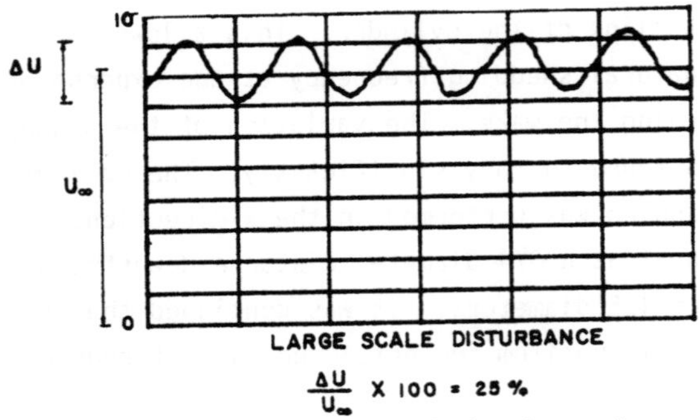

Fig. 10 Example of time variation of the free stream.

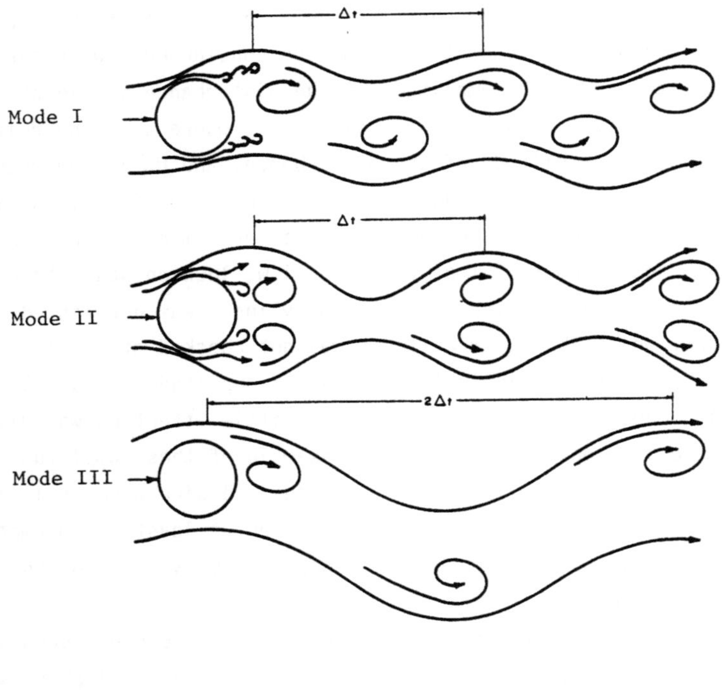

Fig. 11 Schematic representation of vortex shedding patterns.

simultaneously. In Mode III the mechanism is dominated by the subharmonic. The envelopes of the wakes are a distinct characteristic of the shedding pattern as shown in Fig. 11.

To observe the different modes the undisturbed velocity was held fixed as well as all other tunnel parameters, including water temperature and oscillation amplitude. The frequency of oscillation of the oncoming stream was then varied. The lower and upper limits of frequency were set at 0.2 and 1.5 Hz as dictated by difficulties in mechanically generating and measuring periodic disturbances.

Figure 12 shows a set of amplitude spectra that corresponds to the **lock on** process. For no externally imposed oscillations it can be seen in Fig. 12-a that at a Reynolds number of 3000, the shedding frequency is 0.5. The first harmonic is also appearing. When the flow is driven at a frequency much lower than the natural shedding frequency, the

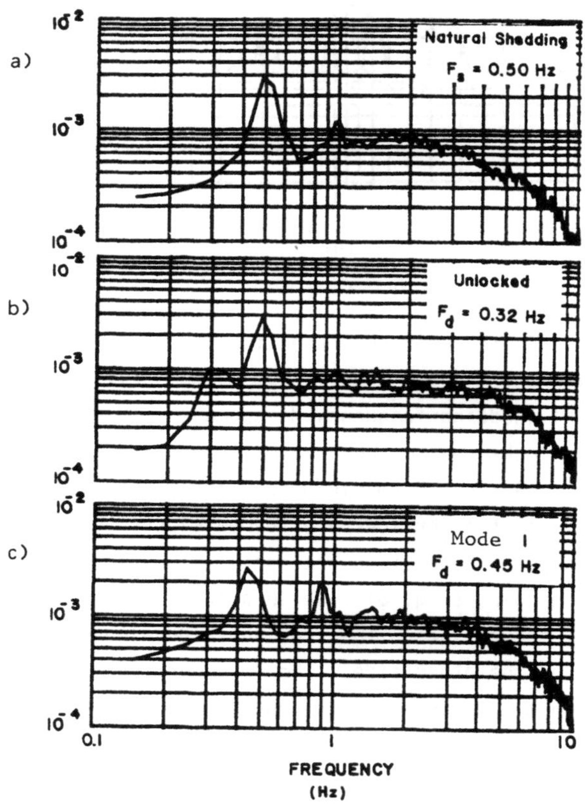

Fig. 12 Amplitude spectra for the driving frequencies that emphasize the three phases of lock-on.

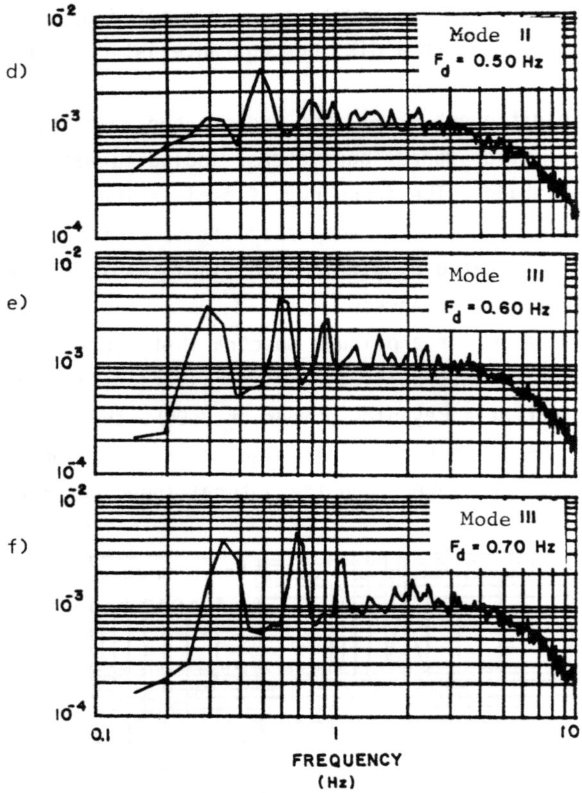

Fig. 12 (cont.)

system is not locked. In the unlocked case, two independent frequency components of the flow are seen in Fig. 12-b. As the driving frequency is increased, the natural shedding frequency disappears. This is because the shedding frequency is locked on to the driving frequency. This is defined as Mode I (Fig. 12-c). A first harmonic often, but not always, will appear in this mode. In Mode I, the vortices are formed closer to the cylinder (Fig. 13) and the points of separation are oscillating more distinctly when compared to natural shedding (Fig. 8). While a large vortex is being formed on one side of the cylinder, the free shear develops the familiar small scale vortices on the other side. The large vortex grows, but essentially remains in the vicinity of the body. Eventually, it is shed in the wake while the shear layer on the opposite side appears to go through a deceleration process. Soon after the free shear rolls up into a new large vortex which starts forming on the side of the decelerating free shear layer.

Mode II occurs only in a narrow frequency band around the natural shedding frequency. Now, two vortex structures are shed simultaneously from both sides of the body, as shown in Fig. 14. As the vortices are forming, the point of separation displaces by almost 20° upstream from θ = 90°. At about the instant of shedding, separation retreats to 15° on the downstream side of the cylinder. This mode is unstable, and quickly jumps to one of the other two modes. We observe in the amplitude spectrum that a subharmonic of one half the driving frequency begins to appear near the natural frequency of shedding, as shown in Fig. 12-d. The small vortices of the free-shear layer are undetectable in this mode as well as Mode III.

With a further increase of the frequency, the phenomenon changes over to Mode III (Fig. 12-e). In this mode the subharmonic becomes dominant (Fig. 15). The first harmonic also grows but its influence is not associated with flow visualization patterns.

The formation of these vortices occur very close to the body, not allowing any space or time for the free shear layer to form small scale instabilities. The point of separation displaces extensively in this mode. As shown in Fig. 15, separation moves from approximately 50° to 160°. This variation indicates that the potential flow is changing directions dramatically. The flow stays attached to one side of the body for much longer distances and thus accumulates more vorticity in the boundary layer. This vorticity is eventually generating a very large and strong vortex which dominates the flow for almost one period of the external oscillation.

To obtain an overall view of the shedding phenomena, the reduced frequencies associated with the vortex shedding are plotted in Fig. 16. The reduced driving frequency is defined as $f_d.D/U_\infty$, where f_d is the driving frequency, D is the cylinder diameter and U_∞ is the free stream velocity.

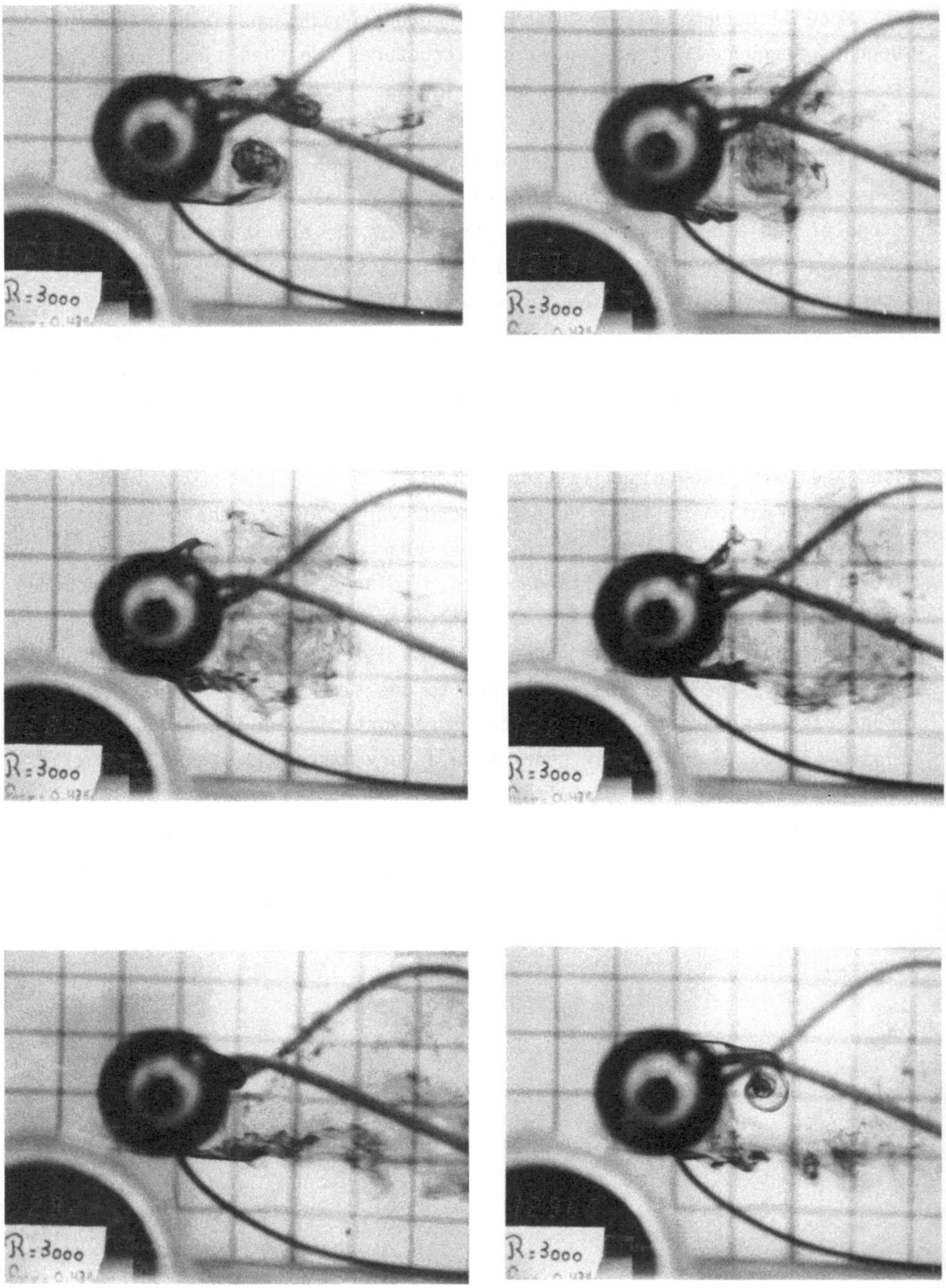

Fig. 13 Instantaneous flow patterns for Mode I.

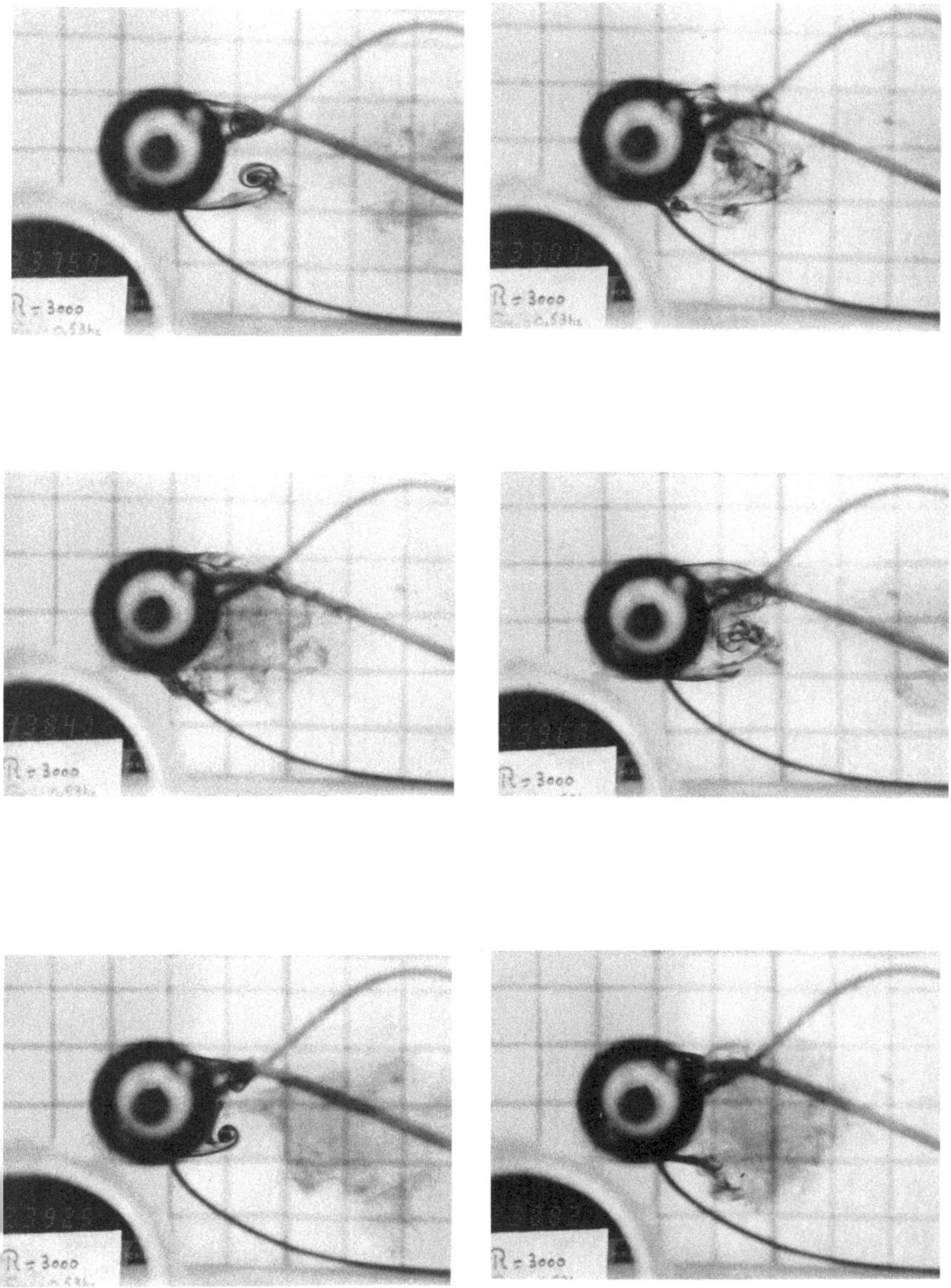

Fig. 14 Instantaneous flow patterns of Mode II.

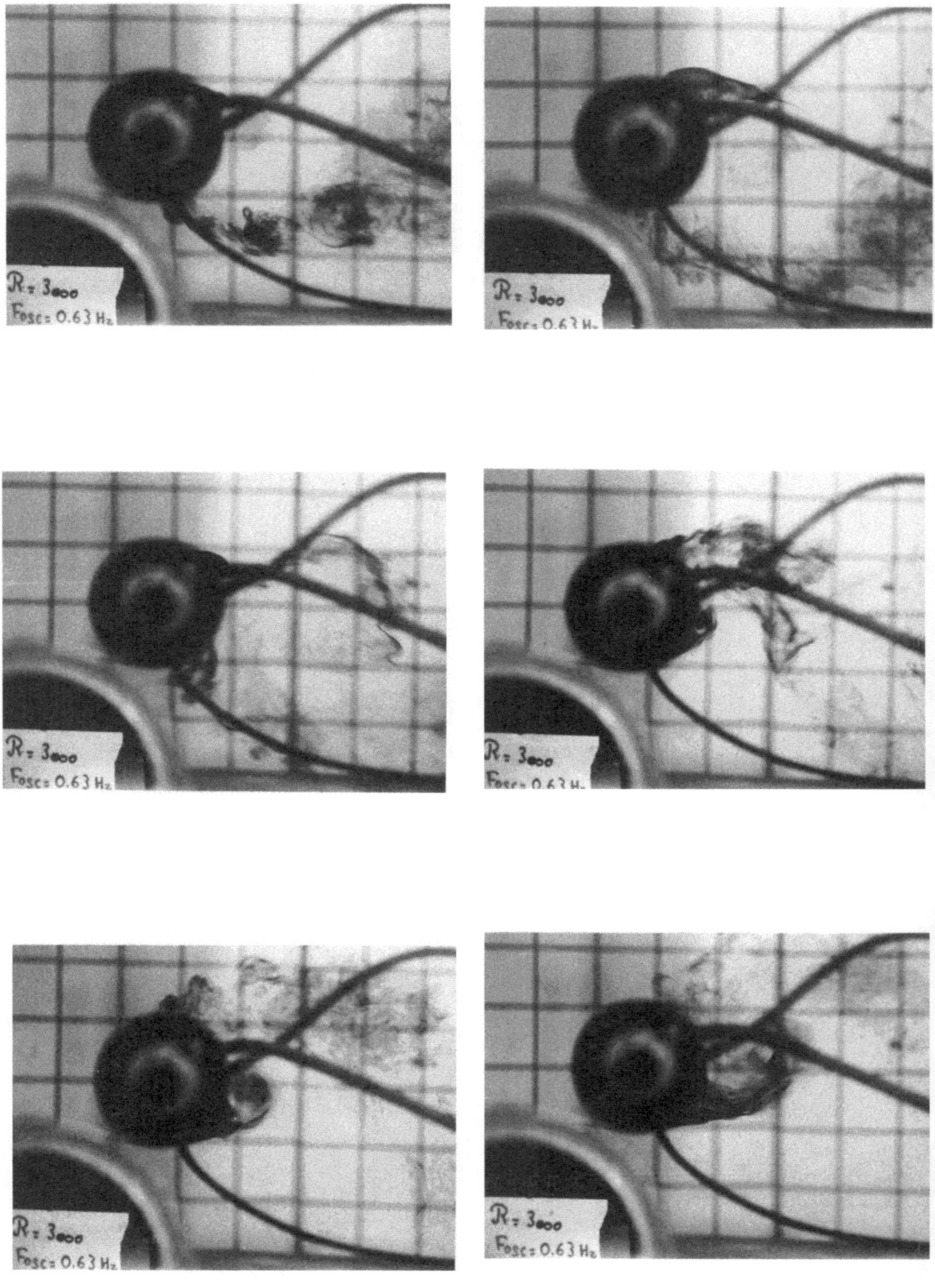

Fig. 15 Instantaneous flow patterns of Mode III.

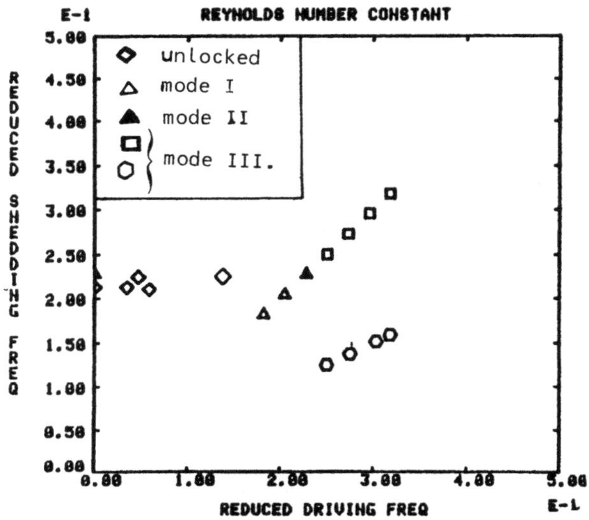

Fig. 16 The reduced shedding frequency, $f_S D/U_\infty$, versus the reduced driving frequency, $f_D D/U_\infty$.

Conclusions

The problems of (a) a cylinder oscillating in a steady stream and (b) a stream oscillating over a fixed cylinder are considerably different due to the apparent mass which is driven with the cylinder in the first case. The first problem has been investigated extensively. In this paper we studied the second. Our findings indicate that the lock-on properties are qualitatively similar. Namely, the wake locks on the outer flow oscillations at frequencies a little lower than the natural shedding frequency but drops to the subharmonic of the external oscillation for driving frequencies above the natural frequency.

Flow visualizations indicate a more detailed picture of the shedding phenomenon. For the range of Reynolds numbers tested, the separating shear layers develop some instabilities in the form of cat's eyes which may pair. The free shear layers together with their instabilities eventually roll up and generate a discrete large scale vortex. It was found that the frequency of cat's eyes formation is unrelated to the large scale vortex shedding; in other words the mechanism of Strouhal shedding is not triggered by the instabilities of the free shear layer. Ho's[18] results seem to support this fact but his work has not yet been published.

Flow visualization also revealed properties of the flow which could not be detected by any measuring device. In particular, it was found that there are three distinct modes of shedding in the lock-on range. The first mode is similar to natural shedding. The second mode is the most interesting and occurs only in the immediate neighborhood of natural shedding and appears to be symmetric. In this mode vortices are shed simultaneously from both sides of the cylinder. In the third mode the period of shedding is doubled and the wake grows sharply in width.

This investigation is still in progress and with the new rig we have had some difficulties reproducing Mode III. Some measurements are being repeated and will be available very soon. Two points of interest will be investigated more carefully. The first is the behavior of separation and the amplitude of its excursion for natural or forced shedding. The second is the three-dimensionality of the phenomenon. Preliminary data obtained recently indicate that the alternate shedding is accompanied by strong periodic motion in the direction of the axis of the cylinder. The motion of the streaklines in this direction will be studied by observing two color dyes emitted from the front and the back of the cylinder respectively.

References

1. Strouhal, V., 1878, "Uber eine Besondere Art der Tonerregung", Anal. Physik Chemie, Vol. 5, pp. 216-251.

2. Morkovin, M. V., "Flow Around Circular Cylinder - A Kaleidoscope of Challenging Fluid Phenomena", Symposium on Fully Separated Flows, Hansen, A. G., ed., ASME, New York, 1964, pp. 102-118.

3. Berger, E. and Wille, R., 1972, "Periodic Flow Phenomena", Annual Review of Fluid Mechanics, Annual Reviews Inc., Palo Alto, CA, Vol. 4, 1972, pp. 313-340.

4. Savkar, S. D., 1976, "A Survey of Flow Induced Vibrations on Cylindrical Arrays in Cross-Flow", ASME Paper 76-WA/FE-21.

5. McCroskey, W. J., 1977, "Some Current Research in Unsteady Fluid Dynamics", ASME Journal of Fluid Engineering, Vol. 99, pp. 8-39.

6. Sarpkaya, T., 1979, "Vortex-Induced Oscillations - A Selective Review", J. Applied Mechanics, Vol. 46, pp. 241-258.

7. Griffin, O. M. and Ramberg, S. E., 1976, "Vortex Shedding from a Circular Cylinder Vibrating in Line with an Incident Uniform Flow", J. Fluid Mechanics, Vol. 75, pp. 257-271.

8. Tanida, Y., Okajima, A. and Watanabe, Y., "Stability of a Circular Cylinder Oscillating in Uniform Flow or in a Wake", J. Fluid Mech., 1973, Vol. 61, part 4, pp. 769-784.

9. Maresca, C., Favier, D. and Rebont, J., 1978, "Effets Instationnaires sur un Profil en Decrochage Dans un Ecoulement Pulse; Comparison Avec le Mouvement de Tamis", 15e Colloque d'Aerodgnamique Appliquée, AAF.

10. Maresca, C., Favier, D., Rebont, Jones, G. and Telionis, D., 1979, "Measurement and Visualization of a Stalling Airfoil in Translational Oscillation", AIAA 12th Fluid & Plasma Dynamics Conference, Williamsburg, VA, July 23-25, 1979.

11. Morkovin, M. V., 1964, "Aerodynamic Loads on Bluff Bodies at Low Speeds", AIAA Journal, Vol. 2, No. 11, Technical Comment., pp. 2058-2060.

12. Michalke, A., 1972, "The Instability of Free Shear Layers", Progress in Aerospace Sciences, ed. D. Kuchemann, Vol. 12, pp. 213-239.

13. Telionis, D. P. and Koromilas, C. P., 1978, Nonsteady Fluid Dynamics, eds. Crow, D. E. and Miller, J. A., pp. 21-32.

14. Koromilas, C. A. and Telionis, D. P., 1980, "Unsteady Laminar Separation: An Experimental Study", J. Fluid Mechanics, Vol. 97, pp. 347-384.

15. Mezaris, T. B. and Telionis, D. P., 1980, "Visualization and Measurement of Separating Oscillatory Laminar Flow", AIAA Paper No. 80-1420.

16. Jones, G. S., Telionis, D. P. and Barbi, C., 1981, "Separation and Wake Interaction of a Pulsating Laminar Flow", AIAA 19th Aerospace Sciences Meeting, St. Louis, MO.

17. Stansby, P. K., 1976, "The Locking-on of Vortex Shedding Due to the Cross-Section Vibration of Circular Cylinders in Uniform and Shear Flows", J. Fluid Mech., Vol. 74, Part 4, pp. 641-665.

18. Ho, Chih-Ming, private communication.

Vortex Shedding from a Circular Cylinder in Oscillatory Flow

C. BARBI ; D. FAVIER ; C. MARESCA

Institut de Mécanique des Fluides
1, rue Honnorat - 13003 Marseille, FRANCF

NOTATIONS

A $= \frac{\Delta V}{2\pi F}$: amplitude of longitudinal oscillating cylinders

C_D : drag coefficient

C_L : lift coefficient

d : diameter of the cylinder

F : frequency of flow oscillation

F_E : frequency of strongest peak in power spectra

F_S : frequency of vortex shedding

F_{So} : " of natural vortex shedding

Re : Reynolds number : $\frac{V_\infty . d}{\nu}$

S : Strouhal number : $\frac{F_S . d}{V_\infty}$

T : period of the oscillating flow

T_S : " of vortex shedding

t : time

V_∞ : velocity of incident flow

ΔV : amplitude of velocity fluctuations

V_n : mean velocity

θ : angular position on the cylinder

λ $= \frac{\Delta V}{V_\infty}$ reduced amplitude.

INTRODUCTION

Since Strouhal demonstrated, one century ago, that circular cylinders gave rise to vortices shedding with a dimensionless frequency $S = F_{So} d/V_\infty$ of constant value over a wide range of conditions, a lot of work relative to aerodynamics of bluff bodies has been done. The main results concerning cylinders in cross-flow are described in detail in the reviews by Morkovin[1].

Bergen and Wille[2] and more recently by Mc Croskey[3]. In particular, it has been shown that the stagnation and separation points, boundary layer and wake of the cylinder are oscillating even if the on coming flow is uniform. Hence, the cylinder experiences fluctuating lift and drag forces with frequency F_{So} and $2F_{So}$ respectively. When the structure is not perfectly rigid the cylinder is excited in an oscillatory motion. Moreover, when the cylinder is forced to oscillate at a frequency close to the natural vortex shedding frequency, the resulting vortex shedding may be captured at the driving frequency of the cylinder. This "locking-on" of the vortex shedding to the cylinder frequency can occur over a range of flow velocity related to the amplitude of cylinder oscillation and Reynolds number.The study of this phenomenon is of special interest in basic understanding of fluid mechanics as well as for aeroelasticity. The effects of wind and water induced oscillations on chimneys, cables and floating offshore vessels respectively are well known.

As experiments are difficult to interpret when performed on spring-mounted cylinder, it is usually prefered to drive a rigid cylinder in order to investigate more easily the "locking-on" problem.

Most recent researches of vortex-induced oscillations and their effects onbodies have concentrated upon vibrations transverse to the flow. However, vortex-induced oscillations occur in the direction in line with the upstream flow. The results given in reference (4) show clearly that longitudinal oscillations generate large lifting forces. So the coupling between longitudinal and transverse degrees of freedom presents a great practical interest.

The litterature presents a lack of experiments on steady cylinders in oscillatory freestream velocity, and whether or not lock-in will also occur in this case is not clear. The results obtained by Hatfield and Morkovin[5] in 1973 appeared to give a negative conclusion, at least in their experimental conditions.

The present paper aims to shed some light on the

"locking-on" regime of fixed cylinders in oscillatory airstream. The work reported here is part of a cooperative effort between two laboratories at Virginia Polytechnic Institute of Blacksburg where the experiments are conducted in water tunnel, and Institut de Mécanique des Fluides de Marseille. At I.M.F.M. the tests have been performed in a subsonic wind-tunnel in the case of a fixed cylinder set in an oscillatory airstream of amplitude velocity ΔV and frequency F. The field of investigation is very wide and our study has been bounded to the subcritical flow regime ($150/300 < Re < 10^5/1.3\ 10^5$). The ΔV and F parameter have been varied in such a way that the quantity $\Delta V/\omega d$ is kept at a constant value. This parameter is equivalent to the A/d parameter of oscillating in line cylinders in steady flow, where A represents the amplitude of the cylinder fluctuation.

The locking-on has been obtained for an oscillatory airstream of $\Delta V/\omega d = 0.2$. The investigations have been carried out by means of several measuring techniques suitable for unsteady flow analysis : torsion dynamometers for measurement of unsteady drag and lift, skin friction gauge at the surface of the cylinder, smoke-filaments visualization, hot wire anemometer probes for velocity profile. Power spectrum analysis of wake velocity, drag, lift, and skin friction fluctuations have been realised by an FFT numerical method.

Stagnation and separation point motions are pointed out in both stationnary and oscillatory flow. Variations of unsteady lift and drag below and in the locking-on regime are discussed with reference to power spectrum. The variation with frequency of vortex wake patterns is studied and peculiar steaklines are shown when locking-on occurs.

2. EXPERIMENTAL FACILITIES AND PROCEDURE

Experiments were performed in the I.M.F.M. low turbulence open circuit wind-tunnel. The rectangular test section ($0.5 \times 1\ m^2$) is 3 m long. Upstream of the nozzle, seven high solidity screens reduce the free stream turbulence level to 0.2 %. The airstream velocity, under steady flow conditions can be varied from 2.5 to 20 m/s. The sinusoidal variation of the speed in the test section is generated by the motion of two flaps pivoted on the floor and the ceiling on each side of the tunnel wall, and by-passing the flow into the settling chamber via a secondary system of ducts (See Figure 1).

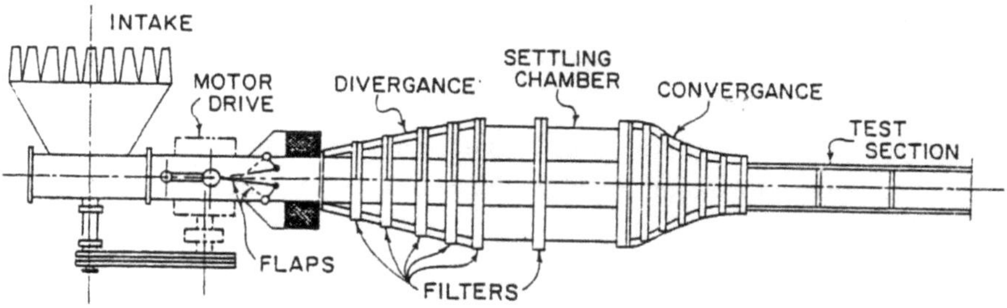

Fig. 1. The I.M.F.M. wind tunnel

In one extreme position, the flaps are extensions of the tunnel walls and in the other, which can be adjusted, they contribute to maximum blockage of the tunnel. The dashed lines in Fig. 1 show schematically the extreme position corresponding to full blockage. The frequency of the generated oscillatory flow is varied by changing the speed of the A.C. motor which drives two rods operating the flaps. The limit frequency of this device is about 7 Hertz. The cylinder first tested was 0.15 m in diameter and generated a natural vortex shedding frequency of 5.38 Hz in our test conditions. This frequency compared to the maximum attainable frequency of the oscillating stream was not able to allow a wide frequency investigation of the locking-on region. Hence, it was decided to test a larger cylinder (0.2 m in diameter) leading to a natural frequency of 3.65 Hz. The cylinders, mounted on a torsion dynamometer, spanned the entire test section and were able to rotate. The angular position θ was read on a scaled ring with an error less than 0.5°. The torsion dynamometer, described in detail in reference(6) delivers an electric output generated by the variation of an electromagnetic field created by the displacement under the action of aerodynamic forces of a small bar supporting the cylinder. The output signal of this dynamically calibrated dynamometer was filtered and amplified. Except when measuring forces, the top of cylinder was screwed to the upper wall of the test-section.

A heated-film skin friction gauge (I.M.F.M. production) has been mounted flush on the surface of the cylinder. The qualitative responses of the gauge were used to determine stagnation and zero-skin friction point motions.

Hot-wire measurements of the oscillating velocity were performed both upstream and in the wake by a Disa probe 55R Series supplied by a Disa 55M Series anemometer and a 55D10 linearizer. This anemometer also processed the heated-film skin friction signal.

The time data domain of velocity, drag, lift and skin friction were reduced and digitally stored through a 800 channels recorder to a HP 9845 B mini-computer. The computer code included also the calibrations of the sensors.

Concerning the power spectra, the results have been deduced from a single cycle time domain record over 19.7 s and 512 data points via a Fast Fourier Transform numerical method. For steady flow, no trigger signal was used and one cycle only was recorded based on Strouhal frequency, whereas in oscillatory flow a trigger signal delivered by the rotation of the motor drive allowed to average the measurements over 20 cycles in case of time domain representations.

Moreover, the wake patterns were revealed by a visualization technique based on upstream emission of white smoke filaments from a seven teeth comb.

3. RESULTS

3.1. Frequency of vortex shedding

Stationary flow

The frequency of vortex shedding was given by the power spectrum of velocity measured by a hot wire probe immersed in the wake. The measurements were repeated for $2.5\ 10^4 < Re < 10^5$ with two cylinders of diameter 0.15 m and 0.2 m and for airstream velocity varying from 2.5 m/s to 8 m/s. The test-section wall blockage ratio ranged from 15 to 20 %. The Strouhal number of vortex shedding was corrected as described in reference (7) and the obtained values of S are about 0.2 which agreeswith the values given in the litterature.

Oscillatory flow

The airstream of the wind-tunnel has been pulsed by means of the upstream device described in section 2. The frequency of oscillation may range from 1.25 Hz to 6.75 Hz with step sizes of 0.25 Hz and 0.5 Hz for d =0.15 and 0.20 m respectively. Typical velocity profiles of the incident flow obtained by hot wire anemometry are shown on figure (2) and compared to perfect sine curves (dotted lines). The Fourier analysis of these profiles attests that the flow generated is nearly sinusoidal.

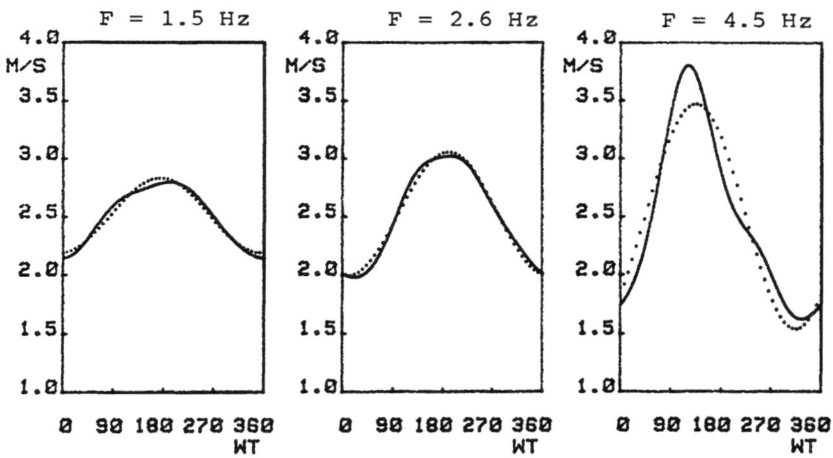

Fig. 2. Velocity profile of the oscillatory flows

As shown by Griffin and Ramberg[(8)], locking-on is sensitive to parameter A/d in the case of oscillating cylinders. It is easy to show that in the case of a pure sinusoidal oscillating flow, the corresponding parameter becomes : $\frac{\Delta V}{\omega . d}$. The litterature reports good locking-on in longitudinal cylinder oscillation for $A/d \simeq 0.2$. In the present experiments, the ratio $\frac{\Delta V}{\omega . d}$ has been also kept at constant value of 0.2, when the frequency was varied. The frequency of corresponding vortex shedding was reduced in the same way that in stationary flow previously described.

The results for the two cylinders are presented on figure 3.

Reduced vortex shedding frequencies F_S/F_{So} are plotted versus reduced oscillating airstream frequencies F/F_{So} (F_{So}=5.38 and 3.65 Hz for d =0.15 and 0.2 m respectively).

When the flow frequency is well below the natural frequency ($F/F_{So} \ll 1$), the shift between F_S and F_{So} is slight, then it increases rapidly for flow frequency close to F_{So}. When locking-on occurs at $F \simeq F_{So}$, the value of the vortex

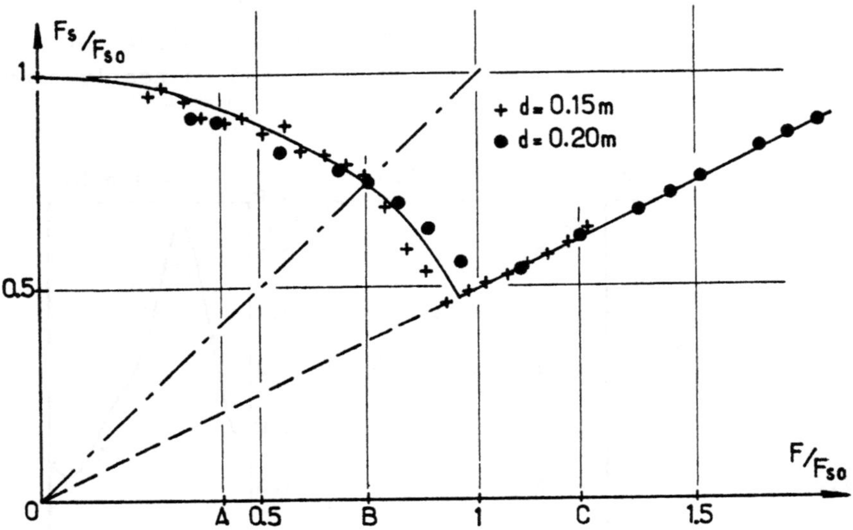

Fig.3. Vortex shedding frequency versus oscillatory flow frequency

shedding frequency is half the flow imposed frequency values. In the locking-on region ($F/F_{So} > 1$), the display of experimental results for both d = 0.15 and 0.2 m gives a straight line of slope 0.5, indicating that the vortex shedding remains captured at half the frequency of the oscillating flow all along the explored frequencies above F_{So}. One can notice on one hand that time records and power spectra exhibit a very good signal over noise ratio as soon as locking-on region is reached, revealing a better organized flowfield than for $0 < F < F_{So}$. On the other hand, higher harmonicsare always present on power spectra.

It is also worthy of note that there is no discontinuity

in the curve of the figure (3) as it may be expected from results presented in references (4) and (5). This unexpected behaviour can be explained with the help of figure(4) on which are plotted the frequency corresponding to the strongest peak obtained on power spectra. These peaks do not always correspond to vortex shedding, and it can be observed on figure(4) that in this case the curve presents discontinuities.

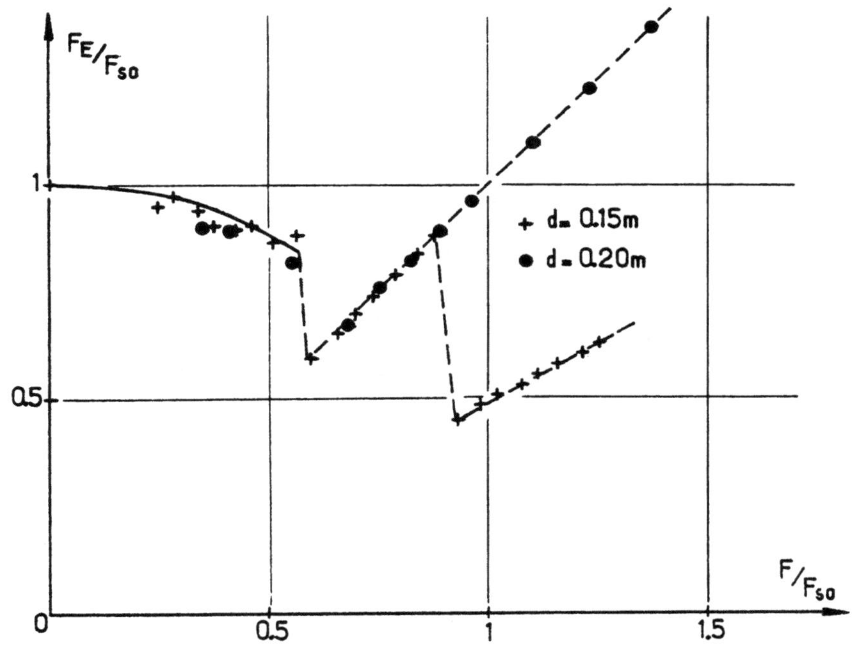

Fig. 4. Frequency of maximum peak versus Frequency of oscillatory flow.

3.2. Heated-film skin friction measurements

The experiments described in the following paragraphs are relative to the cylinder 0.2 m in diameter. No calibration of the skin friction gauge has been done ; therefore the present results allowed only qualitative analysis. Figures 5 and 6 are representations of time domain records, over 5 periods of the vortex shedding, at different angular positions from 0 to 110°. The output signal U has been reduced by its averaged value and the quantity $\frac{U-U_{mean}}{U_{mean}}$ has been plotted. Only one side of the cylinder was scanned.

Fig.5a concerns the stationary flow and Fig. 5b,6a,6b are related to oscillatory flow of frequency 1.5 ;2.6 and 4.5 Hz respectively.

Stationary flow

Because of the lack of trigger there is no phase information from one gauge location to the other.

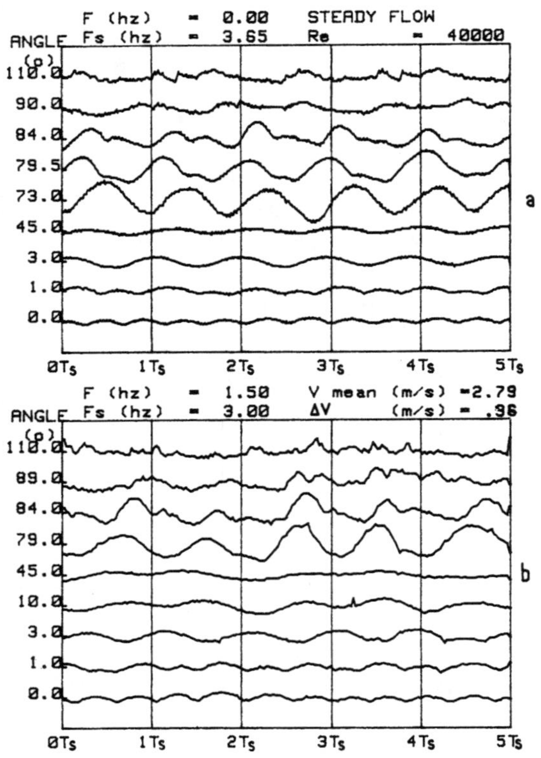

Fig. 5. Skin friction gauge records. a) Stationary flow. b) Oscillatory flow : F= 1.5 Hz.

The records of Fig.5a conduce to define five regions characterising the flowfield. The first one is the stagnation point region from θ = 0 to θ = ± 3°. There is a double peak for θ = 0 which corresponds to the passage of the stagnation point over the gauge, as already mentionned by Dwyer and

Mc Croskey[9]. As the stagnation point passes twice on the gauge during a vortex shedding cycle; the frequency of the output signal seems to be twice the Strouhal frequency. When θ is varied from 0 to 3°, one of the two peaks decreases to disappear for θ > 3°. The slope inversion on the records is indicative of the stagnation point passage on the gauge. We can then infer from these records that the excursion of the stagnation point in steady flow is of about ± 3°.

The second region occurs around θ = 45° where the amplitude of oscillation is very weak. Indeed, in this region of zero pressure coefficient, the curve of mean skin friction variations against angular position presents a plateau as shown in ref. (10). The effect of stagnation point motion results in an incidence variation of ± 3° which induces very small variation of skin friction along the plateau. In the third region where the mean friction curve presents the highest slope[10] at θ ≃ 70°, a weak variation of incidence generates a strong skin friction oscillation shown by the record obtained at θ = 73°. The fourth region characterises, in the same manner that in the first region, the excursion of the rear zero skin friction point which moves between 79.5° and 90° due to the vortex shedding cycle. The fifth region, defined for θ > 90° shows that the boundary layer remains separated from the cylinder during the all vortex shedding cycle with no apparent periodicity in the recorded signal of skin friction (θ = 110°).

Oscillatory flow

The analysis of Fig. 5b relative to F =1.5Hz has indicated quantitative results on stagnation and zero skin friction points excursions similar to those obtained in steady flow. Nevertheless one can note the existence, between 10° and 45°, of an overlap of the signal every two periods.

When the frequency is increased at F = 2.6 Hz, the stagnation motion increases in amplitude (θ = ± 5° as it can be deduced from Fig. 6a. The maximum amplitude of skin friction variation is reached at the same angular position than in steady flow, but the rear zero skin friction moves from 80°

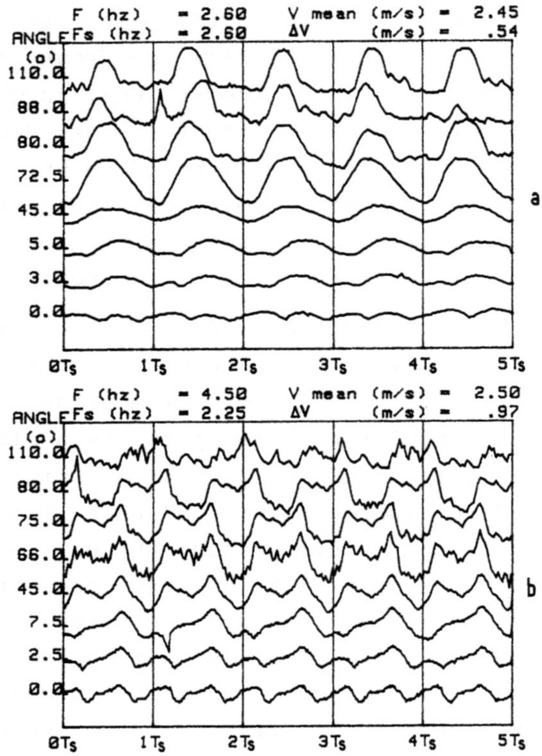

Fig.6. Skin friction gauge records. Oscillatory flow
a) F = 2.6 Hz ; b= F = 4.5 Hz.

up to 110° where very well organized structures of the flow can still be observed.

In locking-on regime (F = 4.5 Hz, Fig. 6b), we can observe a stagnation point excursion as high as $\pm$ 7,5°. The output signal, although periodic, becomes quite different of a sine wave as revealed by the record at θ = 45° where a second harmonic appears. A transition zone characterised by a higher level of turbulence on the output signals is shown up at θ = 66°. The rear zero skin friction is more difficult to define in this case. It may be bounded by θ =75° and θ = 97°.

3.3. Aerodynamic forces measurements

At the selected frequencies A,B,C on Fig. 3, lift and

drag, have been measured both in stationary and oscillatory flows. Power spectra analysis of wake velocity, lift and drag are plotted on Fig. 7 with time domain of lift and drag.

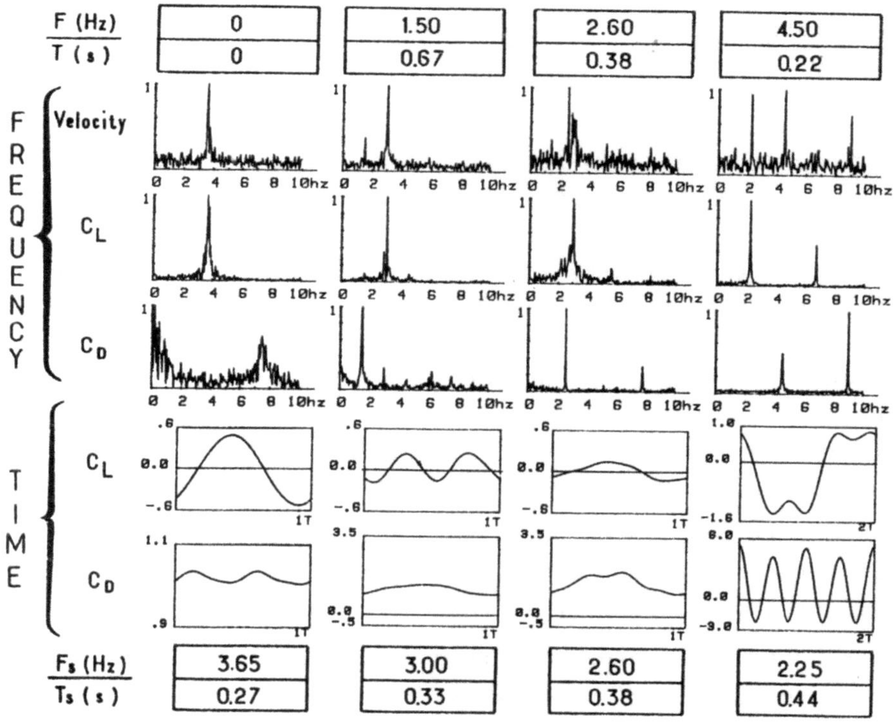

Fig. 7. Power spectra of velocity, lift and drag fluctuations and variations with time.

Each value of the power spectrum has been normalised by the maximum value of each plot. T is the period of the oscillatory flow, in each A,B,C case. Fig. 7 shows that lift behaves like the vortex shedding at frequency F_S, whereas drag combines the influence of the variation of speed at frequency F, especially when ΔV is large, and the influence of the double of the frequency F_S when $\Delta V = 0$. This result is better understood by help of an analogy between the cylinder and the pitching airfoil.

The motion of the stagnation point is then compared to the change of angles of attack of the airfoil. During a period of incidence variation, a marching point on a lift or drag versus incidence curve would cover it two times. Because this curves are odd and even functions respectively, a time domaine representation of lift or drag shows one or two periods respectively.

4. CONCLUSION

The vortex shedding from a cylinder in oscillatory flow at mean Reynolds number of 4.10^4 has been studied by measuring the skin friction on the surface of the cylinder, the lift and drag forces, the velocity in the wake and deducing the power spectrum of these fluctuating quantities. The locking-on phenomenon has occured in the range of frequency experimented. It has been characterised by a vortex shedding at half frequency of the oscillatory flow. Stagnation point exhibits large motion specially when locking-on occurs ($\pm$ 7.5°), whereas the motion of the rear zero skin friction is difficult to bound.

The lift oscillates both in stationary and oscillatory flows at the frequency of the vortex shedding with high amplitude in steady flow and at locking-on regime.

The drag oscillates at a frequency twice the vortex shedding frequency in stationary flow, and at the frequency of oscillatory flow even in the locking-on regime where a strong second harmonic appears.

REFERENCES

1 - Morkovin, M.V.,"Flow around a circular cylinder", A.S.M.E. Symposium on fully Separated Flows, May 1954, pp.102-118.

2 - Berger, E. and Willie, R.,"Periodic flow phenomena", Annual Review of Fluid Mechanics, Vol.14, 1972, pp.313-340.

3 - Mc Croskey, W.J., "Some current research in unsteady fluid dynamics". The 1976 Freeman Scholar lecture - Journal of Fluid Engineering, March 1977, Vol.99.

4 - Tanida, Y., Okajima, A. and Watanabe, Y., "Stability of a circular cylinder oscillating in uniform flow or in a wake" J. Fluid Mech. (1973), Vol.61, part 4, pp.769-784.

5 - Hatfield, H.M. and Morkovin, M.V., "Effect of an oscillating free stream on the unsteady pressure on a circular cylinder". J. of Fluid Engineering, June 1973, pp.249-254.

6 - Valensi, J., Rebont, J., "Efforts aérodynamiques sur un profil d'aile animé d'un mouvement harmonique parallèle à l'écoulement" In Proc.AGARD-FDP Meeting Aerody.Rotary Wing-Marseille, 1972.

7 - Stansby, P.K., "The locking-on of vortex shedding due to cross - stream vibration of circular cylinders in uniform and shear flows". J. of Fluid Mechanics, Vol.74, Pt.4,1976, pp.641-665.

8 - Griffin, O.M. and Ramberg, S.E., "Vortex shedding from a cylinder vibrating in line with an incident uniform flow". J. Fluid Mechanics (1976, Pt.2, pp.257-271).

9 - Dwyer, H.A. and Mc Croskey, W.J., "Oscillating flow over a cylinder at large Reynolds number". J. Fluid Mechanics (1973), Vol.61, Pt.4, pp.753-767.

10- Achenbach, E., "Distribution of local pressure and skin friction around a circular cylinder in cross-flow up to Re = 5 x 10^6". J. Fluid Mechanics (1968), Vol.34, Pt.4, pp.625-639.

Some Characteristics of the Unsteady Wake Flow Past a Circular Cylinder

H.C. BOISSON - P. CHASSAING - H. HA MINH - A. SEVRAIN

Institut de Mécanique des Fluides de l'Institut National Polytechnique de Toulouse - Laboratoire associé au C.N.R.S.
2, rue Charles Camichel - 31071 TOULOUSE CEDEX - FRANCE

Summary

Some properties of the coherent structures in the turbulent wake behind a circular cylinder (R_e = 52300) are analyzed in two downstream sections (X/D = 2 ; X/D = 4) using a phase averaging technique for the velocity fluctuations and the intermittency signal. A comparison with the Von Karman flow (R_e = 100) given by a direct numerical solution of Navier Stokes equations provides some practical information about the relevance of the double decomposition taking into consideration the two dimensional deterministic structures and the "fine scale" three dimensional phase random motions.

1. Introduction

During the last ten years, growing interest have been paid to the study of coherent structures in turbulent flows as revealed by the works of many experimentalists reviewed in the papers of Laufer [1] and Roshko [2] around 1974.

Experiments have been carried out both in natural free turbulent shear flows, mainly shear layers [3] and jets, [4] and in excited flows. As pointed out by Hussain and Zaman [5], excited flows are particularly suitable when getting more stable structures in order to observe their evolution more easily. However, it is well known that, at least for some turbulent regime, natural wakes of two dimensional bodies do exhibit strongly coherent structures responsible for vibrations and eolian tones. Thus, in these natural flows, it is also possible to find some of the advantages of the excited flows as far as one is concerned with the study of coeherent structures. Works of Cantwell [6], [7] and Davies [8] seem to support this idea which will be emphasized in this paper.

In a first part, an experimental method, based on a phase information obtained outside of the turbulent wake, is used to enhance the coherent pattern of the turbulent flow behind a circular cylinder at a Reynolds number of 52300.

In a second part the relevance of this averaged flow to that predicted by a simple numerical model is examined. So far, no exact theoretical approach is able to predict such a type of flow. As suggested by Hussain and Zaman [5], the instantaneous values of the dynamical parameters are splitted into two parts, one corresponding to the unsteady coherent motion and the other to the random phase fluctuation. The coherent flow is then governed by the unsteady Navier Stokes equations (U.N.S.) with an additional term equivalent to the Reynolds stresses. A simulation of the laminar two-dimensional flow is performed for a Reynolds number of 100, based on the numerical solution of the U.N.S. equations. Analogies and discrepancies between the results in both cases are discussed in order (i) to appreciate to what extent such a numerical method can be used when dealing with the turbulent case and (ii) to state on the question of the two dimensionality of the averaged flow field interacting with the three dimensional "fine scale" motions.

2. The experimental study of the turbulent wake

2.1. The experimental set up and data processing

The experiments are conducted in a closed low turbulence wind tunnel. An aluminium smooth cylinder (diameter D = 42 mm) vertically spans the test section (60 cm x 70 cm) corresponding to a blockage ratio of 7% and an aspect ratio of 16.7. This situation does not cause excessive distorsion of the flow according to the results of West and Apelt [9]. In the entry section (7D upstream) a uniform velocity profile ($\overline{U}_0$ = 18.7 m/s) is maintained with a turbulence level lower than 0.2 %.

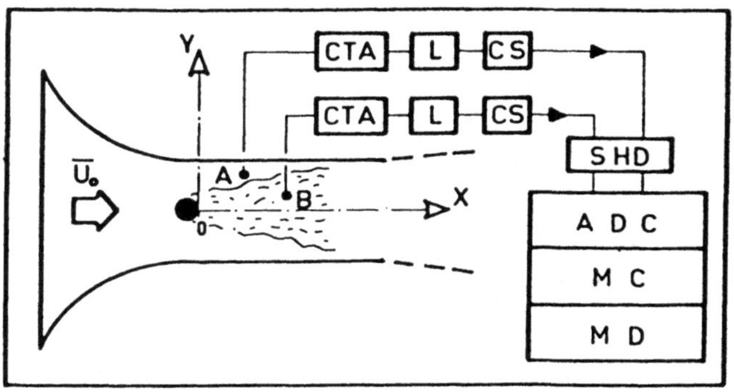

CTA : Constant temperature anemometer ; L : Linearizer ;
CS : Conditioning system ; SHD : Sample and hold device
ADC : Analog to digital converter ;
MC : Minicomputer Plurimat 400 ; MD : Magnetic disk

Fig.1. Schematic diagram of the experimental set up.

The instrumentation is depicted in Fig.1. Two single wire probes are used, the wires being parallel to the cylinder. (A) is the fixed reference probe and (B) is the movable analysis probe. Both signals are identically processed before storage on a magnetic disk. They are low pass filtered at 5 KHz and sampled simultaneously at a frequency of 10 KHz. Further treatments are processed numerically on synchronous samples of 128 x 2048 points (or 26 seconds).

2.2. The measurements

The region which is investigated (X/D = 2 and 4) corresponds to the very near wake of the cylinder. The flow in both sections is highly turbulent and accurate measurements of the velocity require either laser doppler anemometer or flying hot wire as used by Cantwell [6]. Such an apparatus was not available here. Nevertheless, it can be inferred that, since the wire is normal of the plane of the wake, the effective cooling velocity is roughly representative of the modulus of the velocity vector and weakly sensitive to angular variations of this vector in this plane. The signal of such a probe is referred to hereafter as the velocity signal U(t).

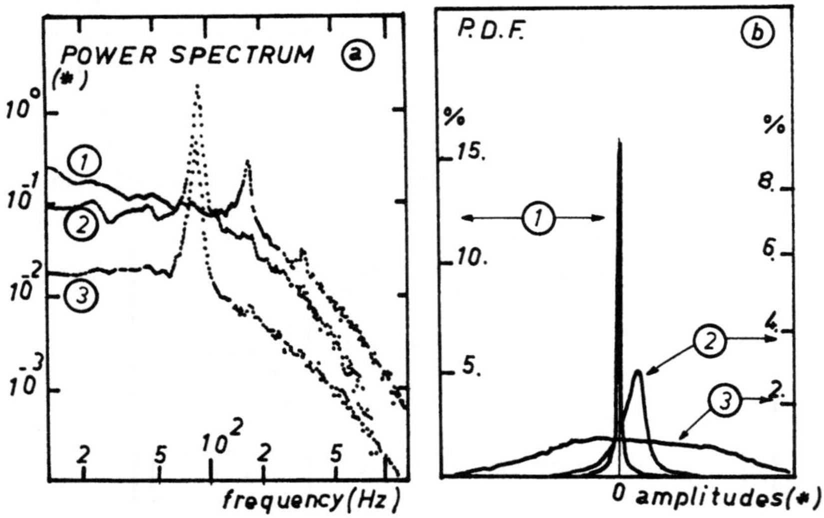

Fig.2. Conventional statistical description of the flow
2a. Power spectra of the velocity at X/D = 4
(1) Y/D = 0.0 ; (2) Y/D = 0.67 ; (3) Y/D = 1.62
2b. Probability density function of the velocity at X/D = 2 ; Y/D = 0.91
(1) Velocity 2nd time derivative ; (2) Velocity time derivative ; (3) Velocity

* Arbitrary units

The power spectra of the velocity fluctuations (Fig.2a) are dominated by a principal peak at a Strouhal number $f_s D/\overline{U}_o$ of 0.198 (f_s being the vortex shedding frequency) a classical result since Roshko (1954) [10]. Moving towards the axis, a first harmonic component becomes more and more important, while the turbulence level is considerably increased.

However this increase of the turbulence level cannot be properly interpreted without taking into account the intermittency of the flow which is very marked all over this region. The intermittency function is calculated from the second time derivative of the velocity signal. It is observed that, for the non turbulent periods, this function is lower than a fixed level which is determined from the values outside the wake. The probability density function (P.D.F.) of this derivative (Fig.2b) exhibits a peak value at the origin corresponding to the smooth part of the velocity signal and is symmetrical with respect to zero. Thus this function appears to be more suitable than the signal itself or its first time derivative for detecting the presence of turbulence. A constant threshold corresponding to 2 to 3% of the maximum value is choosen in a given section. In order to avoid misleading effects of the digital treatment [11], [12], a validation technique is introduced in which a minimum amount of time is alloted for both turbulent and non turbulent intervals (0.2 ms and 1.5 ms respectively for a vortex shedding period of about 11.35 ms and a cut off frequency of 5KHz for the signal).

The properties of the organized motion emerge from the conditional analysis of the velocity fluctuations. Conditional or phase average is defined according to the proposals of Blackwelder [13] or Antonia [14] :

$$\Phi(\tau) = < \phi(t) > = \frac{1}{N} \sum_{n=1}^{n=N} C_n(t)\phi(t) = \frac{1}{N} \sum_{n=1}^{n=N} \phi(t_n + \tau) \qquad (1)$$

$C_n(t)$ is the weighting function

t_n are the reference time stations corresponding to the maxima of the velocity signal of the fixed probe (A) located in the potential flow

τ stands for the phase angle $2\pi\, \tau\, f_s$ ($\tau = 0$ is the phase reference)

$\tau^{*} = \tau\, f_s$ is the non dimensional value of τ

N is the number of periods used to perform the phase average (approximatively equal to 2300).

Three different phase averaged parameters are considered : (i) the phase average of the velocity $<U>$ representing the periodic component, (ii) the coherent correlation of the turbulent fluctuation $<u^2> = <U^2> - <U>^2$ and (iii) the phase average of the intermittency function denoted by $\Gamma(\tau)$ and corresponding to the probability of the presence of turbulence at a given phase angle.

2.3. The Results

2.3.1. The average flow pattern in the plane of the wake.

The phase averages of the flow parameters are given in Figs. 3, 4 and 5. Within the range of accuracy of the method, all these distributions can be considered as nearly periodic with respect to the phase angle τ^{*}. It can be noticed that, owing to the periodicity of these results, it has not been necessary to take into account occasional jitters, the effects of which seem to be of secondary importance unlike in other natural flows, as noticed by Blackwelder [13] for instance.

Considering the phase average of the velocity, a first harmonic component, which is dominant on the axis, is apparent from Y/D = 0.67 and 1.14 at X/D = 2 and 4 respectively, in good agreement with the conventional power spectra of Fig.2a. Similarly the intermittency plots exhibit a double distribution over a period at the same locations. It is deduced that the flow pattern is such that the potential flow from one side can cross over the axis forming large size non turbulent pockets.

From the comparison between the phase average of the velocity and the intermittency, it is found that non turbulent sequences coincide with accelerated parts of the flow, the turbulent breakdown occuring near the maximum of the averaged velocity.

Considering the spatial variations of the intermittency plots (Fig.5), two principal observations can be made : firstly, near the axis, the phase averaged intermittency factor is always significantly different from zero owing to losses in phase lock and to long turbulent periods ; secondly, the central zone involving a double distribution and high values of the intermittency factor is wider when moving downstream. This suggests an entrainment mechanism in which the turbulent kinetic energy is transferred from the vortices towards the external fluid, mainly near the axis where the interaction between the two vortex rows is stronger.

The coherent correlation of the turbulent background velocity exhibits a distribution quite similar to that of the intermittency. Thus this intensity seems to be more dependent on the amount of time during which the flow is turbulent than on the intrinsic variations of the local turbulence intensity.

As a consequence of these properties, the structure of the turbulent vortices

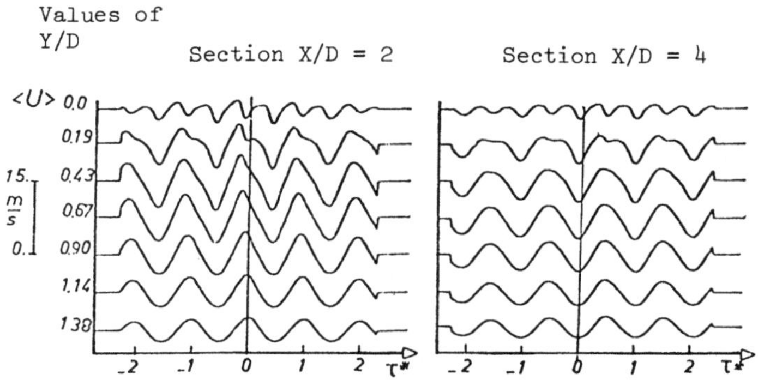

Fig.3. Phase average of the velocity

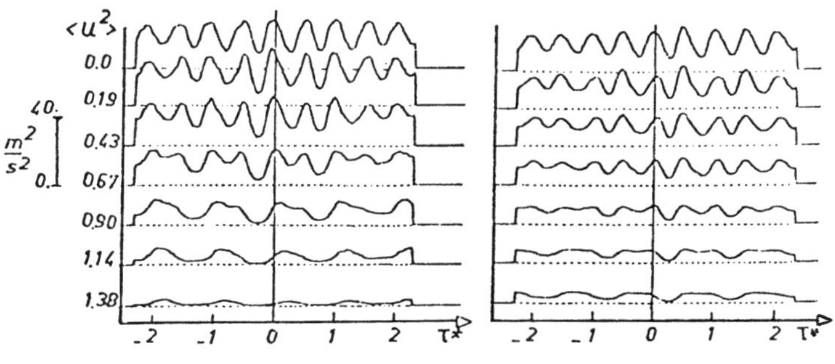

Fig.4. Coherent correlation

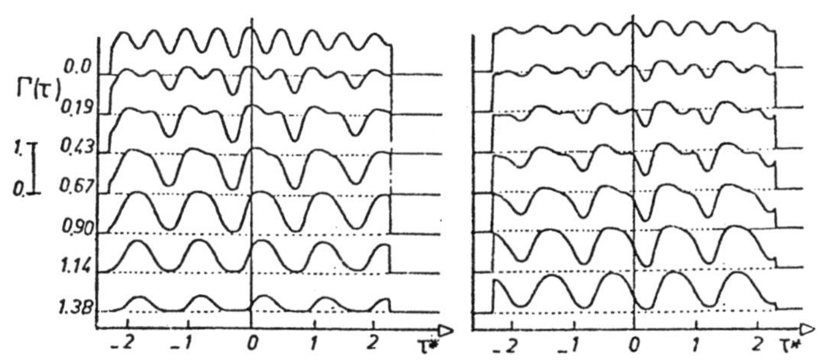

Fig.5. Phase average of the intermittency

appears to be completely controlled by the vortex shedding mechanism. An average convection velocity is obtained from the phase relationship between the two sections and is found equal to 0.82 $\overline{U}_o$, a value corresponding to a wavelength of about 4.2 D which agrees well with the results of Cantwell [6] or Davies [8] amongst others.

2.3.2. Spanwise variations

The analogy between the average flow field just described and the numerical simulation (§ 3) is valuable only if the two dimensional character of the coherent motion is verified. The spanwise variations of the phase average of the velocity and the intermittency are given in Fig.6, the distance between the movable probe and the reference plane (Z) varying from 0 to 4 diameters.

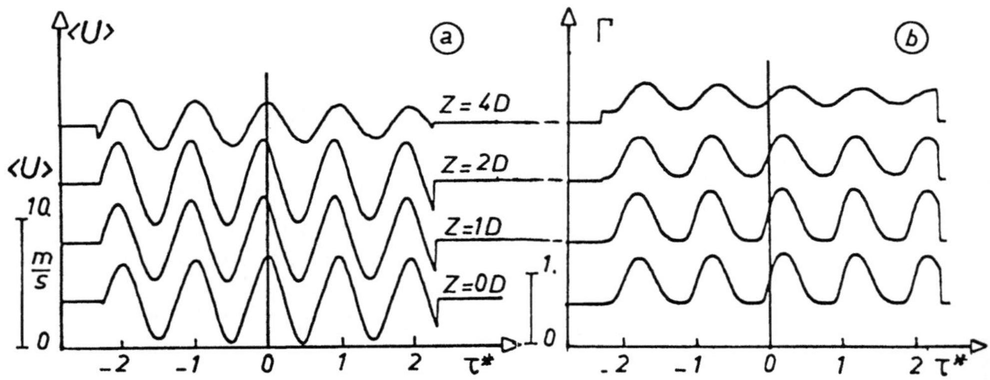

Fig. 6. Spanwise variations of the phase averaged parameters
6a. Velocity ; 6b. Intermittency
X/D = 2 ; Y/D = 1

These results clearly show that there are no-significant differences in the measured parameters at least for distances less or equal to two diameters. As the distance from the reference plane is increased, it is not possible to conclude because the phase average is more sensitive to jitters and the measurement procedure has to be modified. Nevertheless, it seems reasonable to consider that the averaged flow pattern is two dimensional at the analyzed point and this justifies the following analogy.

3. Numerical simulation of the fiscous wake

3.1. The governing equations

Using phase averaging it is convenient to describe unsteady coherent flows by the vorticity transport equation obtained by Hussain and Zaman [5] :

$$\frac{\partial}{\partial t}\langle\Omega_i\rangle+\langle U_j\rangle\frac{\partial}{\partial x_j}\langle\Omega_i\rangle = \langle\Omega_j\rangle \frac{\partial}{\partial x_j} \langle U_i\rangle + \nu \frac{\partial^2\langle\Omega_i\rangle}{\partial x_j\partial x_j} + \underbrace{\frac{\partial}{\partial x_j}(\langle\omega_j u_i\rangle-\langle u_j\omega_i\rangle)}_{\star} \qquad (2)$$

where Ω_i stands for the i component of the vorticity vector. Thus, the unsteady turbulent flow is governed by the same equation as the laminar case except for and added term ($\star$) standing for the turbulent effects.

When an eddy viscosity concept is used to express the "Reynolds" stresses, the equation (2) can be written as :

$$\frac{\partial}{\partial t} \langle\Omega_i\rangle + \langle U_j\rangle \frac{\partial}{\partial x_j} \langle\Omega_i\rangle = \langle\Omega_j\rangle \frac{\partial}{\partial x_j} \langle U_i\rangle + \nu_t \frac{\partial^2\langle\Omega_i\rangle}{\partial x_j\partial x_j} + S_\Omega \qquad (3)$$

where S_Ω is a source terme containing derivatives of the turbulent viscosity ν_t.

In a previous work, it was shown that in steady flows, the source terme S_Ω has a minor importance. Furthermore, if we assume that in a turbulent flow the eddy viscosity is of a higher order of magnitude than the molecular one, it is perfectly legitimate to consider that the actual Reynolds number U_oD/ν_t is in a low range.

Although significant differences between laminar and turbulent flows may be not negligible, since the eddy viscosity is not constant in space and in time, it is yet interesting to compute the laminar case for a Reynolds number corresponding to the same Strouhal number. For this purpose, equation (3) (where ν_t = constant and $S_\Omega = 0$) is simultaneously solved with the Poisson equation for the stream function :

$$\Delta(\psi) = \Omega \qquad (4)$$

for a two-dimensional flow.

The initial conditions correspond to a flow at rest and the boundary conditions are those of no slip on the wall and a mixed Neuman Dirichlet condition at infinity, allowing vortices to go freely out of the downstream boundary [15].

A passive scalar transport equation can be adjoined, the temperature, for instance, being of some interest here for a slightly heated cylinder.

3.2. Numerical code

Equations (3) and (4), transformed in a logarithmic polar coordinate system, are approximated by centered space finite difference schemes. An Alternating

Direction Implicit method is used for the transport equations and the Poisson's equation is solved at each time step by an A.D.I. optimized method. Detailed description of the procedure is given by Martinez [16].

3.3. The results

The results are given for a Reynolds number of 100 since it was checked that the global characteristics were then in agreement with the experimental measurements.

In Fig.7 the velocity field is plotted. The vortices are only apparent when they are attached to the cylinder, otherwise the convection velocity is such that only an oscillation of the flow is depicted.

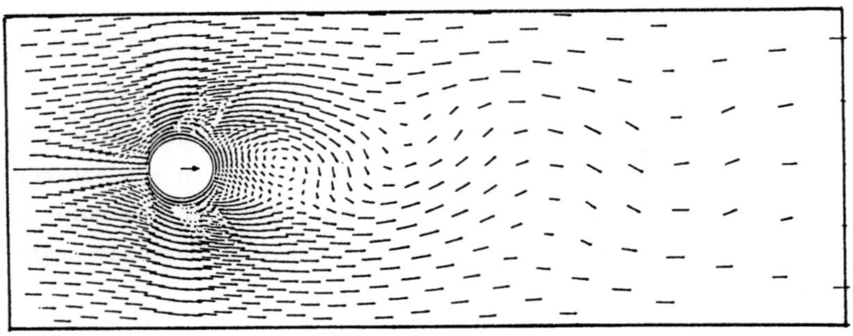

Fig.7. Velocity field R_e = 100

The temperature variations (Fig.8) are more suitable for picturing the coherent vortices. Temperature, as a passive scalar, has been currently used for the evaluation of the intermittency function and is considered as a marker of the flow. It can be observed that the flow pattern exhibits the same unsteady features that have been measured previously in the turbulent wake. However, even though some gross features of the turbulent flow appear in qualitative agreement with the laminar model, some serious discrepancies are to be expected specially in the region near the axis of the wake.

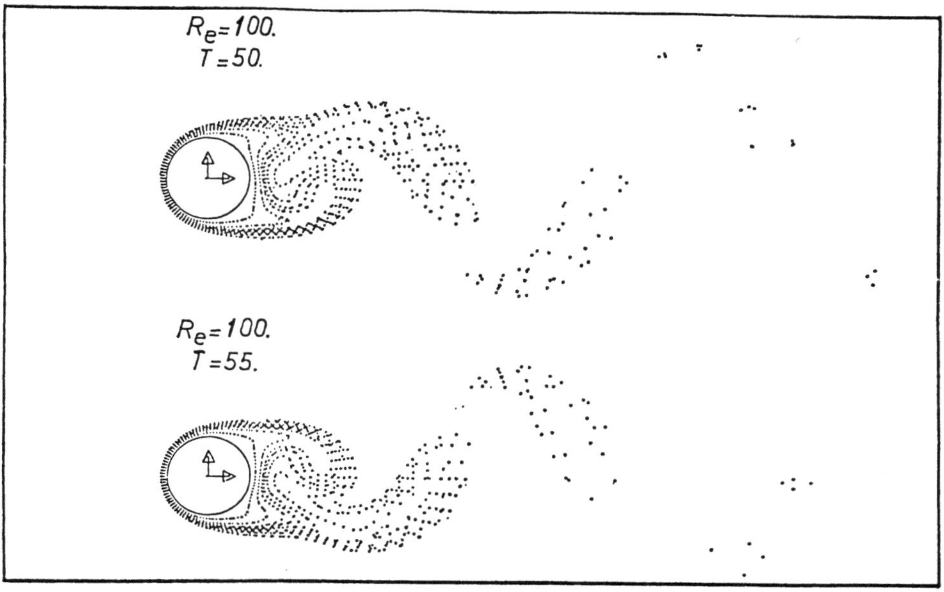

Fig. 8. Isothermal curves R_e = 100. T = dimensionless time.

4. Conclusion

On a qualitative basis, the comparison proved that the approach of the organized wake motion by an unsteady two-dimensional method is valuable, even in the presence of fine scale three dimensional motion. The phase averaged parameters of the flow are actually two dimensional and reflect the periodic structure.

However it seems that a constant eddy viscosity model would probably be unsuitable for predicting the turbulent flow everywhere and a large amount of work is still necessary both in the experimental and numerical fields in order to improve the previous results.

References

1. Laufer, J. New trends in experimental turbulence research. Ann. Rew. of Fluid Mech (1975) 7, pp.307-326

2. Roshko, A. Structure of turbulent shear flows : a new look. AIAA J. (1976) vol.14 (10)

3. Winant, C.D. and Browand F.K. Vortex pairing : The mechanism of mixing layer growth at moderate Reynolds numbers. J. of Fluid Mech. (1974), vol. 63, part.2, pp. 237-255

4. Bruun, H.H. A time domain analysis of the large scale flow structure in a circular jet. Part.1 : Moderate Reynolds Number. J. Fluid Mech. vol.83 part 4, pp. 641-671 (1977)

5. Hussain, A.K.M.F. and Zaman, K.B.M.Q. Vortex pairing in a circular jet under controlled excitation. Part.2 : Coherent structure dynamics. J. Fluid Mech. (1980) vol.101, part 3, pp. 493-544

6. Cantwell, B.J. A flying hot wire study of the turbulent near wake of a circular cylinder at a Reynolds number of 140000. Californian Institute of Technology, Ph.D. Thesis (1975)

7. Cantwell, B.J. Organized motion in turbulent flow. Ann. Rev. Fluid Mech. (1981), 13, pp.457-515.

8. Davies, M.E. A comparison of the wake structure of a stationary and oscillating bluff body using a conditional averaging technique. J. Fluid Mech. (1976), vol.75, part.2, pp.209-231

9. West, G.S. and Apelt, A.J. The effects of tunnel blockage and aspect ratio on the mean flow past a circular cylinder in the range $10^4 < R < 10^5$. (To be published in J. Fluid Mech.)

10. Roshko, A. On the development of turbulent wakes from vortex streets. NACA Rept 1191 (1954)

11. Boisson, H.C., Sevrain, A. and Braza, M. Statistiques sur les durées des épisodes turbulents au passage des tourbillons émis par un cylindre. Euromech colloquium, (Juin 1980) pp. IV3a to IV3c.

12. Boisson, H.C., Chassaing, P. and Ha Minh, H. Conditional analysis of intermittency in the near wake of a circular cylinder. Rapport IMFT (1981) M3-39

13. Blackwelder, R. On the role of phase information in conditional sampling Phys. of Fluids (1977) vol.20, n°10, part II, pp. S232-S242

14. Antonia, R.A. Conditional sampling in turbulence measurement. Ann. Rew. Fluid Mech. (1981) 12, pp 131-156

15. Ha Minh H., Boisson H.C., Martinez G. Unsteady mixed convection heat transfer around a circular cylinder. ASME winter annual meeting (1980) HTD vol.13, pp.35-44

16. Martinez, G. Caractéristiques dynamiques et thermiques de l'écoulement autour d'un cylindre circulaire à nombres de Reynolds modérés. Thèse Docteur Ingénieur, INP Toulouse (1979) n° 48

17. Ha Minh, H., Chassaing, P. Some numerical predictions of incompressible turbulent flows. First Int. Conf. of "Numerical methods in Laminar and Turbulent Flow". Swansea, U.K. (July 1978) pp.287-300

Experimental Analysis of the Wake Behind an Isolated Cambered Airfoil

A. Cenedese

Istituto di Aerodinamica, Università di Roma

G. Cerri

Istituto di Macchine e Tecnologie Meccaniche, Università di Roma

S. Iannetta

Istituto di Idraulica, Università dell'Aquila

Summary

The incompressible flow field in the wake of a cambered airfoil has been experimentally analyzed by means of a LDV technique. Roshko number is given versus incidence angle for different Reynolds number. Measurements include mean velocity and turbulence intensities in streamwise and normal directions for two different incidence angles at different location downstream of the trailing edge. The structure of the wake flow field is presented, the decay of the mean velocity defect has been analyzed and of the RMS velocity distributions too. Some important features of the wake structure are discussed. The measurements reveal that the camber angle influences the growth and the spreading of the wake.

Introduction

The knowledge of the structure of an airfoil or a cascade wake is fundamental for establishing improved aerodynamic design criteria for the surfaces of turbomachinery blades and of helicopter rotors. In fact information on the mean velocity and turbulent intensities in the wake behind an airfoil are useful for predicting noise level, vibration characteristics and loss of efficiency in turbomachinery.

While both laminar and turbulent wakes behind a flat plate immerged in a parallel stream have been experimentally and theoretically widely investigated [1÷7], few investigations exist on airfoil or cascades even though the incidence angle and the geometric characteristics of the profile play an important role in the phenomena [8÷14]. Most experimental and theoretical studies on isolated airfoil have been performed on symmetric profiled ones.

As the shape of the airfoil influences the decay of the velocity defect of the wake and its diffusion downstream, chiefly in the near wake region [3,8, 13,14], experimental investigations were performed on a cambered airfoil immerged in an incompressible flow. Measurements of velocity and turbulence intensity level were carried out in the wake region behind the airfoil at various axial and transverse locations and spectra of the velocity fluctuations were obtained. All the values have been compared with the upstream ones.

Experimental equipment

The incompressible flow field was generated in a water tunnel [14], the test section of which had a square section ($100. \times 100.\ mm^2$) constructed in Plexiglas; behind the airfoil there was a nearly constant pressure flow. Two tanks permit the variation of the free stream velocity to obtain different Reynolds numbers. The cambered airfoil, the characteristics of which are given in fig. 1, was put in the water tunnel and the incidence angle could be adjusted from outside. The measurements were made by means of a Laser Doppler Velocimeter (L.D.V.), the optical arrangement consisted of a 5. mW He-Ne Laser, an optical unit with a Bragg cell and a photomultiplier placed in forward scattering mode (dual beam method). The photomultiplier signal was processed by means of a frequency tracker and then by means of an integrator voltmeter, a RMS voltmeter and a realtime spectrum analyzer (fig. 2).

General remarks on the wake flow

The wake behind a body immerged in a flowing stream is the region where the two flows coming from the opposite surfaces of the profile interact and the flow field is influenced by the presence of the airfoil. Due to the boundary layers on the upper surface and on the lower one, the wake is characterized by:

- high gradients of the mean velocity and of the turbulent energy;
- defect of the mean velocity in the streamwise direction;

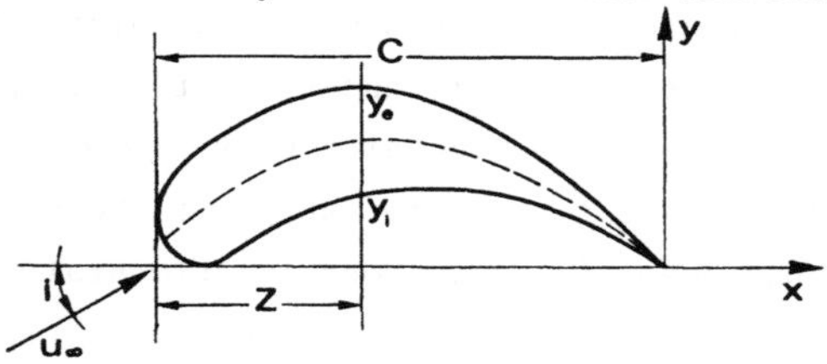

z/C	.0	.1	.2	.3	.4	.5	.6	.7	.8	.9	1.
y_i/C	.072	.022	.084	.125	.159	.166	.159	.147	.113	.059	.0
y_e/C	.072	.284	.344	.375	.378	.356	.313	.263	.194	.109	.0

Fig. 1 - Airfoil geometry.

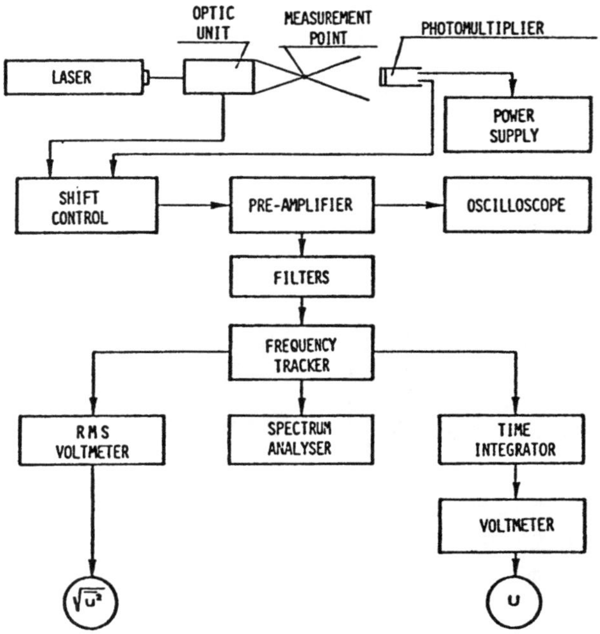

Fig. 2 - Scheme of LDV equipment.

- the wake width;
- different mean turbulent intensities - RMS - of the component of the velocity.

The decay of the mean velocity defect depends on the mass and momentum interchange on either side of the wake which is high in the "near wake" due to its structure that presents high velocity gradient which is an unstable flow condition. The spreading of the wake and the mixing with the outer flow cause a lower gradient in the "far wake".

As separation occurs vortex shedding causes a "quasi-periodic flow". The predominant frequency (f) can be expressed by the dimensionless Strouhal number ($St = f\,C/U_\infty$ referred to the chord) or Roshko number ($Ro = f\,C^2/\nu$).

Results and discussion

The experiments were carried out on an isolated cambered airfoil at different flow conditions (fig. 1).

Spectra of the velocity (fig. 3) were taken far upstream of the airfoil and downstream in the wake region where a peak in the frequency distribution shows the vortex shedding frequency connected with a "quasi-periodic flow".

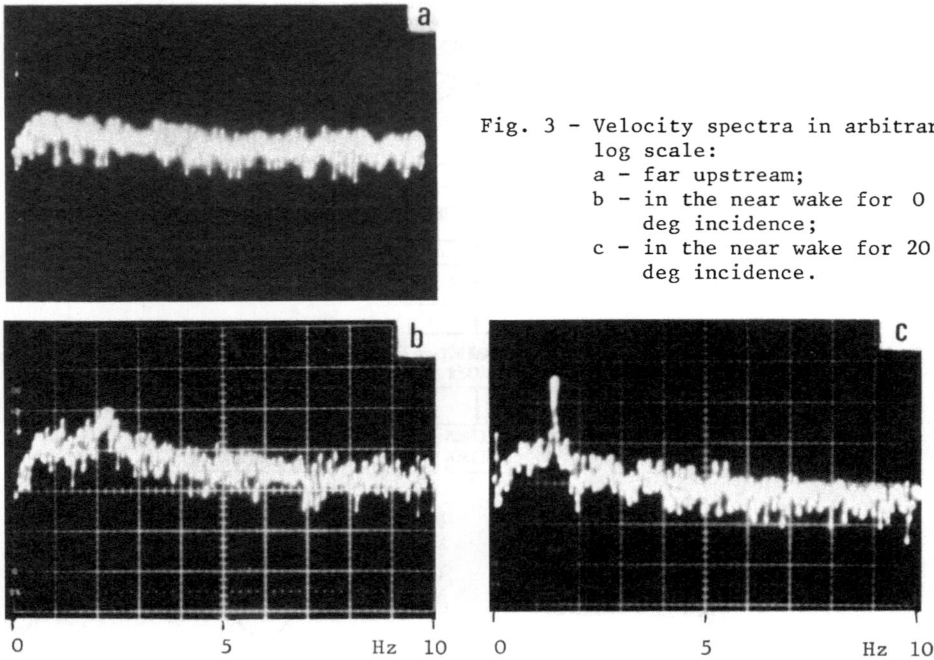

Fig. 3 - Velocity spectra in arbitrary log scale:
a - far upstream;
b - in the near wake for 0 deg incidence;
c - in the near wake for 20 deg incidence.

The peak frequency expressed in terms of Roshko number versus incidence angle and for different Reynolds numbers is given in fig. 4. Measurements were made in different points of the wake region to better observe the phenomenon.

Roshko number is nearly constant at high positive and negative incidence. In these flow conditions the airfoil seems to behave as a bluff body, i.e. the wake width is nearly independent of the incidence angle. When the incidence angle is 90 deg the shedding vortex frequency is lower than at -90 deg . This fact could depend on the shape of the airfoil that presents a curved camber line, a rounded leading edge and a sharp trailing edge, consequently the wake width should be different at the two above mentioned incidence angles. Close to zero incidence angle the vortex shedding frequency reaches a maximum. It is worth noticing that in the neighbourhood of zero deg incidence angle, chiefly for negative ones, the velocity spectra does not present a marked peak but only a maximum (fig. 4, dashed line). This should be related to the lower pressure gradient than for other angles. Consequently the separation point moves toward the trailing edge so the reversed-flow region presents a lower extent and a proper vortex shedding frequency does not seem to exist, but the velocity fluctuates in a range of

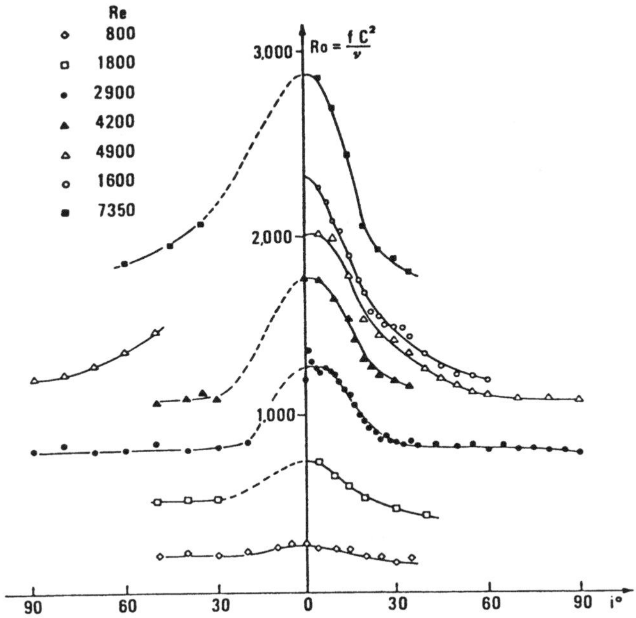

Fig 4 - Roshko number versus incidence angle for various Reynolds numbers.

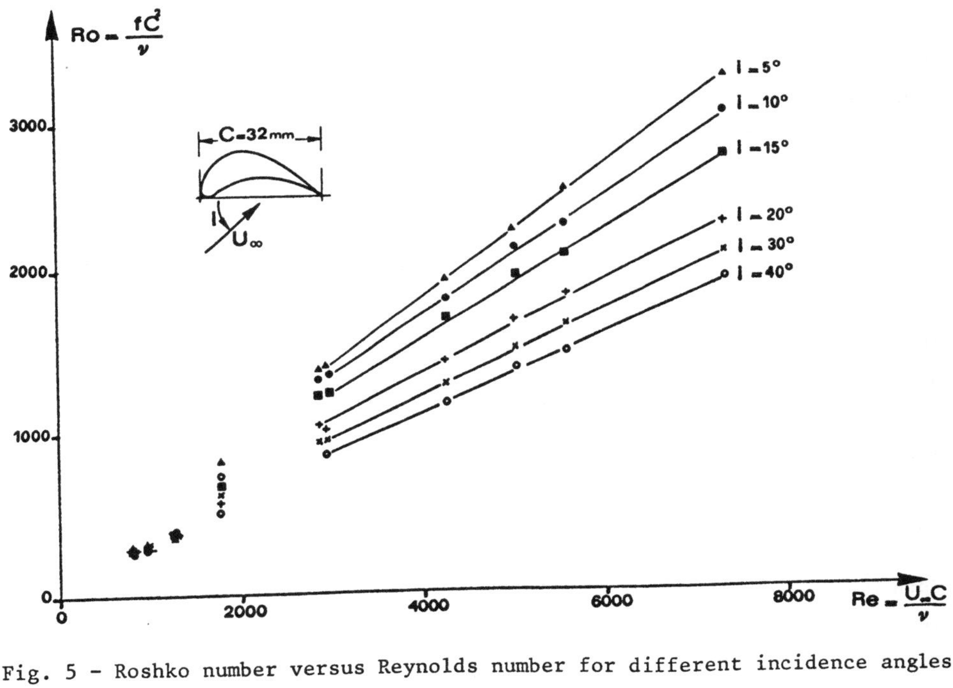

Fig. 5 - Roshko number versus Reynolds number for different incidence angles.

frequency for which the spectrum presents a bump with a maximum. When the incidence angle of the flow makes the separation point move toward the leading edge - the pressure gradient on the suction surface tends to increase - vortex shedding occurs at a definite frequency (peak frequency of the spectrum).

In fig. 5 Roshko number versus Reynolds number at constant incidence angles between 5. and 40. degrees is shown. In agreement with other studies a linear relationship exists between Ro and Re numbers when the incidence angle is constant and for Reynolds numbers higher than 3000. Non-linearity exists below Re = 3000.

In fig. 6 mean velocity distributions and RMS ones of the components of the velocity in x and y directions are given for zero deg of incidence angle and for Reynolds number of 5000. Measurements were taken in the wake region at various distances from the trailing edge. The mean velocity and RMS values are normalized by the far upstream ones. The mean velocity distributions present boundary layer characteristics and chiefly in the near wake the velocity at the edge overshoots the free-stream one. At the trailing section and on the suction side only the mean velocity becomes negative, owing to the reversed-flow region downstream of the separation point. The reversed-flow region disappears before the axial distance of $x/C = 0.66$. The distributions of the turbulence level - RMS - of the U and V velocity components present two maxima due to the interaction of the flows coming from the boundary layers on the suction and pressure sides. These maxima starting from a certain section in the wake reduce until disappearing at $x/C = 3$. Distributions of $\sqrt{\overline{u'^2}} / \sqrt{U^2 + V^2}$ and $\sqrt{\overline{v'^2}} / \sqrt{U^2 + V^2}$ (that is, of the turbulence intensities referred to the local velocity), present a maximum if the sections downstream of the back-flow region are taken into account.

The distributions of the same quantities for an incidence angle of 20 deg and for the same Reynolds number are given in fig. 7. In agreement with the vortex shedding frequency measurements the reversed-flow region has a larger extent than for zero deg. In fact there are negative velocities in the wake beyond the section at $x/C = 1$. The turbulence intensities $\sqrt{\overline{u'^2}}$ and $\sqrt{\overline{v'^2}}$ show two peaks higher than for zero incidence.

Both mean velocity distributions for 0 and 20 deg incidence angles seem to follow the Gauss' function that were derived theoretically for the two dimensional wake profile from a flat plate. The velocity similarity profiles

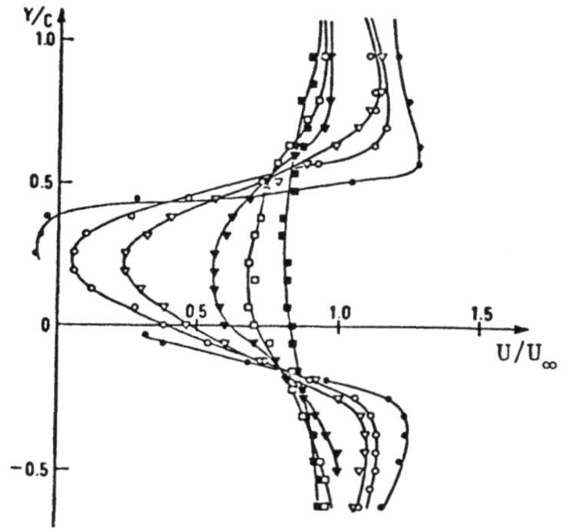

Fig. 6 - Mean and turbulent velocity distributions at different locations downstream from the trailing edge for 0 deg incidence angle.

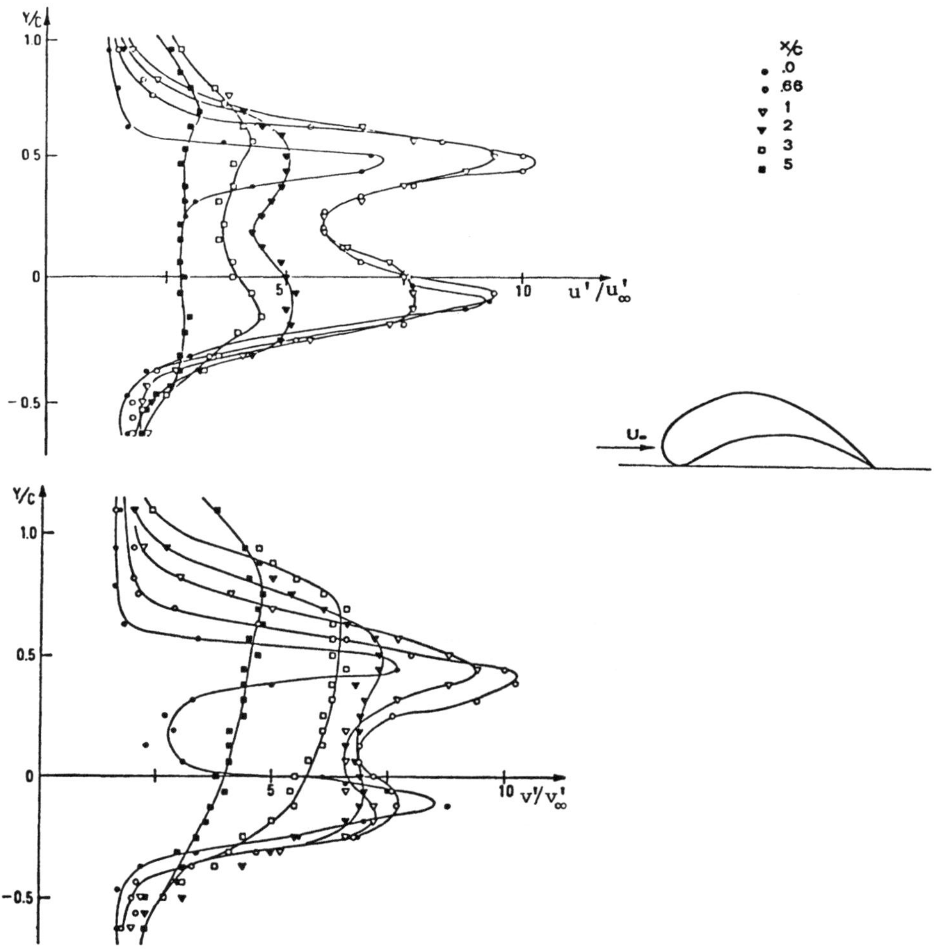

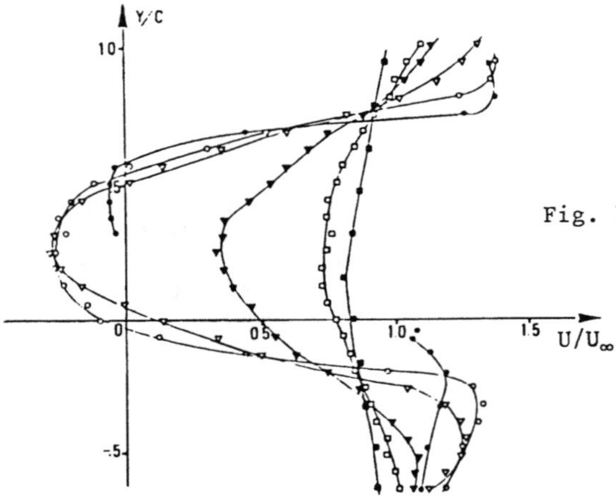

Fig. 7 - Mean and turbulent velocity distributions at different locations downstream from the trailing edge for 20 deg incidence angle.

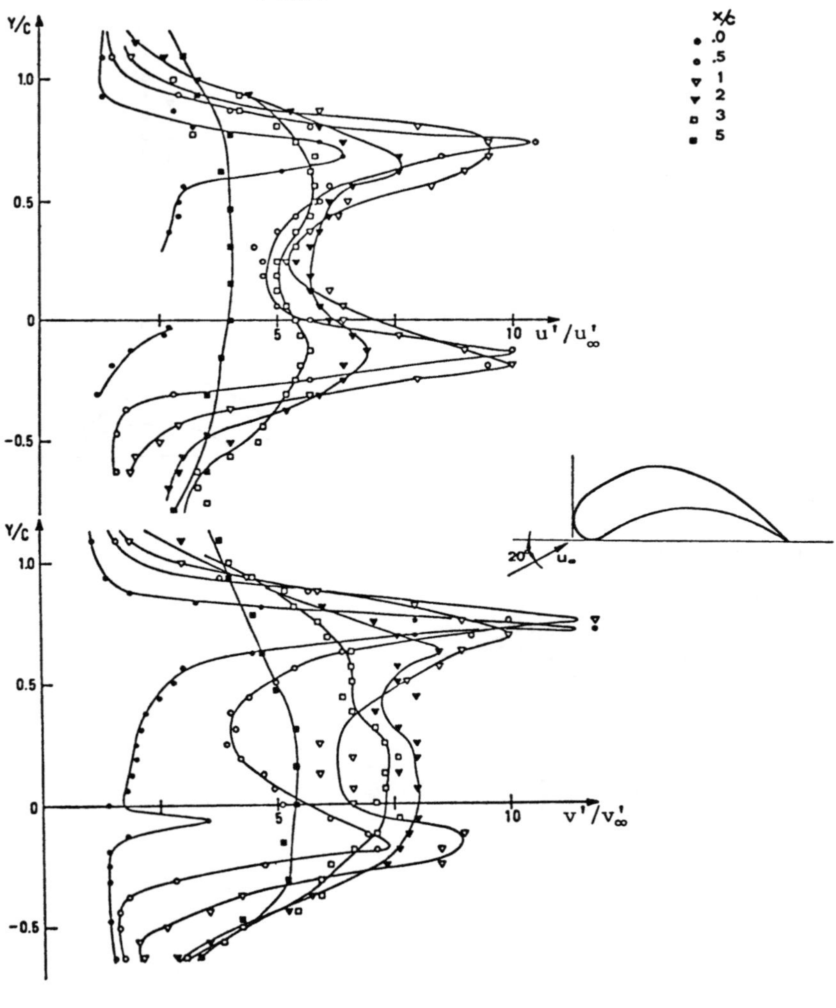

are plotted in fig. 8. As a nondimensionalising velocity scale for the mean velocity distributions the maximum velocity difference was used. As characteristic length scale the distance between the wake centerline (defined as the locus of the minimum points of the mean velocity distribution curves in the wake) and the location where the mean velocity defect was half of the maximum value was used. Due to the asymmetry of the velocity distribution two values have to be used for both suction and pressure side. So, defining the variables $\eta = \frac{y - y_c}{y_{0.5} - y_c}$ and $\xi = \frac{U_e - U}{U_e - U_c}$ where: y_c is the center line y coordinate; $y_{0.5}$ is the half maximum velocity defect value y coordinate; U_e is the mean velocity at the wake edge; U_c is the centerline mean velocity, the similarity rule is:

$$\xi = e^{-0.693\,\eta^2}$$

Similarity is evidenced by the rather good agreement with experimental results, there is a little discrepancy in the region near the wake edge.

Concluding remarks

The experimental results here given agree qualitatively with the ones reported by other Authors whose experiments were carried out chiefly on symmetrical uncambered airfoils or on thin flat plates. For isolated cambered airfoil like the one we experienced, the curvature of camber line influences the spreading of the wake. In fig. 9 the decay of the centerline velocity defect is shown. In the region where the flow is reversed the velocity defect is higher than 1. Just downstream of this region - in the near wake - the decay rate of the velocity defect is high and increases with incidence angle. In the far wake, for $x/C > 3.$, the decay rate becomes independent of the incidence angle.

The width of the wake, given in fig. 10, tends to become the same for both 0 deg and 20 deg incidence angles in the far wake ($x/C > 3$).

The axial distribution of the maximum values of the RMS velocity, fig. 11, shows that for both u' and v' the maximum rises up to reach a maximum value, just on the border of the reversed-flow region, and then decreases as the wake spreads.

It is worth observing that Roshko number versus incidence angle has a profile complementary to the drag coefficient. When Roshko number reaches the maximum value, the drag coefficient is minimum. The difficulties in finding a proper frequency in the neighbourhood of zero deg incidence

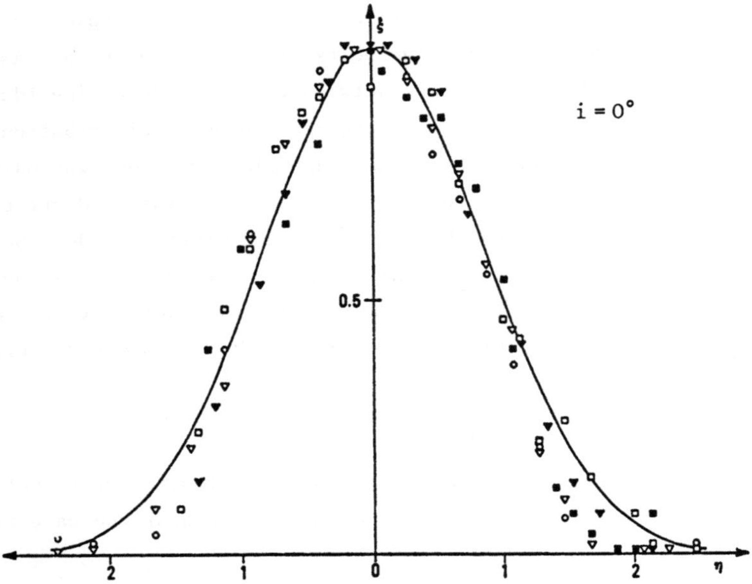

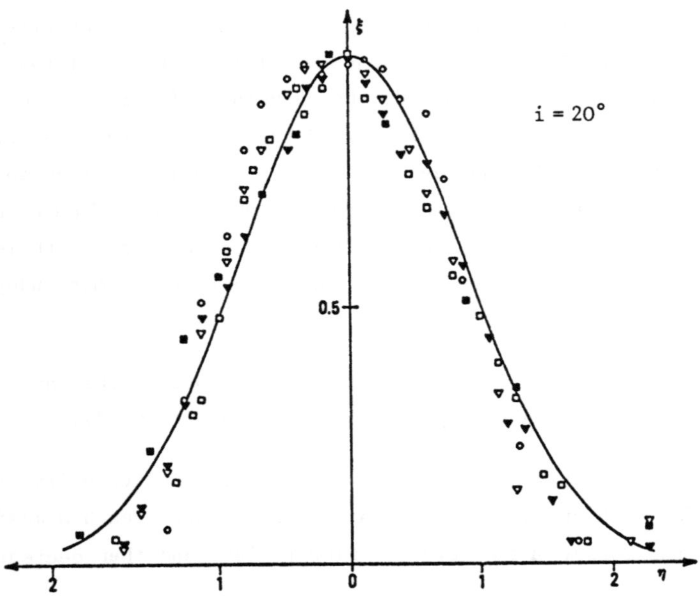

Fig. 8 - Non dimensional streamwise velocity profiles.

angle - that should be close to the one for which the drag coefficient is minimum - connected with a bump in the velocity spectrum (without a marked peak as has been discussed above), denotes the absence of well-organized large structures that shed with a fixed frequency. The influence of the walls in the phenomena should be taken into account, so a new water tunnel with a higher test section is being built. With this new tunnel measurements on cascades will be feasible too.

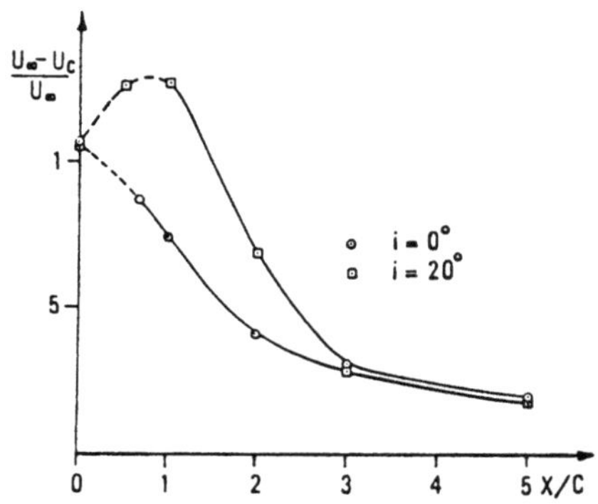

Fig. 9 - Decay of centerline velocity defect.

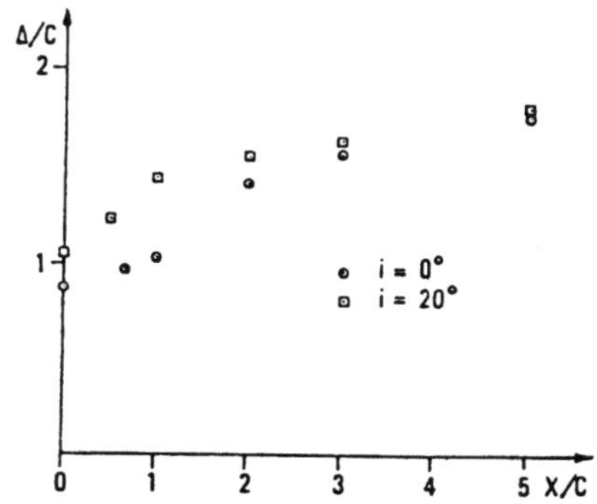

Fig. 10 - Wake width variation downstream the airfoil.

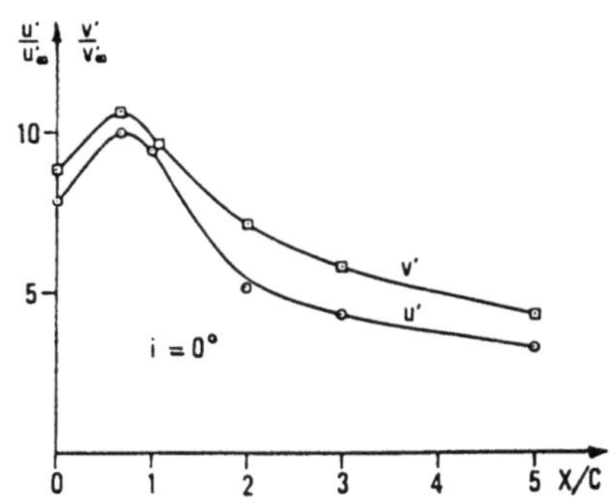

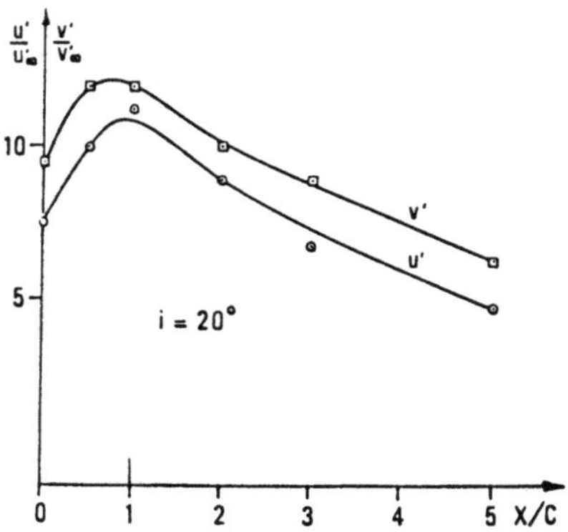

Fig. 11 - Maximum values of RMS velocity distributions along the wake.

References

1. Chevray, R., Kovaznay, L.S.G., "Turbulence measurements in the wake of a thin flat plate", AIAA Journal (1969), vol. 7, n. 8, pp. 1641-1642.
2. Toyoda, K., Hirayama, N., "Turbulent near wake of a flat plate - Part I - Incompressible flow", Bull. of the JSME, vol. 17, n. 108 (1974), pp. 707-712.
3. Toyoda, K., Hirayama, N., "Turbulent near wake of a flat plate - Part II - Effects of boundary layer profile and compressibility", Bull. of the JSME, vol. 18, n. 120 (1975).
4. Ray, R., Lakshminarayana, B., "Three dimensional characteristics of turbulent wakes behind rotors of axial flow turbomachinery", Journal of Engineering for Power, vol. 98 (1976), pp. 218-228.
5. Nishioka, N., Miyagi, T., "Measurements of velocity distributions in the laminar wake behind of a flat plate", Journal of Fluid Mechanics, vol. 84 (1978), part 4, pp. 705-715.
6. Irwin E. Alber, "Turbulent wake of a thin, flat plate", AIAA Journal, vol. 18 (1979), n. 9, pp. 1044-1049.
7. Agrawal, M.L., Pande, P.K., Prakask, R., "Study of the turbulent near wake of a flat plate", AIAA Journal, vol. 15 (1977), n. 5, pp. 740-743.
8. Ray, R., Lakshminarayana, B., "Characteristics of the wake behind a cascade of airfoils", Journal of Fluid Mechanics, vol. 61 (1973), pp. 707-730.
9. Gustafson, W.A., Davis, D.W.Jr., Deffenbangh F.D., "Analysis of the turbulent wake of a cascade airfoil", Journal of Aircraft, vol. 14 (1977), n. 4, pp. 350-356.
10. Satyanarayana, B., "Unsteady wake measurements of airfoil and cascades", AIAA Journal, vol. 15 (1977), n. 5, pp. 613-618.
11. Hill, P.G., Schaub, U.W., Senoo, Y., "Turbulent wakes in pressure gradients", Trans. of the ASME, Journal of Applied Mechanics, (1973), pp. 518-524.
12. Johnson, D.A., Bachals, W.D., "Transonic flow past a symmetrical airfoil - Inviscid and turbulent flow properties", AIAA Journal, vol. 18 (1980), n. 1, pp. 16-24.
13. Lakshminarayana, B., Davino, R., "Mean velocity and decay characteristics of the guidevane and stator blade wake of an axial flow compressor", Trans. of the ASME, Journal of Engineering for Power, vol. 102 (1980), pp. 51-60.
14. Cenedese, A., Cerri, G., Iannetta, S., "Analisi mediante Anemometria Laser Doppler della scia prodotta da un profilo alare", Proceedings of the 5th AIMETA Congress, Palermo (Italy), 23-25 October 1980.

Influence of Wall Vibrations on a Flow With Boundary-Layer Separation at a Convex Edge

E.P. SUTTON, G.P. EVANS, M.D. McGUINNESS and K.M. SVEHLA

Cambridge University Engineering Department

Summary

Experiments on flow with a separation bubble at a sharp-edged inlet to a pipe from a plane wall are described. The shear layer bounding the bubble was thin and laminar at separation but turbulent over most of its length. Small-amplitude vibrations of the separation edge reduced the length of the bubble by as much as a third, by influencing the development of the coherent structures in the shear layer. These and other observations draw attention to the importance of the large-scale unsteady features of separated and reattaching flows with turbulent shear layers.

Introduction

Flow with boundary layer separation and reattachment containing a more-or-less two-dimensional recirculating region or 'separation bubble', can usefully be described in terms of its time-mean characteristics (Figure 1a), neglecting the large temporal fluctuations which are often present. The shear layer bounding

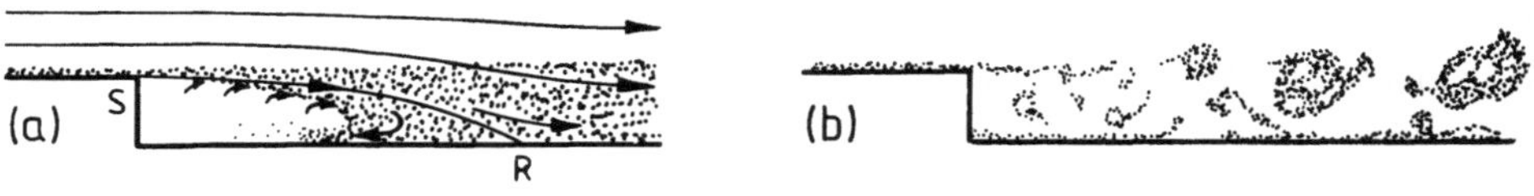

Fig.1. Flow with separation bubble: (a) Time-mean, (b) Instantaneous

the separation bubble entrains fluid from the interior of the bubble; there is no net flow across a dividing streamline which extends from the separation point at S to the reattachment point R, so the fluid entrained must be replaced by a return flow from near R. In the upstream half of the bubble, where velocities are low, the static pressure is approximately constant. Towards the downstream end there must be a pressure gradient, positive towards R, to drive the return flow into the bubble. Influences which increase the entrainment rate of the shear layer cause the length of the bubble to decrease.

This description is a valuable one, but flow visualization reveals that the instantaneous flow rarely bears any close resemblance to it, except at rather low Reynolds numbers. Figure 1b is an attempt at an instantaneous picture of the 'separation-bubble' region when the shear layer is turbulent over its whole length. Large coherent structures develop, as in a shear layer bounding a jet. They appear to dominate the flow, and they certainly play a major role in the entrainment process. They interact with one another and with the wall as they increase in size and pass downstream. They continue past the time-mean reattachment position, which is *not* a conspicuous feature of the instantaneous flow pattern.

It seems unlikely that flow with separation and reattachment at high Reynolds numbers will be adequately understood without full account being taken of the role played by these large-scale unsteady features. The main part of this paper concerns the effects of influencing them by periodic excitation, which produced large changes in the time-mean flow. Some other relevant experimental results are also briefly mentioned.

Axisymmetric flow with a separation bubble: Apparatus

An annular separation bubble was formed at a sharp-edged inlet to a circular pipe from the plane end wall of a cylindrical settling chamber (Figure 2)[1]. The settling chamber was supplied with air from a centrifugal blower. It had wire-mesh screens and a honeycomb, and the boundary layer was removed from its cylindrical wall downstream of the honeycomb. The diameter of the pipe was 144 mm, and the area contraction ratio was approximately 36:1.

To apply periodic disturbances to the shear layer, the separation edge itself was made to vibrate in an axial direction[2], with frequency up to 275 Hz and amplitude up to about ± 0.08 mm. A flexible diaphragm was incorporated in the plane end wall of the settling chamber and driven electromagnetically by means of an AC winding attached to it, in the field of a fixed DC winding, as shown diagrammatically on the right of Figure 2.

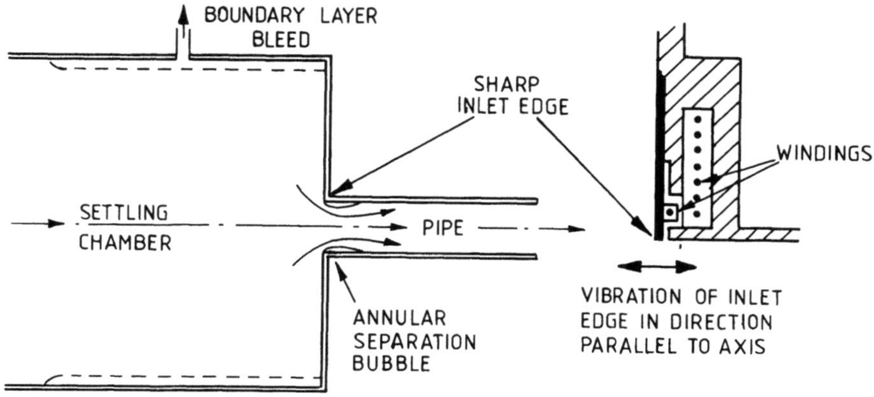

Fig.2. Apparatus for axisymmetric-flow experiments

The flow without excitation[1,2]

At Reynolds number $Re_D = 10^5$ (based on the mean velocity in the pipe and the pipe diameter) the momentum thickness of the laminar boundary layer at separation was about 0.03 mm. Transition followed very quickly, and the separated shear layer was turbulent over most of its length. The length of the separation bubble was about 125 mm, according to surface oil-flow observations, uniform within ± 2 mm and approximately constant over the Reynolds number range from 10^5 to 2.1×10^5. Its maximum (radial) height was about 12 mm.

As far downstream as the maximum bubble height, the flow closely resembled an unconfined contracting jet from a sharp edged orifice. The streamlines of the potential flow coincided with the streamlines of such a jet as given by free-streamline theory. The rate of growth of the shear layer, and the spectral characteristics of hot-wire signals from the potential flow adjacent to the shear layer, were as expected for the shear layer of an unconfined jet from a nozzle; the thickness increased linearly with axial distance, and the spectra indicated the existence of quasi-periodic coherent structures with associated characteristic frequency decreasing inversely with axial distance from the inlet plane, in quantitative agreement with published data for jets.

A reference static pressure p_1 measured at the pipe wall in the approximately constant-pressure region at the upstream end of the bubble was used together with the stagnation pressure in the settling chamber to calculate a reference velocity U_1, an approximation to the velocity in the potential flow just outside the shear layer at the edge of the contracting jet. At Reynolds number 10^5, U_1 was about 17 m/s.

The effects of excitation by vibration of the inlet edge[2]

Figure 3 shows the effects of vibration on the length L of the separation bubble as determined by surface oil-flow observations, over a range of Reynolds number and dimensionless frequency of excitation. The dimensionless excitation frequency n_E is defined

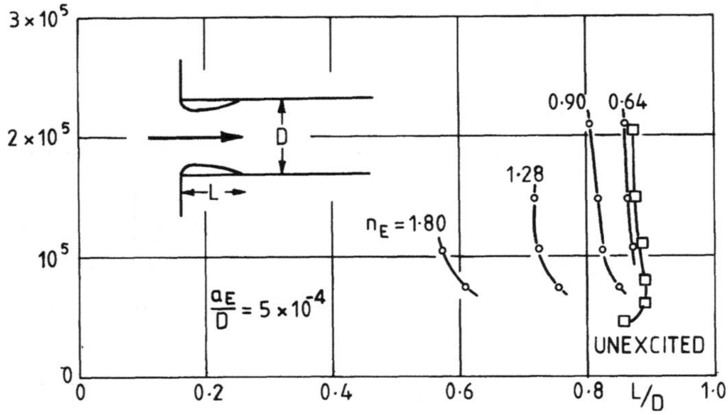

Fig.3. Variation of bubble length with dimensionless excitation frequency and Reynolds number

as $f_E D/U_1$, where f_E is the excitation frequency, D is the pipe diameter, and U_1 is the reference velocity defined above. The dimensionless amplitude of vibration was $a_E/D = \pm 5 \times 10^{-4}$.

At dimensionless frequencies n_E less than about 0.5, vibration had little effect on the bubble length. With increasing n_E, the length of the bubble decreased steadily. At n_E up to about 1.0 the effect remained fairly small, but at $n_E = 1.8$ with $Re_D = 10^5$ the bubble length was reduced by more than one third.

Figure 4 shows how the excitation affected the rate of growth of the shear layer. The locus of points at which the axial

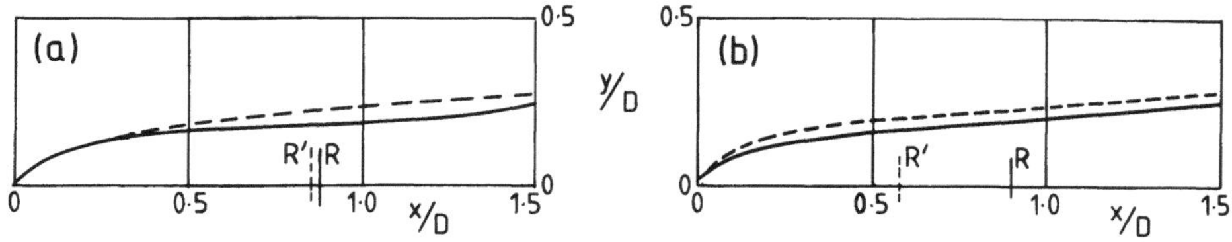

Fig.4. Shear-layer growth: Locus of turbulence intensity 5%. (a) n_E = 0.9, Re_D = 2.1 × 10^5 (b) n_E = 1.8, Re_D = 1.05 × 10^5
———— unexcited, - - - - - excited

component u' of turbulence intensity was 5% of U_1 on the potential-flow side of the shear layer is shown for flows with and without excitation. Excitation at n_E = 0.9 (Figure 4a) increased the rate of growth of the shear layer from about half-way along the bubble. Excitation at n_E = 1.8 (Figure 4b) increased the rate of growth almost from the beginning.

It appears that the main effects of periodic excitation at the separation edge were delayed to a distance downstream at which the shear layer was able to respond. Comparison with the results of the spectral observations made on the unexcited flow showed that the main response ocurred somewhat upstream of the position at which the characteristic frequency associated with the coherent structures in the unexcited shear layer was equal to n_E. The higher the excitation frequency the further upstream the shear layer responded. The lack of response when n_E was less than about 0.5 can be accounted for by the intervention of reattachment well before the characteristic frequency of the shear layer would have fallen as low as 0.5 in the absence of reattachment.

An increase in the rate of growth of the shear layer would be expected if the excitation organised and enhanced the pairing of the coherent structures. The observed effects on the length of the separation bubble can be accounted for by the increase in entrainment rate corresponding to the increase in rate of growth of the shear layer, or more directly to the enhancement of pairing of coherent structures.

Whether it is more meaningful to say that the bubble shortens because of the increase in entrainment rate or to say that it shortens because the more rapidly growing shear layer approaches and interacts with the wall sooner is not obvious. What does seem clear is that progress towards a thorough understanding of the reattachment process is likely to require a more detailed study of the interactions of the coherent structures with the wall.

The effect of shortening the reattachment surface

Figure 5a shows how the static-pressure distribution along the pipe wall changed when the pipe was progressively shortened (in the absence of excitation). Reducing the pipe length to

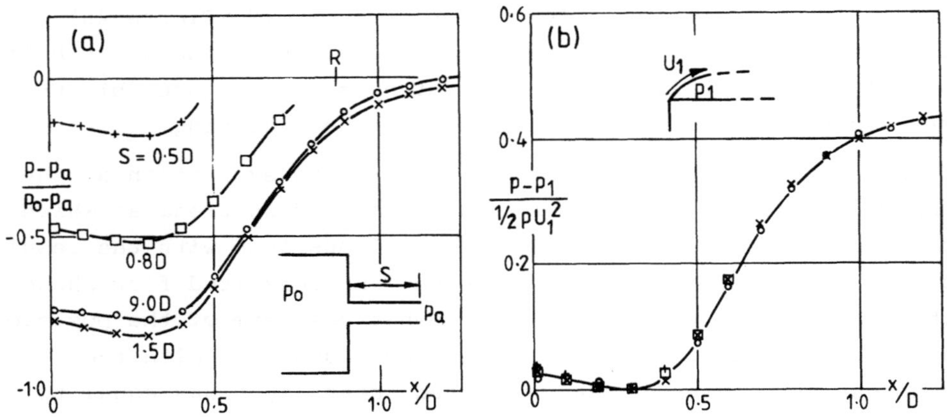

Fig.5. Wall pressure distribution along pipes of various lengths, (a) relative to constant outlet pressure p_a, non-dimensionalised with (p_o-p_a); (b) relative to minimum pressure p_1, non-dimensionalised with $\frac{1}{2}\rho U_1^2$ ($Re_D = 2.2 \times 10^5$)

less than that for needed for time-mean reattachment curtailed the pressure rise associated with reattachment. The outlet pressure p_A at the end of the pipe was always the same, so the result was an increase in the general level of pressure downstream of separation.

In Figure 5b the same pressure distributions have been referred to p_1 as datum and non-dimensionalised with $\frac{1}{2}\rho U_1^2$. (In this case p_1 was the minimum pressure at the wall). The pressure distributions in these terms are identical. It seems that the return flow into the bubble from near the time-mean reattachment position R plays a less specific part than the time-mean description of the flow suggests, and that more attention should be paid to l o c a l interactions between the coherent structures and the wall.

Similar conclusions have been reached by Simpson et al[3] from experimental observations of separated turbulent boundary layers at considerably higher Reynolds numbers.

Further investigations are in hand of effects of periodic excitation when the pipe is shortened.

<u>Further experiments to investigate the unsteady features of separation-bubble shear layers</u>

The effects of periodic excitation of flows with separation bubbles have drawn attention to the importance of the unsteady features of such flows whether or not artificially excited. Figure 6 shows some preliminary results of an attempt to derive some information about them from steady three-dimensional-flow experiments, with which an approximate analogy may be expected to exist.

A long thin strut of uniform cross-section, shown in Figure 6c, was mounted across the working section of a wind tunnel at zero incidence and 45° sweep. The flat "leading edge" of the strut was roughened by covering it with sandpaper, in order to apply more-or-less random small-scale perturbations to the shear layers which separated from its otherwise sharp spanwise edges.

Figure 6a shows the pattern of limiting streamlines on one of the side faces of the strut, as deduced from flow visualisation by the surface oil-flow technique, over a small part of the span well away from the tunnel walls. The shear layer from the separation edge at the nose seems to have rolled up piecewise into concentrated 'vortices', (Figure 6b), at first on a smaller scale and then on a larger scale as the smaller vortices combined. A spanwise-mean reattachment line RR could be located approximately, though the attachment position was not everywhere easy to identify. At intervals along the span, the larger vortices trailed away downstream and were replaced by new vortices from upstream (Figure 6b).

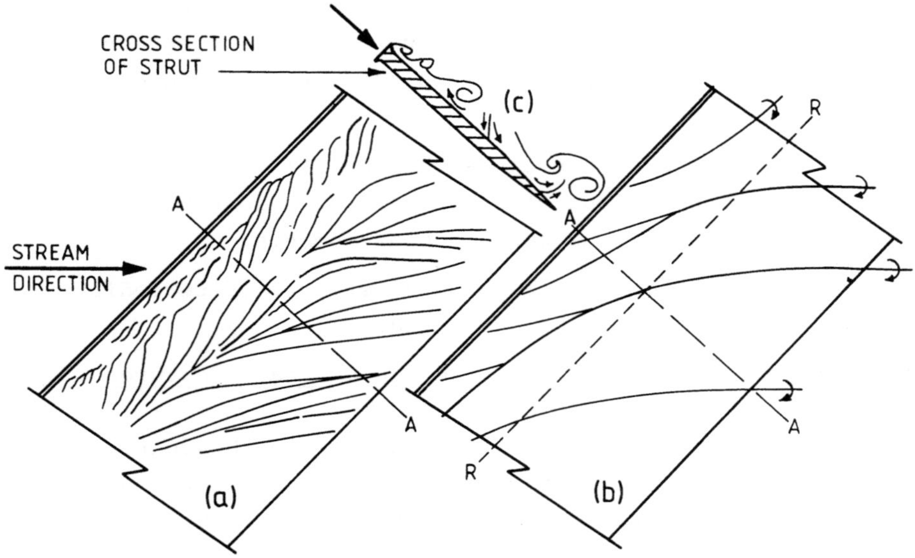

Fig.6. Three-dimensional flow past a swept strut with separation at the leading edge. (a) Limiting streamlines at surface. (b) Paths of vortices. (c) Flow in a plane through AA normal to the span. RR is an estimate of the spanwise-mean reattachment position.

The sketch in Figure 6c illustrates features of the flow in a cross-section plane normal to the span at AA, deduced from Figures 6a and 6b. It resembles an instantaneous view of a two-dimensional flow about an unswept strut of the same cross-section, with typical coherent structures in the separated shear layer. The spanwise development of the flow in planes normal to the span should bear some resemblance to the time-wise development of the two-dimensional flow, including the interaction of the coherent structures in the shear layer with the reattachment surface.

Further experiments will include some in which spanwise-periodic variations in the shape of the leading edge of the swept strut represent periodic excitation of the two-dimensional flow.

Concluding remarks

The relevance of the experiments on the influence of wall

vibrations on a flow with a separation bubble to studies of the unsteady-flow mechanics of turbulent shear-layer reattachment has been sufficiently emphasised already.

Their practical significance must also be mentioned. Very small structural vibrations may have a large effect on flows with separation bubbles, not only in practical engineering situations but also in experimental research. The same is true of excitation by noise[2]. As regards experimental research, it should be noted that effects of vibration or noise can be responsible for spurious apparent effects of variation of Reynolds number.

References

1. Evans, G.P. Separation bubble at pipe entrance. Ph.D. Dissertation, Cambridge University Engineering Department, 1973.

2. McGuinness, M.D. Flow with a separation bubble: Steady and unsteady aspects. Ph.D. Disseration, Cambridge University Engineering Department, 1978.

3. Simpson, R.L., Shivaprasad, B.G., and Chew, Y.-T. Some features of unsteady separated turbulent boundary layers. I.U.T.A.M. Symposium on Unsteady Turbulent Shear Flows, Toulouse, 1981.

Turbulent Pulsating Flow in the Entrance Region of a Pipe

H. MAINARDI and P.K.PANDAY

Laboratoire d'Energétique et de Mécanique Appliquée
Ecole Supérieure de Transport, d'Energie et de Propulsion
Université d'Orléans, 45046 ORLEANS-CEDEX.

Summary

The effect of flow pulsations on the characteristics of turbulent flow in the entrance region of a cylindrical pipe is presented. A velocity wave which causes local friction in the turbulent boundary layer is observed in the core flow for relative amplitude of flow fluctuations of 15%. It is shown that the friction coefficient depends on the mode of acoustic resonance of the installation and that the radial distribution of static pressure can not be considered constant in the entrance region. ($x \leqslant 50D$)

Introduction

During last ten years many research workers have studied the development of boundary layer on flat plate in turbulent pulsating flow. But such stadies in the case of pipe flow are non existent. The influence of flow fluctuations on the internal flow in a cylindrical pipe with profiled entry is presented here [1] . The mean values of longitudinal velocity, static pressure and shear stress in pulsating flow are compared with those obtained from a two dimensional numérical analysis of stationary turbulent flow. The measured values of periodic fluctuations of static pressure and longitudinal velocity are compared with the results of a one dimensional analysis which requires the introduction of a coefficient of friction. It is shown that the coefficient of friction depends on the mode of resonance of the installation. The flow fluctuations create a longitudianl velocity wave in the core flow in the entrance region. The distribution of the shear stress shows that the pressure gradient in the radial direction is not negligible.

Stationary Flow

A two dimensional model of turbulent flow is used to determine the longitudinal and radial velocities, the shear stress and the static pressure from the equations : of state

$$\bar{p}\,\bar{\rho}^{-\gamma} = \text{const.} \qquad (1)$$

of continuity

$$\frac{\partial}{\partial x}\left(\bar{\rho}\bar{u}\right) + \frac{1}{r}\frac{\partial}{\partial r}\left(\bar{\rho} r \bar{v}\right) = 0 \tag{2}$$

of momentum

$$\bar{\rho}\left(\bar{u}\frac{\partial \bar{u}}{\partial x} + \bar{v}\frac{\partial \bar{u}}{\partial r}\right) = -\frac{\partial \bar{p}}{\partial x} + \frac{\partial}{\partial x}\left(\mu_{eff}\frac{\partial u}{\partial x}\right) + \frac{1}{r}\frac{\partial}{\partial r}\left(r\mu_{eff}\frac{\partial u}{\partial r}\right) \tag{3}$$

where

$$\mu_{eff} = \mu + \bar{\rho}\,\ell^2 \left|\frac{\partial u}{\partial r}\right|$$

the mixing length is that given by Van DRIEST [2]

$$\ell = 0.41\,y \text{ for } \ell < 0.14\,\delta \quad \text{and} \quad \ell = 0.14\,\delta \text{ for all other } y$$

The calculated values are compared with experimental values.
Figure 1 shows the longitudinal velocity profile as fonction of y/R for some discrete planes.

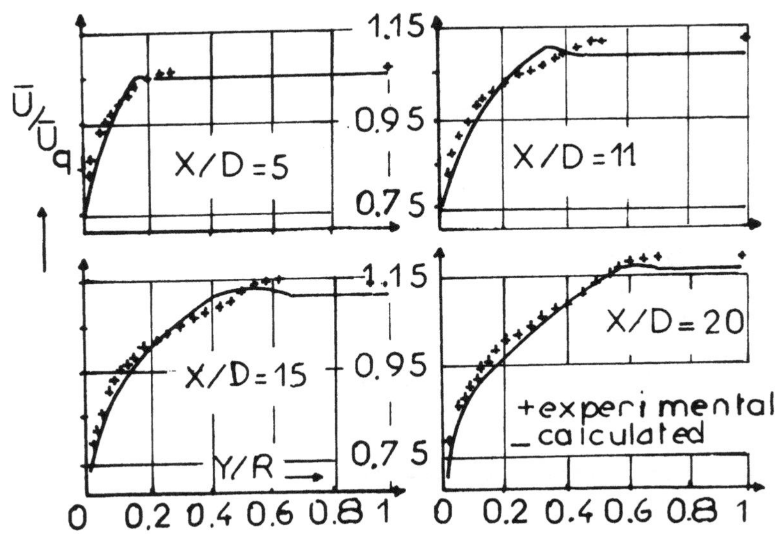

Fig.1. Mean longitudinal velocity profile in stationary flow

Figure 2 gives the static pressure distribution as fonction of x/D.

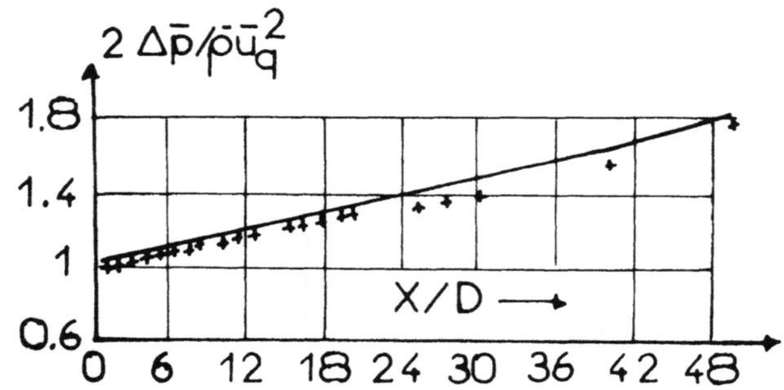

Fig.2. Static pressure distribution in stationary flow

Figure 3 shows the wall coefficient of friction calculated from the following expression

$$C_f = 2\left[\frac{d\theta}{dx} + \frac{\theta}{\bar{u}_x}(H+2)\frac{d\bar{u}_x}{dx} - \frac{1}{\bar{u}_x^2}\cdot\frac{c}{\bar{\rho}}\int_0^{\delta}\tau_r\,dy\right]$$

where C is a constant, C = 3.5 for x $\leq$ 30 D and C = 1 for x $>$ 30 D

θ is the momentum thickness and H the form factor which in a fonction of the displacement thickness δ_1

$$\delta_1 = R\int_0^1\left(1-\frac{\bar{u}}{\bar{u}_x}\right)\frac{y}{R}\,d\left(\frac{y}{R}\right)$$

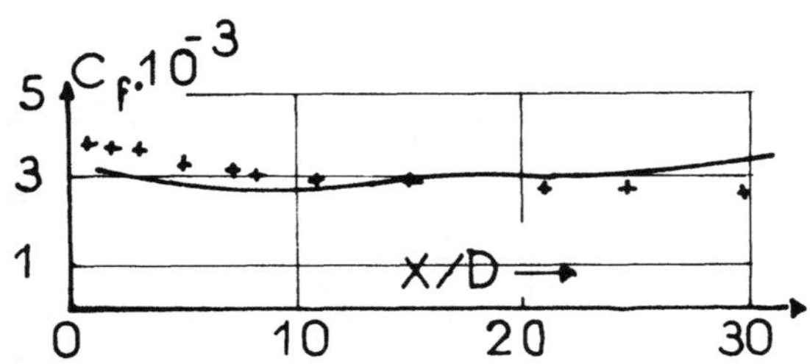

Fig.3. Wall coefficient of friction in stationary flow

Turbulent pulsating flow-Fluctuating components

A one dimensional analysis has been used for determining the fluctuations of flow velocity and static pressure at the wall. [3]

The equations gouverning the flow are :

state

$$\tilde{p}\,\bar{\rho}^{-\gamma} = \text{const.}$$

continuity

$$\frac{\partial \tilde{\rho}}{\partial t} + \bar{\rho}\,\frac{\partial \tilde{u}_q}{\partial x} + \bar{u}_q \frac{\partial \tilde{\rho}}{\partial x} = 0$$

momentum

$$\frac{\partial \tilde{u}_q}{\partial t} + \frac{1}{\bar{\rho}}\,\frac{\partial \tilde{p}}{\partial x} + \bar{u}_q\,\frac{\partial \tilde{u}_q}{\partial x} + f\,\tilde{u}_q^{\,2} = 0$$

Figure 4 gives the calculated values of these fluctuations compared to those obtained experimentaly. The superposition of calculated and experimental values is obtained by giving a value dependent on the mode of resonance to the coefficient f. (fundamental mode f =0.088, 2nd and 3rd mode f=0.175 in the region $x \leq 25D$ and f =0.125 for $x > 25D$)

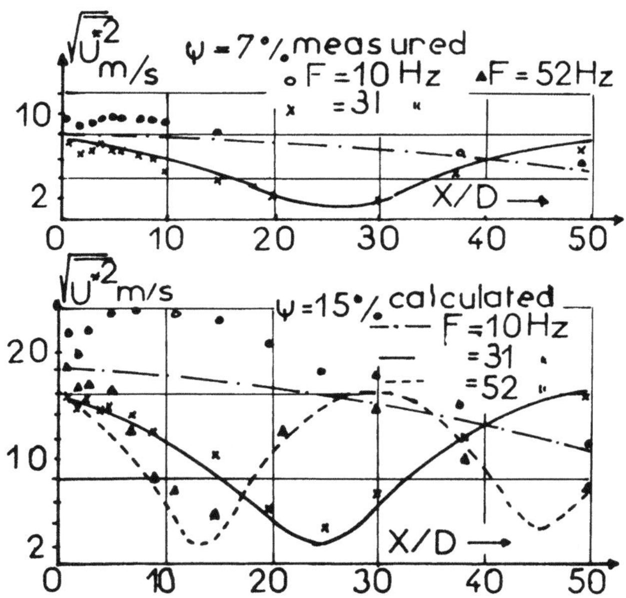

Fig.4.1. Fluctuations of flow velocity.

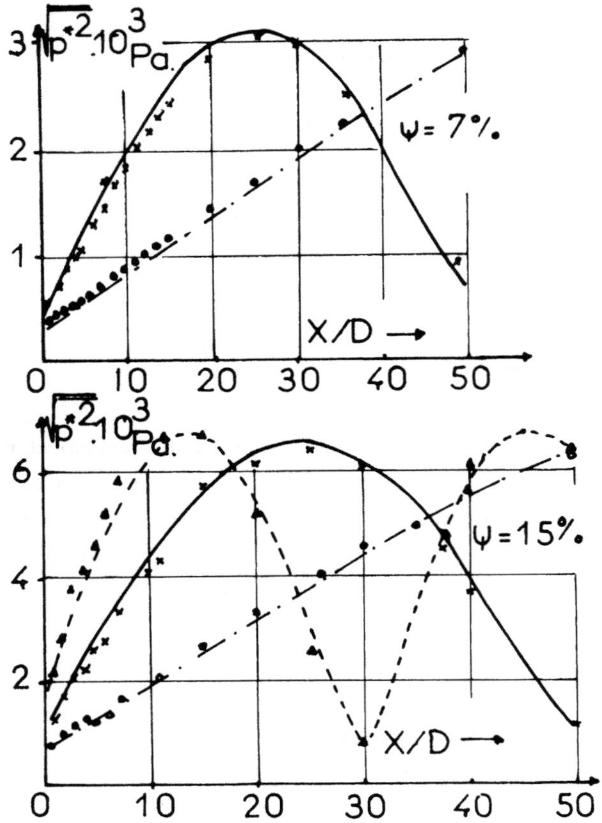

Fig. 4-2. Fluctuations of static pressure at the wall

Turbulent pulsating flow-Mean values

The evolution of the mean longitudinal velocity profile of the turbulent pulsating flow for relative amplitude of flow fluctuations of 15 % is given in figure 5 for 0,10,31 and 52 Hz. These frequencies correspond to stationary flow and 1st, 2nd and 3rd mode of acoustic resonance. It is seen that these distributions of longitudinal velocity depend on the mode of resonance and that there exists a longitudinal wave in the core region different from the standing wave. This wave effect is dependent on the development of the boundary layer whose thickness is defined by the condition $\bar{u}_r = 0.99\bar{u}_x$ (figure 6)

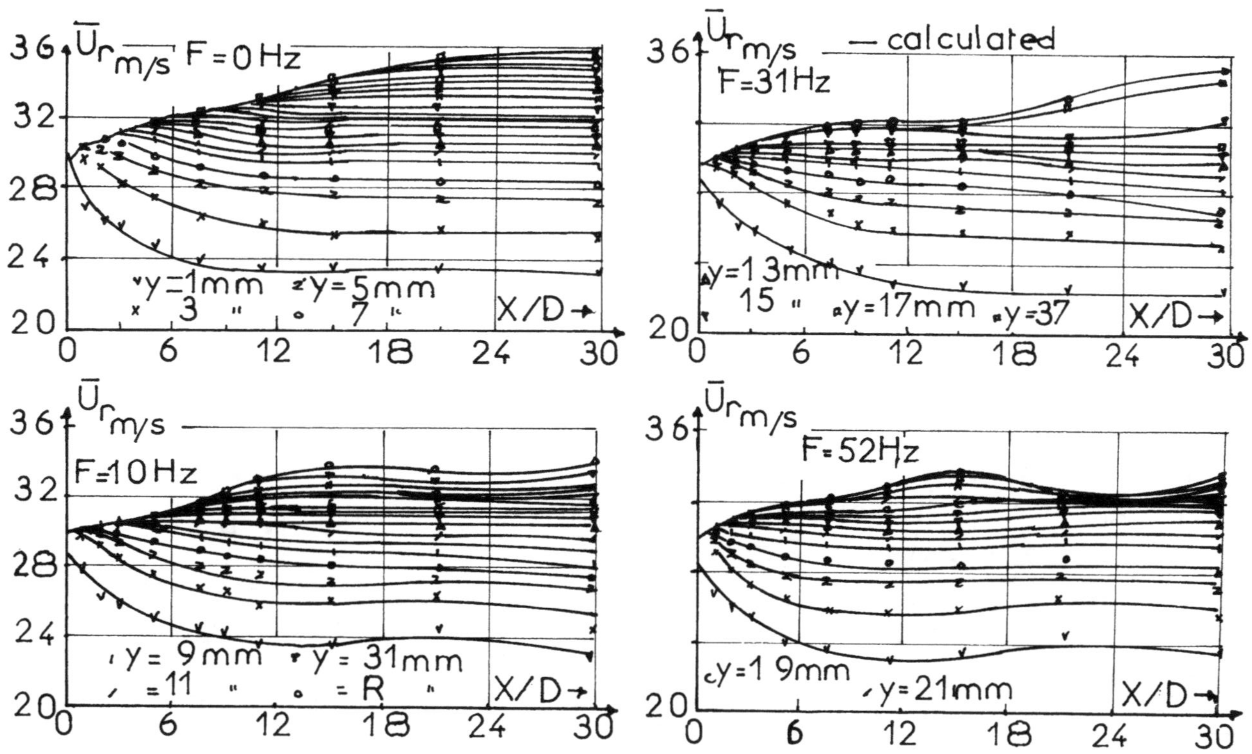

Fig.5. Mean longitudinal velocity profile in pulsating flow

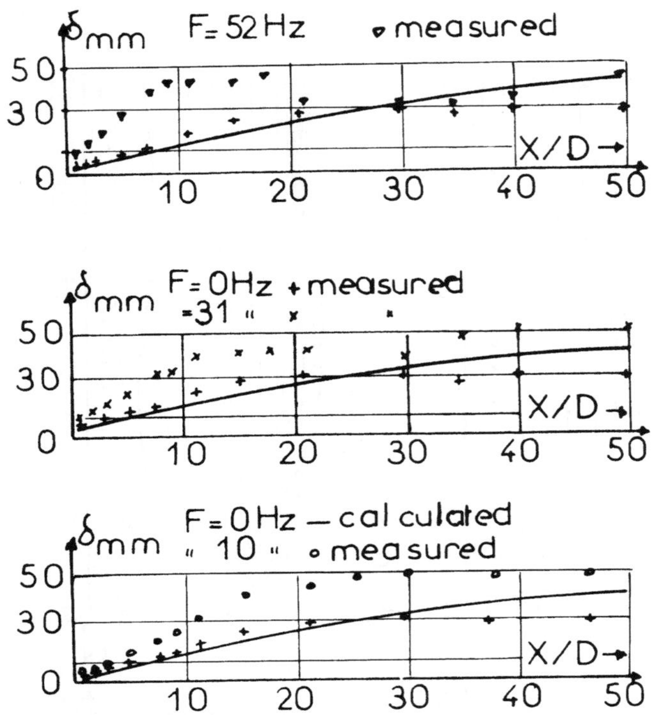

Fig.6. Development of the boundary layer

The mean pressure loss ($\bar{p}_F$) for pulsating flow is compared with the corresponding value for stationary flow in figure 7, both having the same mean flow velocity.

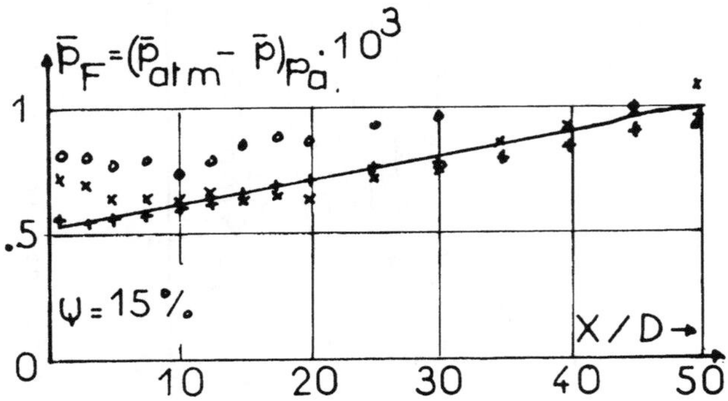

Fig.7. Mean pressure loss

Our experience shows that the pressure loss in pulsating flow can be evaluated from the expression

$$\bar{p}_F = \bar{p}_{F=0} + \frac{\bar{\rho}}{2} \overline{u_x^{*2}}$$

where u_x^* is the fluctuation of the longitudinal velocity which can be obtained either experimentaly or from one dimensional calculations. The increase in the pressure loss ($\bar{p}_F - \bar{p}_{F=0}$) is also given in figure 8

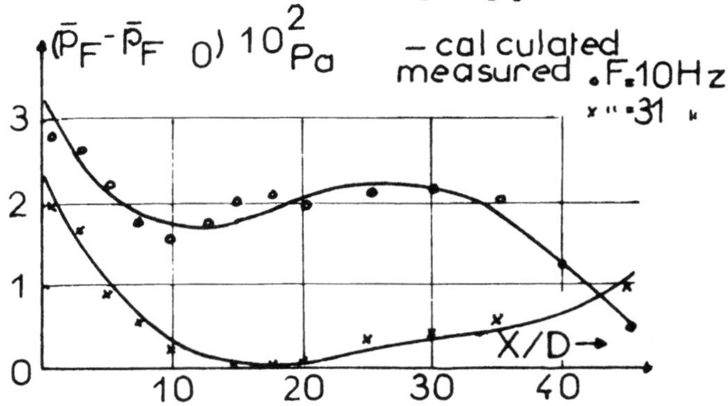

Fig.8. Pressure loss

The experimental values are compared with the values calculated from the following polynomial expressions

$$\bar{p}_{F=0} = 3.067 \ln x - 0.9563 x^5 + 7.0155 x^4 - 15.8 x^3 + 7.5627 x^2 + 76.057 x + 522$$

$$\bar{p}_{F=10} = \bar{p}_{F=0} + 0.602 x^5 - 2.82 x^4 - 28.5 x^3 + 187 x^2 - 312 x + 333$$

$$\bar{p}_{F=31} = \bar{p}_{F=0} + 0.904 x^5 - 7.11 x^4 + 2.28 x^3 + 11.3 x^2 - 30.8 x + 235$$

A velocity wave which causes local friction in the turbulent boundary layer is observed in the core flow for relative amplitude of flow fluctuations of 15 %.

The velocity wave in the core region creates a distribution of the shear stress which depends on the mode of acoustic resonance of the installation. The figure 9 gives the measured values of shear stress and it is seen that local shear stress in pulsating flow compared to that in turbulent stationary flow is much higher.

The local shear stress in pulsating flow can be evaluated from the expression : $\tau = \overline{u^* v'}$

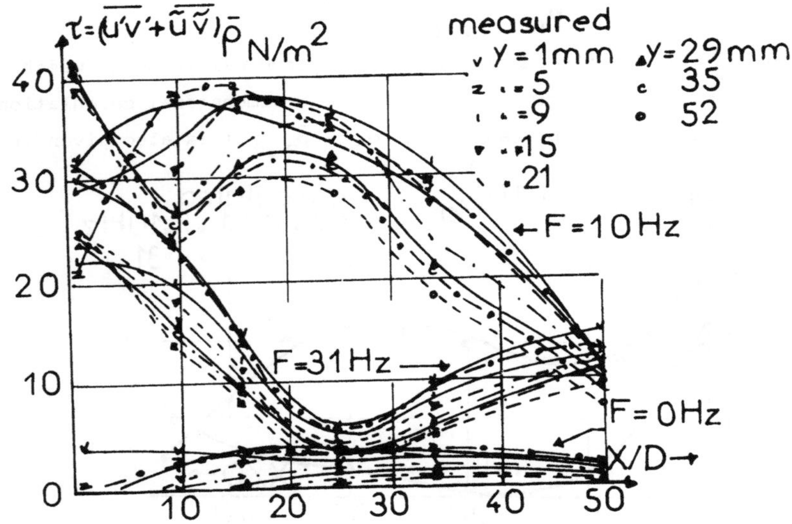

Fig.9. Shear stress-Longitudinal distribution of shear stress

This distribution of the shear stress shows that in the case of pulsating flow with relative amplitude of flow fluctuations of 15 % the radial distribution of static pressure can not be considered as constant. We have therefore calculated the local coefficient of static pressure [4] defined by

$$C_{p_r} = \frac{\bar{p}_r - \bar{p}_F}{\bar{\rho}\bar{u}_r^2/2} = -\left[\frac{\partial}{\partial x}\int_R^r \left(\bar{u}_r\bar{v}_r + \overline{u_r^* v_r^*} + r\frac{\partial \bar{u}_r}{\partial r}\right) dr + \overline{v^{*2}}\right] \cdot \frac{1}{u_r^2/2}$$

The figure 10 gives the variation of C_{pr} as fonction of x/D. We see that for the first two modes of acoustic resonance this coefficient presents a maximum depending on the mode. The maximum of the curve seems to correspond to the thickness of the boudary layer which fills the pipe at a distance of 25D for the first mode and at 35D for the 2nd mode.

Conclusion.

The mean static pressure of the turbulent pulsating flow can be evaluated from fluctuations of the longitudinal velocity. A velocity wave different from the stationary wave exists in the core region of the pipe which modifies the development of the boundary layer. This wave creates a local friction which is much more important than the corresponding value in stationary flow. As a consequence the radial pressure distribution can no more considered negligible.

10

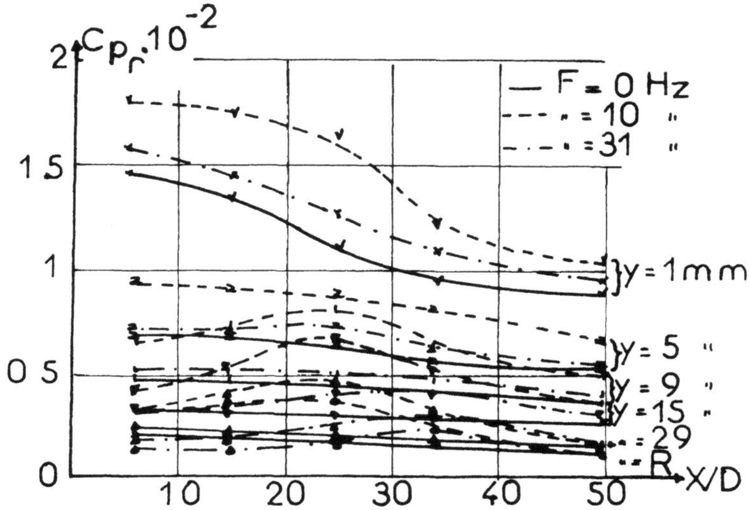

Fig.10. Variation of static pressure coefficient.

References.

[1] MAINARDI, H: Etude des paramètres de l'écoulement turbulent pulsé dans une canalisation cylindrique en régime d'établissement. Thes.Doct.ès-Sci.Univ.Orléans.27 Janv.1981.

[2] Van DRIEST : On turbulent flow near a wall.Journ.of the Aero.Sci. nov.1007-1011 (1956)

[3] THOMAS,JM : Etude théorique et expérimentale des écoulements pulsés en canalisation circulaire.Thes.Doct.ès-Sci.Univ.Orléans.11 June 1974.

[4] LAWS,E.M.,LIM,E.H,LIVESEY,J.L. : Turbulent pipe flows in development and decay.2° Symposium Turbulence shear flow.2/4 july.London (1979)

Notation.

C Coefficient

C_f Friction coefficient

D pipe diameter (D = 2 R)

f friction coefficient defined by $2C_f/D$

p static pressure

Q rate of flow

r radial coordinate

u longitudinal velocity

u_q mean flow velocity defined by $4\,\bar{Q}/\pi D^2$

v radial velocity

x distance along pipe axis mesured from entry

y radial distance measured from the wall

greek letters

δ boundary layer thickness

ρ fluid density

μ dynamic viscosity

τ shear stress

ψ relative amplitude of flow fluctuation defined by $(\bar{Q}_{max} - \bar{Q}_{min})/(\bar{Q}_{max} + \bar{Q}_{min})$

superscripts

$\bar{}$ time averaged value

$\sim$ pulsating component

$'$ turbulent fluctuation

$*$ instantaneous value

Unsteady Turbulent Shear Flow in Shock Tube Discontinuities

Joseph A. Johnson III, Raghu Ramaiah, and Lin I

Department of Physics
Rutgers University
New Brunswick, New Jersey

Summary

Using a pressure-ruptured shock tube and an arc driven shock tube, we have studied the evolution of turbulent fluctuations at contact surfaces with $N_2O_4 \rightleftarrows 2NO_2$ mixtures and at ionizing shock fronts in argon. We have focused on point density diagnostics derived from crossed light beam correlations and electric probes. Turbulent bursts are found for which dynamical and spectral analyses suggest a particle-like evolution of fluctuation segments with a unique and characteristic frequency, independent of flow history and overall flow conditions.

In collisional fluids and plasmas, there are several mechanisms through which perturbations in unstable flow may be amplified: (1) Rayleight-Taylor instabilities, arising when a pressure gradient is opposed by a density gradient; (2) rotational convection instabilities, arising when vortices are produced in such a manner as to enhance the spatial dimensionality of the flow; (3) ion acoustic instabilities, arising from the direct coupling of the induced electromagnetic field of the perturbation and local ion dynamics; (4) reaction-diffusion instabilities arising because of an inhibition of dispersion by a temporally competitive non-equilibrium process. The first two are traditional "fluid" mechanisms, the second two being usually associated with highly ionized gases. These natural instabilities can lead to turbulence in boundary layer-like flows which show two kinds of statistical fluctuations [1]: (1) random fluctuations with indeterminate phase relationships between various flow segments; (2) localized fluctuations with deterministic dynamics and fixed phase relationships from flow segment to segment. The second category (i.e., flow with "bursts" or "large coherent structures") has recently grown in importance since it plays a large role in the transport and mixing processes from which turbulence derives much of its crucial significance [2].

There is no complete theory of turbulence for these phenomena [3,4,5]. However, approaches to the problem exist which suggest that (near transition) reaction rates can be distorted [6] and that a boundary layer approximation affords a generalizable quantitative context for measurements on turbulent bursts in unstable contact surfaces [7] and detonation waves [8]. Furthermore, we have just completed a study of bursting instabilities in collisional ionizing shock fronts which has analogous implications [9]. All of the results just cited have treated the instantaneous in-place manifestations of turbulence. As an extension of these interests, we report here on the real time evolution of two kinds of unstable shock tube discontinuities, both of which are boundary layer-like due either to the nature of the velocity discontinuity [10] or to the presence of curvature [11].

The flow properties of the shock tubes which we have used are summarized by the x-t plot in Figure 1. Energy is stored in a driver section: for the arc discharge tube a 14.5 μf capacitator is charged to 18.5 kV; for the pressure ruptured tube, a diaphragm contains a pressure load of between 1 to 3 atm. The energy is released by the sudden discharge ($\lesssim$20 μsec) of the capacitor in the arc-discharge shock tube and by the sudden rupturing of the diaphragm in the pressure loaded shock tube. This release causes a blast wave or, alternatively, a contact surface to propagate downstream into the driven section preceded in both cases, due to the strength of the disturbance, by a shock wave. The details concerning the production of shock waves by shock tubes are discussed in many places [12].

In the pressure ruptured tube, our measurements are performed on the contact surface. A firing to firing variation in flow velocity at the turbulent contact surface is available through

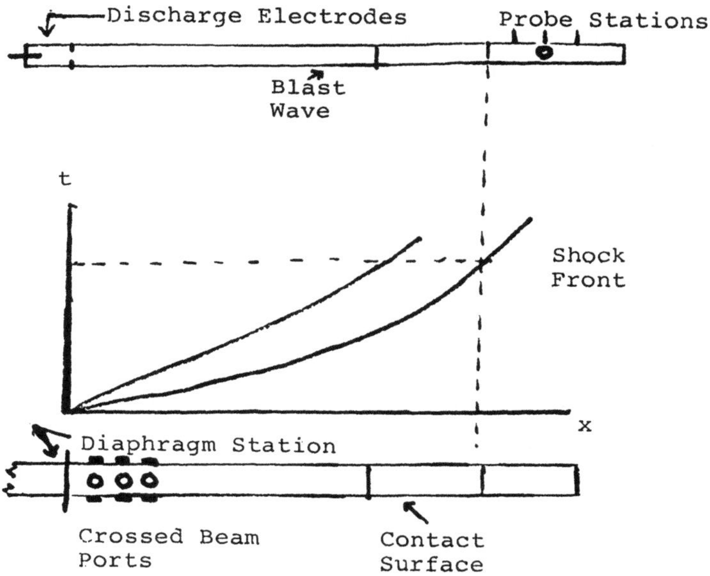

Figure 1. x-t Diagram for Shock Tube Discontinuities. The arc-discharge tube (upper sketch) is 190 cm long and 5 cm in diameter. The test section is 160 cm from the 10 cm long discharge section with inlets for electric probes as indicated above. The 12.7 cm diameter pressure ruptured shock tube (lower sketch) has a 153 cm long driver section and a 336 cm long driven section. The test section is 153 cm long, immediately adjacent to the diaphragm. See references [7],[10], and [13] for details.

the variation in primary shock wave Mach numbers, M_s, $1.5 \le M_s \le 3.0$ producing a range in local Reynolds number [7] of $100 < Re < 1000$. The test section contains two ports separated by 15 cm at which crossed-beam stations are placed. Blue light (4350 Å) absorption of NO_2 is used and a direct measure of the covariance of density fluctuations at the point of intersection is obtained from [13]:

$$F(x,y,z) = \frac{1}{T}\int_0^T \ln i_1(t)\cdot \ln i_2(t)\, dt = I(x,y,x)A$$

where $i_j(t)$ is the light intensity at path (j) with respect to the instantaneous average level $\langle G_j \rangle$ $(i_j(t) = G_j(t)/\langle G_j(t)\rangle)$, T is a convenient averaging interval, A is the cross section at the intersection of the beams, and I(x.y.z) is a measure of the intensity of turbulent density fluctuations. In the arc driven tube, our measurements are performed on the shock front. Over the length of the test section, the shock wave is heavily attentuated,providing a range of Mach numbers during a firing which can vary by as much as 5%/cm. Conventional electric probes [14] operated under ion saturation conditions are the major diagnostics. The probes are separated by 13.2 cm and give a direct measure of ion density throughout the plasma of interest. In both cases, the primary diagnostics are supplemented by the usual assortment of pressure gauges and line-averaged optical sensors.

First, we turn to the question of the dynamics of bursts in these two systems. Direct observations using pressure gauges roughly 3 cm apart have already provided evidence that turbulent bursts in contact surfaces propagate as discrete flow entities with velocity at roughly 90% of the local flow velocity[7]. Similarly, turbulent bursts in collisional ionizing shock fronts measured over distances of about 0.7 cm also move at velocities of roughly 90% of the local velocity; in this case, the velocity measurements are derived from the dispersion relations calculated from the power spectra [9]. These behaviors are generally consistent with the reports on large coherent structures in other manifestations of boundary layer flow [15].

A sample of the contact surface data is given in Figure 2. The driver gas is a N_2O_4 mixture {1.0% ($N_2O_2 \rightleftarrows 2NO_2$); 99% He}. In Figure 2(a), the arrival of the contact surface is indicated by the sudden increase in absorption of blue light at the first and second stations. The lag in the two signals results from the time of flight between the two stations. Figure 2(a) also shows the prominent distortions in what should be otherwise a smooth monotonic change in the light signals. These are the tubulent bursts as they show up in line-averaged blue light aborption. In Figure 2(b), the covariance of each station is displayed. The signatures of the turbulent bursts

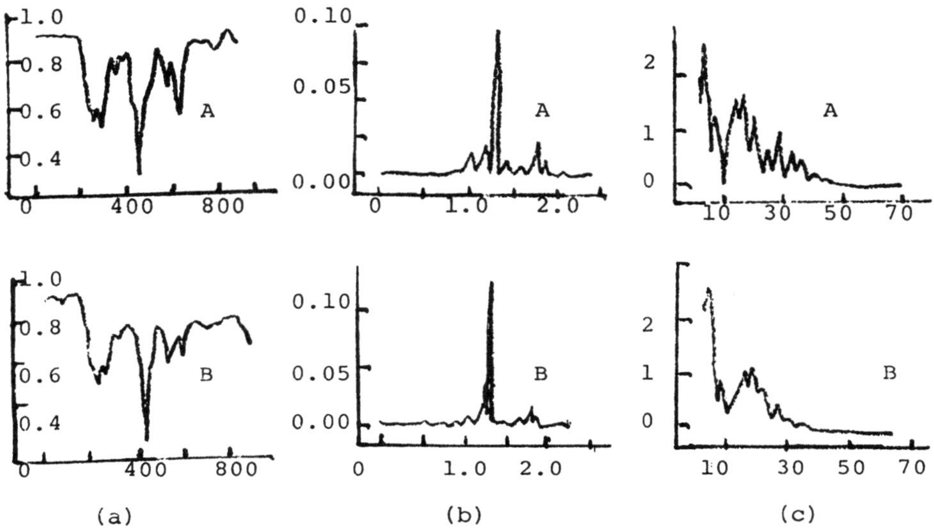

Figure 2. Crossed Beam Data: ($N_2O_4 \rightleftarrows 2NO_2$;He) Mixture for Driver Gas; N_2 for Driven Gas. (a) Light Absorption. The upper trace is the absorption in one of the two crossed beams at the first station (Station A). The lower trace is the absorption in one of the two crossed beams at the second station (Station B) 15 cm downstream. $M_{S,A}=2.1$; $M_{S,B}=2.3$. The vertical axes are in arbitrary (amplitude) units; the horizontal axis is sample number, 4 μs/sample. (b) Normalized Covariance I (x,y,z) for Crossed Beams. This the the measure of turbulent density fluctuations, at the points of intersection, for the two stations. The vertical axes are arbitrary (power) units; the horizontal axes is local particle time in msec. (c) Power Spectra for Crossed Beams. The vertical axes are arbitrary (power) units. The horizontal axis is sample number, 122 Hz (sample).

here are the results in the region of intersection, a volume of approximately $(0.5\ cm)^3$. Power spectra for the covariance functions are performed using standard FFT digital analyses and are displayed for these data in Figure 2(c). These three sets all imply a particle-like evolution of bursts. The velocity of the contact surface increases by more than 10% between stations A and B; however, the power spectra show a substantial repeating of the overall features. The results represented in Figure 2 are found in all cases studied.

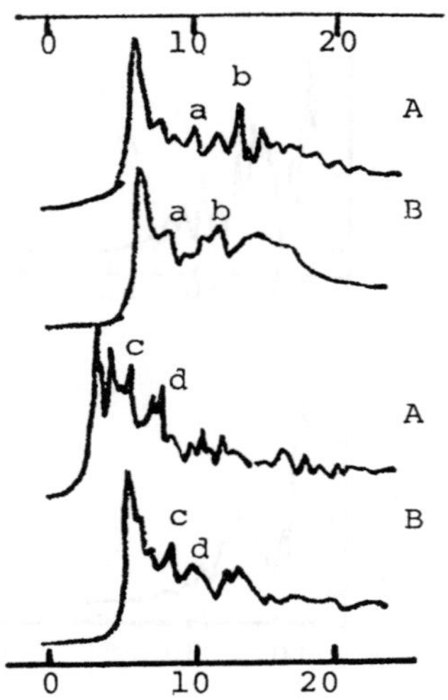

Figure 3. Electric Probe Data: Shock Heated Argon. The upper two traces give relative ion density during a discharge into Argon at initial pressure p_1=40 mTorr; $M_{S,A}$=52, $M_{S,B}$=39. The vertical axes (not shown) would have been arbitrary (amplitude) units; the horizontal axis is in cm; measured in a frame of reference at rest in the local gas. The labels (a,b,c,d) suggest corresponding clumps in the ion density profiles.

A set of ion density profiles from the arc-driven shock tube is shown in Figure 3. In the top two curves, the shock front data from a single firing are given in which the argon pressure prior to discharge was 40 mTorr; the shock wave decelerates between the two stations (with 13.2 cm separation). In both cases, we can identify bursts in the plasma; we can also observe that the disturbances persist over the 13.2 cm distance. Peak to peak correspondence of the clumps can be easily traced as indicated by the assigned labels. This allows us to notice that the bursts seem to catch up with the shock fronts, a behavior of a sort consistent with soliton-like dynamics [16]. The results represented in Figure 3 are found in all our data suggesting a particle like behavior under widely varying environments for the evolution of bursts in turbulent shock fronts.

Secondly, we have examined the evolution of the spectral behaviors of the turbulent bursts. Figure 2(c) suggests a persistence of prominent components of the spectral profile for an accelerating discontinuity. In Figure 4, the power spectra are shown for a firing in which the contact surface decelerates by roughly 20% between the first station and the

Figure 4 (right) Crossed Beam Power Spectra: ($N_2O_4 \rightleftarrows 2NO_2;N_2$) Mixture for Driver Gas; N_2 for Driven Gas. $M_{S,A}=2.35$; $M_{S,B}=2.2$. The scales here are the same as in Figure 2(c).

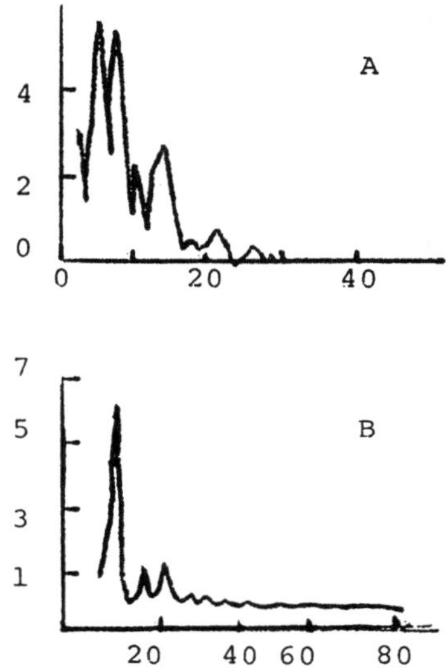

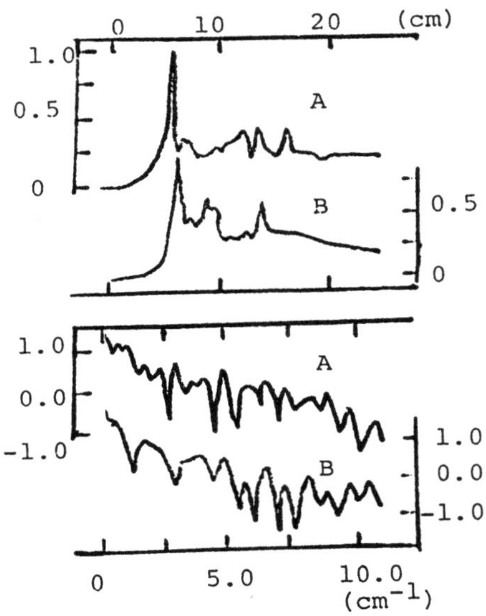

Figure 5 (left). Ion Density Profiles and Associated Power Spectra. p_1= 60 mTorr Argon; $M_{S,A}=33$, $M_{S,B}=25$. The scales for the upper two traces are the same as Figure 3. The lower two traces are the corresponding power spectra. The vertical axes are arbitrary logarithmic units; the horizontal axis is wavenumber K in cm^{-1} from $K = \omega/v_{gas}$.

second. The gross features of the spectral profiles in terms of prominent modes are certainly reproduced in the evolution of the bursts. In firings processed to date, we find this to be the case, including the apparent tendency for a burst during its history to provide more high frequency modes, the higher its velocity.

For bursts in the shock fronts, the same general features are observed. By way of illustration, a firing is displayed in Figure 5 in which the shock front decelerates and the spectral profiles are determined. The power spectra have a clear mode to mode correspondence. The behaviors $S_A \alpha K^{-1.39}$ and $S_B \alpha K^{-1.67}$ indicate a turbulent-like spectra dependence [17]: $S \alpha K^{\beta}$. However the differences in power law trends indicate that the relative strength of the higher wave number modes decreases as the shock front propagates and decelerates.

Finally, we turn to the possibility of an explicit connection between these bursts and other boundary layer flow. In this, we can be guided by the well established consensus with regard to the appropriateness of the treatment of contact surfaces as boundary layers [10] as well as the unique association of bursts in boundary layers [18] and in shock-tube discontinuity [8] with the Orr-Sommerfeld solutions for Falkner-Skan velocity profiles.

In the latter regard, the amplification of fluctuations is expected to follow a predictable serpentine profile, parameterized by a characteristic frequency and a maximum in the turbulent intensity vs Reynolds number plane [19]. This behavior is found for turbulent bursts in a boundary layer [20,18] and for turbulent bursts in a detonation wave [11,8]. Although our earlier results on contact surfaces are suggestive [7]: the pressure gauges used are wall mounted and bandwidth limited; the interferometer provides line-averaged (rather than point) data; and the Mach number range studied does not provide adequate determination of the change in turbulent intensity with increasing Reynolds number. However, our present data are based on point diagnostics, reach higher flow speeds with 4 times the previous bandwidth and permit us to fashion a test of the predictions of Orr-

Sommerfeld-like behaviors. Specifiacally: (a) the strength of the fluctuations in a burst should begin to decrease with increasing Mach number above $M_S \approx 2$ (see figure 12 in reference 7 and associated discussion for the relationship between shock wave Mach number and contact surface Reynolds number); (b) a single dominant frequency should characterize the spectral profiles of fluctuations associated with the turbulent bursts.

Our data for turbulent bursts at the contact surface confirm these expectations. The local Reynolds number is given by Re α ℓ where ℓ is a characteristic length scale; $\ell \alpha t_L$ where t_L is the separation between the burst and discontinuity [8]. Thus, one can map I vs t_L into I vs Re where I is the turbulent intensity. The expectation (a) above means $d^2I/dt^2_L < 0$. Figure 6(a) shows a plot of $\Delta I_x/\Delta t_L$ where, in each covariance with at least two bursts, a computation of $(I_{x,i} - I_{x,j})/t_{L,i} - t_{L,j})$ is performed; averages are determined at a given M_S and the profile $\Delta I_x / \Delta t_L$ vs M_S is achieved as a representation of $\Delta I/\Delta Re$ vs Re. The negative slope is qualitatively consistent with the Orr-Sommerfeld profiles. In addition, in each of the power spectra, an estimate of the fundamental mode, f_o, is made by assuming an integral relationship between it and each prominent spectral component. This gives an estimate of f_o at each firing and in all spectra when we require that a single fundamental frequency should describe all data. The results are given in Figure 6(b). The average value of the fundamental frequency is $f_o \approx 1041$ Hz; the data are not inconsistent with the assumption of a constant value for this frequency over the range of speeds studies.

With these results, we see the first evidence of the possibility of a nonempirical unified treatment of all boundary layer-like unsteady shear flow without regard for flow history. In particular, we see new manifestations of a "constant" characteristic frequency with these turbulent bursts of the sort previously found in conventional boundary-layer flow [21]. The usefulness of the interpretation of turbulent bursts as characterized by Orr-Sommerfeld profiles should therefore be

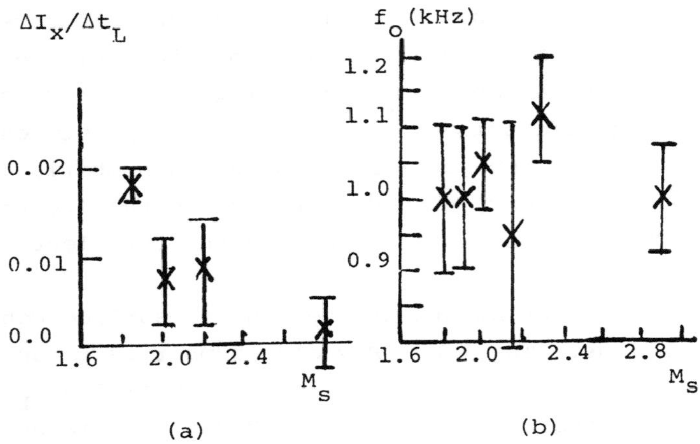

Figure 6. Crossed Beam Data on Boundary Layer-Like Profiles. (a) I_x is the value of I at the peak of an identifiable burst and t_L is the time interval between that peak and the arrival of the constant surface at the same point in a frame of reference at rest with respect to the local flow (see Figure 2). For two peaks in a given covariance profile $\Delta I_x/\Delta t_L = (I_{x,2} - I_{x,1})/(t_{L,2}-t_{L,1})$. Each profile corresponds to a definite value for M_S. All values for $\Delta I_x/\Delta t_L$ obtained a comparable values for M_S are averaged without regard for flow history. (b) Each identifiable peak in a power spectrum, f_m, is interpreted in terms of a relationship $f_{m,i}=n_m f_i$ where $n_m= 1,2,3,\ldots$ and f_i is the estimate of f_o, the fundamental frequency, provided by $f_{m,i}$. The values for f_i obtained from comparable values of M_S are averaged without regard for flow history, $f_o = \langle f_i \rangle$.

tested for the widest possible range of flow environments.

This work was supported in part by NASA grant NSG 3280.

References

1. A.S. Monin, Sov. Phys. Usp. 21, 429 (1978).

2. B.J. Cantwell, Am. Rev. Fluid Mech. 13, 457 (1981).

3. A.M. Yoglom, Am. Rev. Fluid Mech., 11,505 (1979).

4. C. Speziale, Phys. Fluids, 23, 459 (1980).

5. M. Kac, The Boltzmann Equation (New York:Springer) 379 (1973).

6. J.A. Johnson III and S.C. Chen, Phys. Letts., 68A, 141 (1978).

7. J.A. Johnson III, W.R. Jones, and J. Santiago, J. Phys. D: Appl. Phys., 13, 1413 (1980).

8. J.A. Johnson III, Appl. Letts., 37, 275 (1980).

9. L.I., J.A. Johnson III and J.P. Santiago (submitted to Journal of Plasma Physics); also Lin I, Ph.D. dissertation 1981 (unpublished).

10. L. Landau and E. Lifshitz, Fluid Mechanics (reading, Mass: Addison-Wesley) 114 (1959).

11. D.R. White, Phys. Fluids, 4, 465 (1961).

12. J.N. Bradley, Shock Waves in Chemistry and Physics, (London:Methuen) 88 (1962).

13. J.A. Johnson III, R. Ramaiah and J. Santiago, Rev. Sci. Instr., (1981, to be published).

14. R.P. Smy, Adv. in Phys., 25, 517 (1976).

15. L.S.G. Kovasznay, Structure and Mechanism in Turbulence, Vol. I, H. Fiedler (ed) (New York: Springer) 1 (1978).

16. R.C. Davidson, Method in Nonlinear Plasma Theory, (New York: Academic), 15 (1972).

17. V.N. Taytovick Theory of Turbulent Plasma (New York: Consultants Bureau) 1 (1977).

18. S. Tsuge, Phys. Fluids, 17, 22 (1974).

19. M.R. Osborne, SIAMJ Appl. Math 15, 539 (1967).

20. G. Sthubauer and P.S. Klebanoff, NACA Report 1289 (1956).

21. R.A. Antonia, H.Q. Dank, A. Prabhu, Phys. Fluids, 19, 1680 (1976).

Turbulence Structures in the Wake of an Oscillating Airfoil

J. DE RUYCK, Ch. HIRSCH

Vrije Universiteit Brussel, Dept. of Fluid Mechanics,
Pleinlaan 2, 1050 BRUSSELS, Belgium

Summary

An experimental set-up for turbulence measurements in the wake of a 60 cm chord NACA 0012 oscillating airfoil is described. The airfoil oscillates around an axis at 25 % chord distance from the leading edge, with a sinusoidal motion. A slowly moving slanted hot wire anemometer technique combined with a space and time conditional averaging is used to determine all non zero Reynolds stress and velocity wake profiles. Data acquisition is performed through an on-line sampling, digitizing and recording microprocessor controlled system which records the hot wire anemometer signal as well as the probe and wing position as functions of time on digital tape.
Measurements are performed at small and large oscillation amplitudes (1.5° and 5°), at three mean incidences which correspond to situations with no stall, stall onset and deeper stall, and at three frequency coefficients k = 0, 0.37 and 1.06.
Results of time and space dependent Reynolds stress distributions are presented and compared for the different combinations of oscillation parameters.
The turbulence behaviour in presence of the oscillating flow is discussed and compared with the steady state situations.

1. Introduction

Helicopter blade behaviour, turbomachine blade row interactions, marine propeller flows and vertical axis windmills are typical application fields of unsteady phenomena, where viscous effects such as flutter or stall play an important role in the overall flow properties.

Actually computational methods for solving the Navier Stokes equations are subject to a spectacular development and one can expect that these methods will become operational with reasonable computer time costs within the coming decennia.

The comparison of Navier Stokes solutions and reality is however still limited by a lack of experimental data concerning the turbulence input of the basic equations [1]. More particularly, the question of the influence of the unsteadiness on the turbulence structure is still unanswered [2,3] .

At the present time, several experimental studies have been presented about velocity and turbulence structures in boundary layers around fixed wings or plates placed in an oscillating external pressure gradient [4,5, 6, 7, 8] . In the case of an oscillating surface experiments are however axed on unsteady airloads and to some extend on velocity ditributions [9, 10, 11] .

The objective of this research is therefore to provide experimental information on the turbulence structures in wakes and boundary layers of an oscillating airfoil.

In the present paper, the experimental technique is described and results are presented of measurements made at 5 % chord distance downstream of the trailing edge and at various oscillating conditions. The results concern instantaneous wake profiles of velocity and all non zero Reynolds stresses, at each phase of the oscillation period.

Similar results will be available in the near future at other distances from the trailing edge and for identical flow and oscillation conditions.

2. Experimental set-up

A 60 cm chord NACA 0012 profile is used for the experimental set-up. This airfoil is positioned in a 2 m by 1 m test section in one of the VUB wind-tunnels, where speeds of 17 m/s are reached. A Reynolds number of 5.10^5 can be obtained.

The airfoil is excited by a 300 Watt motor reductor which allows for an oscillating frequency up to 8 Hz at ± 5° oscillation amplitude. Higher frequencies can be obtained at smaller amplitudes. Sinusoidal pulsation is obtained using a crank-connecting-rod mechanism, where a long connecting rod allows for a motion which is sinusoidal within 2 % distorsion. Speed is kept constant within one oscillation period by regulation of the motor shaft velocity (0 - 5000 RPM).

A slanted rotating hot wire is mounted on a moving support in the wake of the airfoil. A step motor allows for a slow and constant speed motion of the probe support across the whole wake area. The step motor control delivers the probe position coordinate in digital form for recording purposes.

3. Data acquisition facility

The data acquisition facility of the VUB Fluid Mechanics Department is an essential and unique facility for the present investigation.

The aim of this facility is to sample, digitize and write on tape up to 8 analog or digital input signals. Each analog signal (e.g. hot wire anemometer voltage) is filtered and DC amplified by the system before being sampled, digitized through a 10-bit ADC and recorded on tape. The maximum data transfer rate is about 150 kHz. The system is controlled by a ZILOG Z80 micro processor.

A constant DC offset is subtracted from the hot wire anemometer signal before being DC amplified, in order to have an acceptable resolution for the measurement of the fluctuation amplitudes.

A 1 bit wing position reference signal is combined with a 9 bit probe position coordinate and recorded as one 10 bit signal on the tape drive.

A detailed description of this data acquisition facility can be found in ref [12].

4. Rotating slanted hot wire measurement technique

The technique for obtaining velocity and Reynolds stresses from a rotating slanted hot wire anemometer signal is extensively described in [13] and is only briefly reviewed in this section.

The anemometer output voltage E and the effective cooling velocity Q_{eff} are related by King's law through

$$E^2 = A + B\, Q_{eff}^{\ n} \tag{1}$$

The velocity Q can be reconstructed from this signal through

$$Q_{eff} = Q\, [\cos^2 \psi + k \sin^2 \psi\,]^{1/2} \tag{2}$$

where ψ denotes the angle between velocity and the plane normal to the wire, while k is a correction factor. If U, V and W denote the three velocity components in an arbitrary coordinate system, then

$$Q = \sqrt{U^2 + V^2 + W^2} \tag{3}$$

Hence, expressing ψ as function of U, V and W and differentiation of both sides of eq. (2) yields

$$dQ_{eff} = \frac{\partial Q_{eff}}{\partial U} dU + \frac{\partial Q_{eff}}{\partial V} dV + \frac{\partial Q_{eff}}{\partial W} dW = A_u\, dU + A_v\, dV + A_w\, dW \quad (4)$$

where A_u, A_v and A_w can be expressed as functions of Q, k and ψ.

If the main flow is steady, eq. (4) can be rewritten as

$$dQ_{eff} = A_u\, u + A_v\, v + A_w\, w \quad (5)$$

where u, v and w are turbulent velocity components.

Squaring and time averaging of eq. (5) yields.

$$\overline{(dQ_{eff})^2} = \overline{(A_u\, u)^2} + \overline{(A_v\, v)^2} + \overline{(A_w\, w)^2} + 2\,\overline{A_u A_v\, uv} + 2\overline{A_u A_w\, uw} + 2\overline{A_v A_w\, vw} \quad (6)$$

If the turbulence level is not too high, ($T_u \lesssim 10$ %) A_u, A_v and A_w have small fluctuations and eq. (6) can be approached by :

$$\overline{(dQ_{eff})^2} = A_u^2\, \overline{u^2} + A_v^2\, \overline{v^2} + A_w^2\, \overline{w^2} + 2\, A_u A_v\, \overline{uv} + 2A_u A_w\, \overline{uw} + 2\, A_v A_w\, \overline{vw} \quad (7)$$

where A_u, A_v and A_w are expressed as functions of $\bar{Q}$ and $\bar{\psi}$. The different Reynolds stresses appear in the r-h-s of eq. (7) while $\overline{(dQ_{eff})^2}$ can be obtained from $\overline{(dE)^2}$ using eq. (1) through

$$nB\, Q_{eff}^{n-1}\, dQ_{eff} = 2\, E\, dE$$

$$\overline{(dQ_{eff})^2} = \left\{\frac{2\,\bar{E}}{nB\,\bar{Q}^{n-1}}\right\}^2 \overline{(dE)^2} \quad (8)$$

Hence, measurement of $\bar{E}$ and $\overline{(dE)^2} \simeq \overline{(E - \bar{E})^2}$ at different probe positions allows to estimate the main flow velocity magnitude and direction through eqs (1), (2) and to estimate the Reynolds stresses through eqs (7) and (8).

5. Averaging techniques

The above reasoning is valid for a steady flow while in the present investigation a large scale fluctuation on E and Q is impressed by the periodic motion of the blade. Hence, averages are to be taken at times when main values of E or Q are identical, which can be performed by a periodic sampling technique (ensemble average).

In this way values with overbars such as $\bar{E}$ are to be interpreted as periodic averaged values which vary periodically with time.

The data acquisition system records instantaneous values of $E(t) - E_o$, E_o being a constant offset voltage which is subtracted from E(t) before entering the inlet amplifier and the ADC. Hence $\overline{E}(t)$ can be reconstructed from $\overline{E}(t) = (\overline{E(t) - E_o}) + E_o$, while $\overline{(dE)^2}$ is found from

$$\overline{(dE)^2} = \overline{(E(t) - \overline{E}(t))^2} = \overline{E^2}(t) - \overline{E}^2(t) \tag{9}$$

E(t) being in principle statistically independent of $\overline{E}(t)$.

The probe position is considered to be fixed in the absolute system of the windtunnel or in the blade relative system. The reconstruction of space profiles using a fixed probe means that the measurements are to be repeated at many distrete positions. In order to improve the measurement efficiency, a technique is developed where the probe is moved slowly with time, while the probe position is recorded simultaneously on tape. The probe speed is chosen slow enough with respect to the oscillation speed in order to have a quasi-fixed probe within a few wing oscillations. Averages are taken over about 30 events, spread over a distance of .8 mm, which corresponds to an error of 1 to 2 % maximum with regard to the variations within this distance. This technique represents a considerable gain in time, since in this way the ensemble average delivers space information. On the other hand, continuous, instantaneous velocity and Reynolds stress distributions can be reconstructed from this technique, increasing the amount of experimental information.

6. Results

A short summary of the obtained results is presented. Data are taken at low Mach numbers and at a constant Reynolds number of 3.10^5. Following combinations of oscillation parameters were considered :

i) Oscillation amplitude

Two values of oscillation amplitude are considered : a value where the swept distance at trailing edge is small (1.5°) and a value where this distance is large compared with the steady state (unstalled) wake thickness (5 .

ii) Mean incidence

Three situations are considered when choosing the mean incidence : a situation with no stall (0°), one where stall onset occurs (5°) and one where deeper stall is present over the whole suction side of the oscillating airfoil (9°).

iii) Oscillation frequency

Three situations were considered : steady state, low and high frequency.

The corresponding frequency coefficients are respectively $k = 0$, .37 and 1.06 where $k = \frac{\omega c}{2U_\infty}$, while the frequencies are respectively 0, 2.2 and 6.16 Hz. "Low frequency" means that the passage time from leading to trailing edge of the boundary layer fluid is small compared with the oscillation period. Hence, this situation can be considered as close to consecutive steady state situations, although significant disturbances due to the wing motion are found at the high values of mean incidence (stall). At "high" frequency, the passage time from leading to trailing edge has the same magnitude as the oscillation period and large phase shifts are found between wing oscillation and turbulence response in the wake and in the rearmost part of the wing boundary layers.

Some of the obtained results are selected on figures 2 to 7. On figures 2 to 4 instantaneous wake profiles as seen in the absolute system are shown for the 5° amplitude test cases at several phase angles of the oscillation period, while on figures 5 to 7 comparisons are made between unsteady and steady wake turbulence profiles. The phase angle ϕ is chosen to be zero when the wing incidence is maximum. The transverse space coordinate in the various figures is measured with respect to a reference direction defined by a parallel to the wind-tunnel axis passing through the oscillation point of the wing (see fig 1).

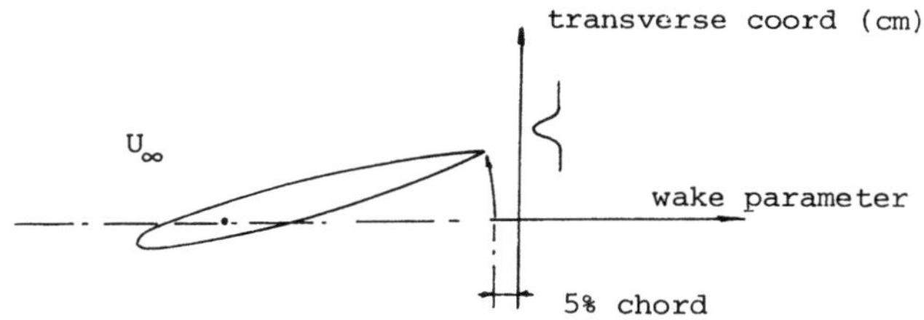

Figure 1 : transverse space coordinate

On figure 2 velocity and non-zero Reynolds stress profiles are shown in the case of 0° mean incidence and at high frequency (k=1.06). U, v and w denote respectively the streamwise, spanwise and pitchwise turbulent velocity components. Since the flow is basically two-dimensional, the $\overline{uv}$ and $\overline{vw}$ stresses are of lower order of magnitude and are not considered. The phase angle varies from O to π (one sweep), while identical profiles are found for phase angles from π to 2π (sweep back, not shown). The $\overline{v^2}$ profiles (not shown) are quite similar to the $\overline{u^2}$ profiles (fig 2.a) but are 8 times smaller in magnitude.

A dip in the $\overline{u^2}$ turbulence profiles is found where the two wing boundary layers meet, while the $\overline{uw}$ profiles obviously change sign at this position (figs 2.a and 2.d). As the wing moves from one side to the other, increased peaks are found in the turbulence profiles at pressure side, especially in the $\overline{uw}$ and $\overline{w^2}$ profiles (figs 2.c and 2.d). A slight asymmetry is observed between 0 and π phase positions which is probably due to slight inaccuracies in the setting of the wing.

On figure 3 profiles are shown at 5° mean incidence and 5° amplitude. Low (figs 3.a and 3.b) and high (figs 3.c and d) frequencies are compared.
As the wing is at its extreme position (ϕ = O), a leading edge separation is observed visually and the stalled flow is detected some time later in the wake, as can be seen from figures 3.a and 3.d. The difference in scale between the figures is to be noted.
At low frequency, the complete suction side region is simultaneously reached by the stalled flow (fig 3.a $\phi = 0$ to $\pi/2$). The maximum turbulence level of about 16 % corresponds nearly to the 9° incidence steady state wake turbulence intensity. The high turbulent flow disappears within a short time (fig 3.a $\phi = .8\pi$ to π) as the wing oscillation amplitude reaches its minimum value -5° corresponding to zero incidence. At increasing incidence (fig 3.b) profiles are found with turbulence levels almost identical to those at 0° mean incidence (fig 1 : 6 to 7 %).
A complete different behaviour is found at the high frequency. The stall disturbs gradually the suction side of the wake (fig 3.c) and the turbulence reaches a maximum when the wing is already halfaway its increasing incidence sweep (fig 3.d : $\phi = 3\pi/2$). Due to the higher oscillation frequency, the separated fluid reattaches on the suction surface before reaching the wake. The stalled region is smaller than in the low frequency case. Important turbulence peaks of 10 % are found in the pressure sides of the wake (fig 3.d).

On figure 4 $\overline{u^2}$ profiles are shown at 9° mean incidence and high frequency (k=1.06). Differences similar to those at 5° mean incidence were found between low (not shown) and high frequency. Turbulence is higher (over 20%) while very important pressure side turbulence peaks are observed (fig 4.a : up to 20% at $\phi = 4\pi/3$). Another observation is that a phase shift of about $\pi/3$ is present between velocity minimum and turbulence maximum in the stalled wake portion, the velocity being minimum at $\phi = 4\pi/3$ (fig 4.b) and the turbulence maximum at $\phi = 5\pi/2$ to 2π (fig 4.a).

On figure 5 a comparison is made between steady and unsteady non-stalled profiles at low and high frequencies for 0° and 5° mean incidences, at $\phi = \pi/2$ position which corresponds to the wing passage of zero amplitude (mean position). A strong similarity is observed between the low frequency profiles at 0° and 5° mean incidence, while both do not differ significantly from the steady state profiles. Higher differences are found at high frequency, with turbulence peaks at pressure sides of the $\overline{uw}$ profiles, as already mentioned.

On figure 6 a similar comparison is made between two data sets taken at approximately the same wing displacement speeds. This is obtained by comparing the high amplitude/low frequency (5°amp/k=.37) and the low amplitude/ high frequency cases (1.5°amp/k=1.06). This is done at mean incidences 0° and 5°. Data are again compared with the steady state profiles at the zero amplitude passage ($\phi = \pi/2$). These two data sets show strong similarity in the turbulence profiles (except that the 5°amplitude/5° mean incidence profile is stalled at decreasing incidence).

On figure 7 comparisons are finally made between 9° mean amplitude steady and unsteady $\overline{u^2}$ profiles. On figure 7.a, the wake hysteresis between the zero amplitude passages is shown for the low frequency (k=.37). A non stalled profile is found at increasing incidence (as already mentioned) while deep stall is detected at decreasing incidence. The pressure side turbulence peaks seen on fig 4.a at high frequency are much higher than the steady state turbulence level seen on fig 7.a. On figure 7.b a comparison is made in the low amplitude test case. In this case the oscillation amplitude is very small compared with the complete stalled wake thickness and the turbulence levels are less affected by the wing motion.

7. Concluding remarks

A sample of the detailed information about the instantaneous turbulence structure in the wake of an oscillating airfoil is presented. Although only one wake traverse at 5% chord distance from the trailing edge is considered, unsteady turbulence properties can be extracted from the present results.

In the non-stalled conditions, very similar wake profiles are found at identical maximum oscillation speeds. The turbulence profiles do not differ greatly from the steady state ones at relatively small speeds while an increase in turbulence is found mainly in the pressure side of the $\overline{uw}$ and $\overline{w^2}$ profiles.

As soon as stall is present, an important influence of the frequency on the wake turbulence behaviour is detected due to different mechanisms of stall propagation. Very high turbulence peaks can be found even at the non-stalled side of the wakes at high oscillation frequency, mean incidence and oscillation amplitude.

More detailed information about the turbulence structure and stall propagation will be obtained by the measurements to be performed in the wing boundary layers.

REFERENCES

[1] Mc CORMACK R.W., "Three Dimensional and Unsteady Separation at High Reynolds Numbers", AGARD LS94, 1978

[2] Mc CROSKEY W.J., "Some Current Research in Unsteady Fluid Dynamics", Trans. ASME, J. of Fluid Eng. 99, 8-39, 1977

[3] TELIONIS D.P., "Unsteady Boundary Layers, Separated & Attached", AGARD Conf. Proc. n° 227, p. 16-1 tot 16-21, 1977

[4] COUSTEIX J., HOUDEVILLE R., DESOPPER A., "Résultats Expérimentaux et Méthodes de Calcul relatifs aux Couches Limites Turbulentes et Ecoulement Instationnaire", AGARD Conf. Proc. n° 227, p. 17-1 to 17-16,1977

[5] PATEL M.H., "On Turbulent Boundary Layers in Oscillatory Flow", Proc. Royal Soc., Vol. A353, pp. 121-144, 1977

[6] SCHACHENMANN A., ROCKWELL D., "Oscillating Turbulent Flow in a Conical Diffuser", J. of Fluid Eng. Vol. 98, pp. 695-702, 1976

[7] SATYANARAYANA B., "Some Aspects of Unsteady Flow Past Airfoils and Cascades", AGARD Conf. Proc. n° 177, Paper n° 25, Sept. 1975

[8] SIMPSON R., "Features of Unsteady Turbulent Boundary Layers as Revealed from Experiments", AGARD Conf. Proc., n° 227, p. 19-1 tot 19-10, 1977

[9] Mc CROSKEY W.J., PUCCI S.L., "Viscous-Inviscid Interaction on Oscillating airfoils", AIAA-81-0051, 1981

[10] Mc CROSKEY W.J., "The Phenomenon of Dynamic Stall", Unsteady Airloads and Aeroelastic Problems in Separated and Transonic Flow, VKI Lecture Series 1981-4

[11] Mc ALISTER K.W., CARR L.W., "Water Tunnel Visualisations of Dynamic Stall", ASME Journal of Fluids Eng., Vol. 101, n° 3, pp. 377-380, 1979

[12] GOOSSENS M., HAVERBEKE A., DEDONCKER H., "A fast Programmable Multichannel Datalogger", Microprocessors and their applications, North-Holland Publ. Comp., p. 155, 1979

[13] KOOL P., "Determination of the Reynolds Stress Tensor with a Single Slanted Hot Wire in Periodically Unsteady Turbomachinery Flow", ASME Paper 79-GT-130

ACKNOWLEDGMENTS

The present research has been made possible through support of the NFWO and of the US Army contract n° DA-JA 37-80-C-0367

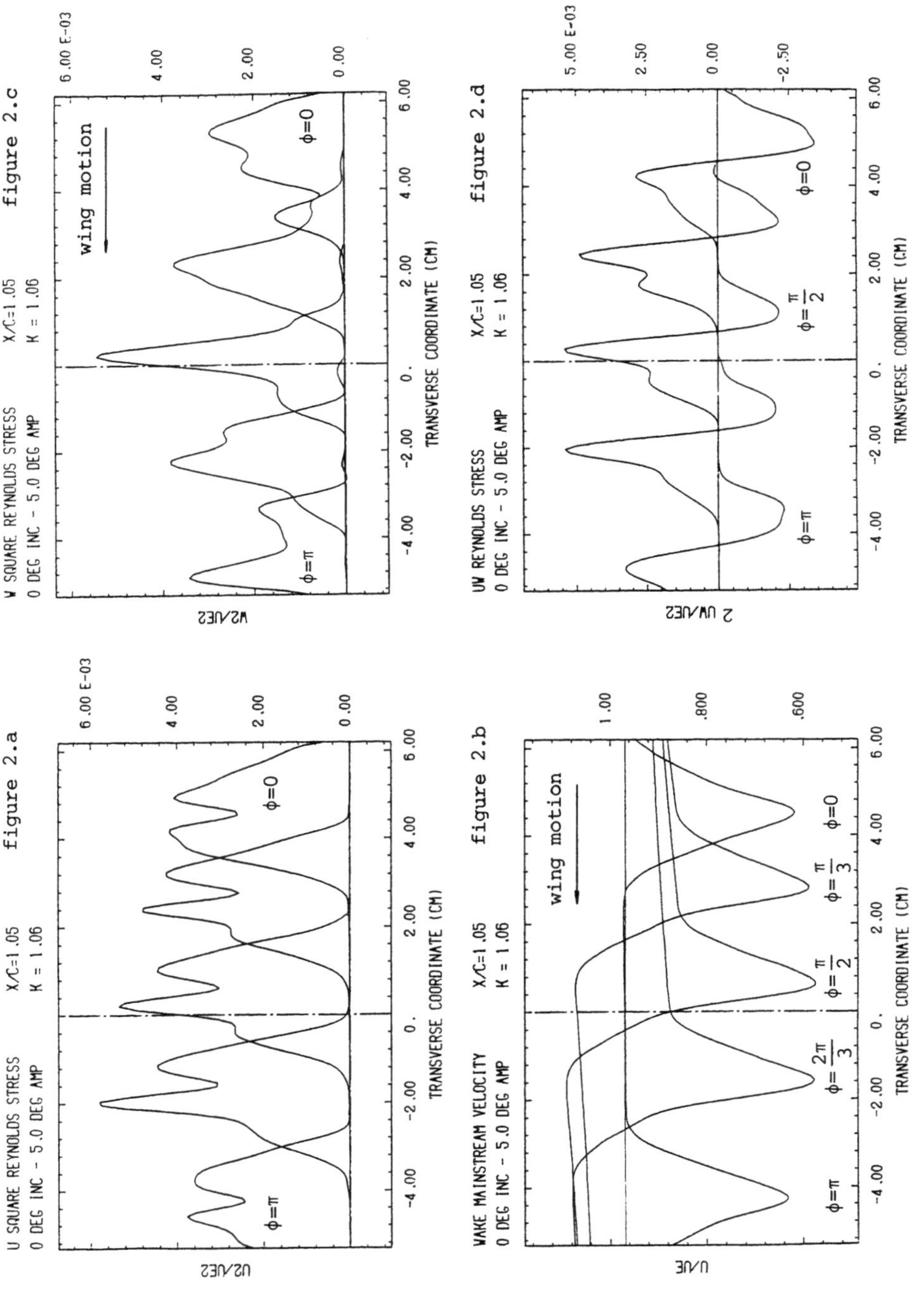
figure 2.a
U SQUARE REYNOLDS STRESS
0 DEG INC - 5.0 DEG AMP
X/C=1.05
K = 1.06
U2/UE2
TRANSVERSE COORDINATE (CM)
figure 2.b
WAKE MAINSTREAM VELOCITY
0 DEG INC - 5.0 DEG AMP
X/C=1.05
K = 1.06
wing motion
U/UE
TRANSVERSE COORDINATE (CM)
figure 2.c
W SQUARE REYNOLDS STRESS
0 DEG INC - 5.0 DEG AMP
X/C=1.05
K = 1.06
wing motion
W2/UE2
TRANSVERSE COORDINATE (CM)
figure 2.d
UW REYNOLDS STRESS
0 DEG INC - 5.0 DEG AMP
X/C=1.05
K = 1.06
2 UW/UE2
TRANSVERSE COORDINATE (CM)

<u>figure 2</u>

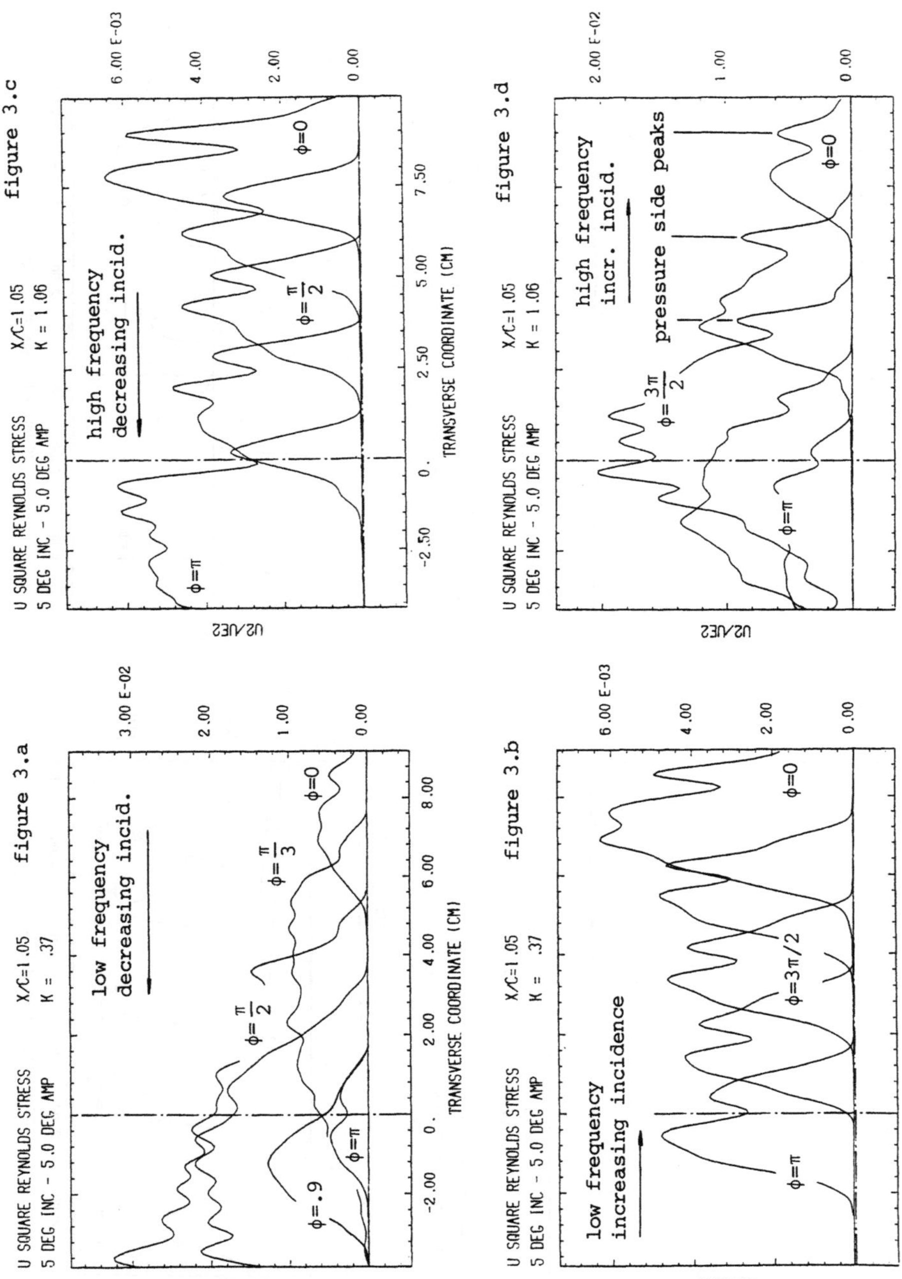
figure 3.a
U SQUARE REYNOLDS STRESS
5 DEG INC - 5.0 DEG AMP
X/C=1.05
K = .37
low frequency
decreasing incid.
U2/UE2
TRANSVERSE COORDINATE (CM)
figure 3.b
U SQUARE REYNOLDS STRESS
5 DEG INC - 5.0 DEG AMP
X/C=1.05
K = .37
low frequency
increasing incidence
U2/UE2
TRANSVERSE COORDINATE (CM)
figure 3.c
U SQUARE REYNOLDS STRESS
5 DEG INC - 5.0 DEG AMP
X/C=1.05
K = 1.06
high frequency
decreasing incid.
U2/UE2
TRANSVERSE COORDINATE (CM)
figure 3.d
U SQUARE REYNOLDS STRESS
5 DEG INC - 5.0 DEG AMP
X/C=1.05
K = 1.06
high frequency
incr. incid.
pressure side peaks
U2/UE2
TRANSVERSE COORDINATE (CM)

figure 3

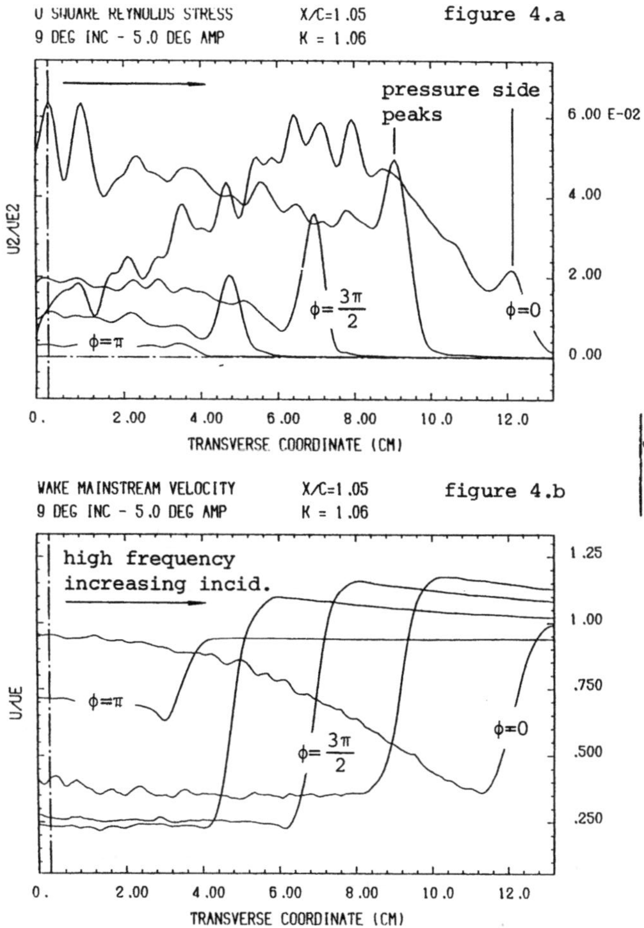

figure 4

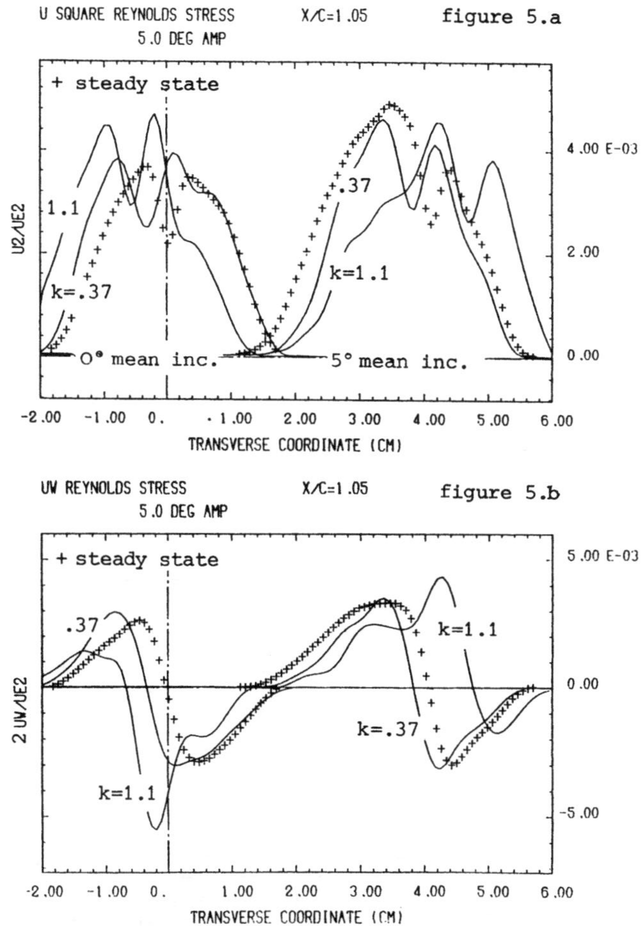

figure 5

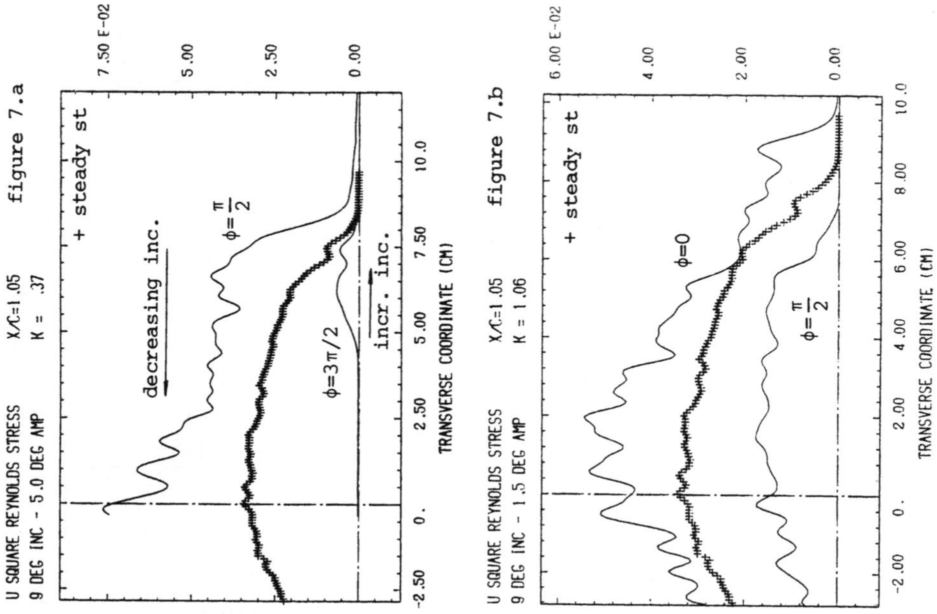

figure 7

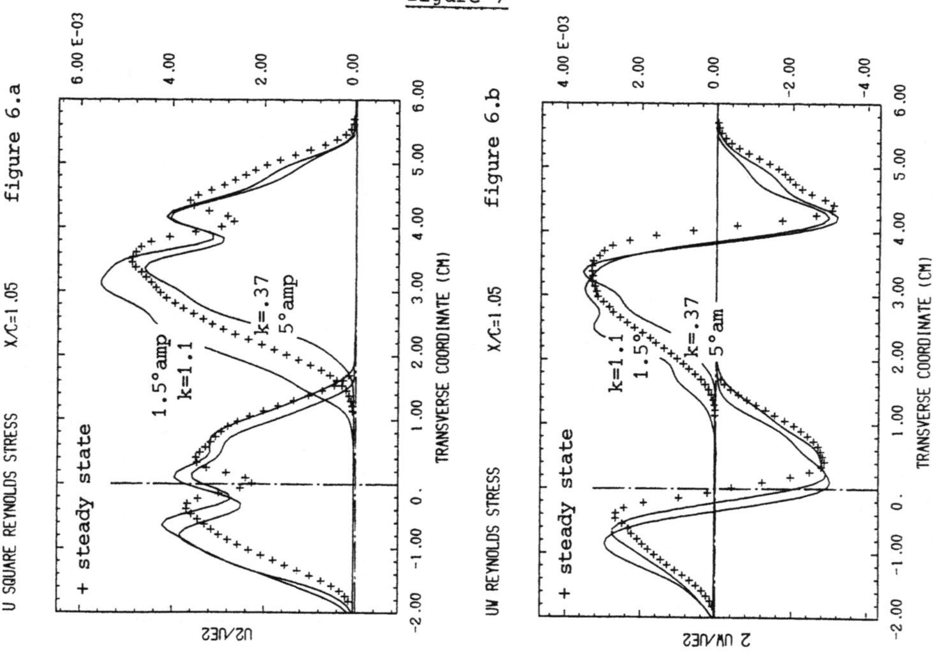

figure 6

An Investigation of Vortex Shedding Below the Keel of a Floating Offshore Vessel in Waves

D.T. BROWN and M.H. PATEL

London Centre for Marine Technology
Department of Mechanical Engineering
University College London
LONDON, U.K.

Abstract

This paper presents a theoretical analysis for modelling the vortex shedding from the submerged keel edges of a flat-bottomed barge. The analysis is capable of computing the resultant added damping forces which can be superimposed on the results from potential flow theory to arrive at a more accurate assessment of barge roll motions in waves. The results of small and large scale tests aimed at verifying the validity of the model are also presented.

1. INTRODUCTION

A current practical problem in the offshore industry is associated with the roll motions of large ocean-going flat-bottomed towed barges used for transporting jacket structures and other large cargoes. These roll motions are lightly damped and the accuracy of predicting this damping and thus the peak roll-angle amplitude at resonance is crucial to the size of cargo fastenings and the risk assessment of costly long-distance towing operations. At present barge motion response to waves are computed using potential-flow wave diffraction theory with very crude empirical estimates of the additional damping due to vortex shedding and other viscous effects. The total damping force on a barge excited by a gravity wave at the free surface can occur due to the combined effect of three energy dissipating flows in the fluid around the vessel keel. As illustrated in Fig 1, these are: (1) the potential flow associated with outwardly radiating gravity waves generated by the vessel, (2) vortex shedding and the generation of turbulence and (3) the viscous boundary layer flow due to relative motion between the vessel submerged surface and the local fluid. It should be noted that the potential flow in (1) is an overall field effect whereas the flow mechanisms in (2) and (3) are highly localised to the barge submerged surface.

This paper presents theoretical and experimental work aimed at improving the accuracy and reliability of the prediction of damping force due to vortex shedding from the submerged keel edges of a flat-bottomed barge. A theoretical analysis based on point vortices in a potential base flow is used to predict vortex movements and the resultant induced pressures. These pressures are integrated over the barge submerged surface to yield estimates of the roll-damping coefficients averaged over one cycle of roll oscillations. The theoretical analysis is supported by experimental data for a self-excited barge in still **water and** for wave-excited barge motions for both small and large scale models.

2. THEORETICAL MODELLING OF VORTEX SHEDDING

2.1 The Basic Model

The theoretical model of vortex shedding is based on the flow below a bluff, plane-based, two-dimensional body having right-angled corners between the sides and lower face. The flow is assumed to remain attached on the side faces and separate at the corners. Point vortices are introduced into the flow at positions close these corners, and the velocity at any vortex position is calculated as the sum of the two-dimensional, irrotational potential-flow velocity around the body and the velocity induced at the vortex position by all the other vortices in the flow. These velocities can be obtained by assuming the body to extend upwards to infinity (see Fig 2) and then using a Schwartz-Christoffel transformation to project the exterior of the body into an upper half plane with the body boundary along the real axis. The transformation is based on work by Clements [1], and is given by

$$z = \frac{-2s}{\pi} [\sin^{-1}(\lambda) + \lambda(1 - \lambda^2)^{\frac{1}{2}}] \tag{1}$$

such that the points $\lambda = \pm 1 + 0i$ in the transformed plane correspond to points $z = \mp s + 0i$ in the physical plane, and s is the half beam of the barge.

A sinusoidally varying base flow is chosen, this being similar to the flow around a floating vessel rolling near to resonance in waves. The base flow must satisfy the boundary conditions that for roll at frequency ω,

$$|\underset{\sim}{u}| \rightarrow 0 \quad \text{as} \quad |\underset{\sim}{z}| \rightarrow \infty , \tag{2}$$

$$\underset{\sim}{u} = (- U_o \sin \omega t, 0) \text{ at } z = 0, \tag{3}$$

where $\underset{\sim}{u}$ is the complex velocity of the base flow and U_o is constant. It follows that the complex potential of the base flow is given by

$$W_1(\lambda) = \frac{4s\, U_o\, \lambda}{\pi} \sin \omega t . \tag{4}$$

Now at each half cycle of roll motion two vortices are introduced into the flow close to the points $\lambda = \pm 1+0i$ to represent vortices being shed just downstream from the corners of the rolling barge. Image vortices of opposite sign and equal strength to those in $\xi > o$ are placed in $\xi < o$ at conjugate positions to the actual vortices so that the condition of zero flow across the body boundary $\xi = o$ is maintained. The contribution to the complex potential due to the vortices is then given by

$$W_2(\lambda) = - \sum_j \frac{ik_j}{2\pi} \log (\lambda - \lambda_j) + \sum_j \frac{ik_j}{2\pi} \log (\lambda - \bar{\lambda}_j), \tag{5}$$

where λ_j, $\bar{\lambda}_j$ are positions of the jth vortex and its image respectively with k_j being their strenghs. Thus the complex potential for the complete flow is given by

$$W(\lambda) = W_1 + W_2. \tag{6}$$

These vortices will convect under the influence of the base flow and the effects of other vortices and images in the field. The complex velocity of a vortex at λ_m, say, is given by

$$\left.\frac{dW}{d\lambda}\right|_{\lambda_m} = \frac{4s\, U_o}{\pi} \sin \omega t - \sum_{\substack{j \\ j\neq m}} \frac{ik_j}{2\pi} \frac{1}{(\lambda_m - \lambda_j)} + \sum_j \frac{ik_j}{2\pi} \frac{1}{(\lambda_m - \bar{\lambda}_j)}, \tag{7}$$

and so using

$$\lambda_m (t + \Delta t) \doteq \lambda_m (t) + \left.\frac{dW}{d\lambda}\right|_{\lambda_m} \Delta t \tag{8}$$

for each of the m vortices the resulting positions $\lambda_m(t + \Delta t)$ at Δt seconds later may be found.

2.2 Predictions of Vortex Motions and Induced Pressures

The expected vortex paths around the barge keel can be predicted by using equation (8), and transforming the vortex positions at each time instant back into the physical plane through equation (1). These paths have been 'sketched' in Fig 3 for a typical case close to roll resonance. The vortices indicated by dotted lines are those introduced following an upward keel edge motion and these tend to 'convect' upwards around the corner of the barge. Similarly the vortices shown by full lines are those introduced following downward keel edge motions and these tend to 'convect' downwards, again rounding the corner of the barge. The validity of this approach has been substantiated by the results of flow-visualisation tests performed on a right-angled corner using both hydrogen bubble and dye injection techniques (see ref 3). A frame by frame analysis of cine films of the vortex shedding at roll resonance indicates qualitatively that the vortices follow similar paths to those predicted by the theory.

The initial positions of the vortices, their strengths and the time instant of vortex formation in the base-flow cycle are all input variables for the theory as it is stepped through time with increments of Δt. However, the use of very simplified boundary layer theory, evidence from the flow-visualisation experiments and numerical tests on the theory allow the identification of physically representative values for all these parameters.

Once the vortex positions have been identified for any time instant during the motion, the pressures induced by the vortices on any element of the barge submerged surface can be evaluated from the current complex velocity and the rate of change of velocity potential at the element. The effects of these pressures, integrated over the barge surface and over one cycle of the barge and base-flow oscillation, can be interpreted as an average energy dissipation per cycle by the barge into the flow. This energy dissipation allows the evaluation of an effective damping value due to the shedding and subsequent movement of the vortices.

3. EXPERIMENTAL VERIFICATION - STILL WATER TESTS

3.1 Test Details

A 1:36 scale model of a wall-sided flat-bottomed barge with dimensions of 2.40m by 0.80m and a draught of 0.10m was used for these tests. The model was designed with three detachable keel edges of different profiles in order to vary the extent of vortex shedding in the flow around the keel - the three edge profiles being a sharp right-angled edge and two with circular quadrants of 30mm and 50mm radii.

The barge model was tested in a large still-water basin of dimensions 50m by 28m with an average depth of 2m. The roll-damping forces were deduced by applying a known inertial oscillating roll moment to the freely floating barge and measuring the amplitude and phase of the resultant oscillating response.

3.2 Measured Roll Response and Roll Damping

The data for measured barge roll response per unit applied roll moment amplitude for the three keel edge profiles is presented in Fig 4 in terms of the non-dimensional values of the damping coefficient computed from the roll response data plotted against excitation frequency. At resonance, the barge roll amplitude is reduced by a factor of 1.8 due to the presence of vortex shedding and this is illustrated in Fig 4 by a corresponding increase in the damping factor. However, there is no observable consistent difference between data for the two rounded keel edges. Another notable feature of the data in Fig 4 is the strong dependence of the damping coefficient with frequency. This is a consequence of the added damping induced by potential-flow gravity wave radiation of energy outwards from the barge. The data also indicate an approximately linear variation of damping with roll amplitude for both the rounded and right-angled keel edge profiles.

These experimental data have been compared with a linearised potential-flow diffraction theory (see ref 5) based on representing the fluid flow by discretised oscillating sources on panels in the barge submerged surface and using a boundary integral formulation to compute the source strengths. Figure 4 indicates the theoretical curve computed with 624 panels in this case. It is observed that agreement between theory which accounts for the gravity-wave radiation only and the rounded keel edge profile data is good indicating that the effects of vortex shedding and skin friction for these profiles is very small. The same cannot be said for the right-angled keel edge profile and the vortex shedding theoretical model presented here is aimed at explaining and predicting this observed discrepancy in damping coefficient by a theoretically representative model of the real fluid flow below the barge keel.

4. EXPERIMENTAL DATA FOR BARGE WAVE INDUCED MOTIONS

4.1 Test Facilities and Barge Models

Wave excited tests on models of a typical flat-bottomed ocean going barge have been carried out at scales of 1:36 and 1:108. The 1:36 scale tests were performed at Hydraulics Research Station, Wallingford, England in an 18m square by 1.5m deep wave basin using the same barge model as for the still water tests. Only two of the keel edge profiles - the right angled edge and the rounded profile of 50mm radius - were used in these tests. All six components of the barge motion were measured in both regular and irregular long-crested gravity waves. The tests in irregular seas were conducted for three wave spectra and three orientations to waves. Similar experiments have also been performed on an identical smaller scale (1:108) barge model at University College London in a wave tank with dimensions of 15m by 2.2m with 1m water depth.

4.2 Motion Responses in Roll

The experimental data at both scales for rounded and sharp keel edges agree closely with each other and with the predictions of potential-flow diffraction theory for all the components of motion with the exception of roll. A typical roll response curve is presented in Fig 5 for the 1:36 scale model test. It is observed that at roll resonant frequency, ω_n, the response for the right -angled keel edge is nearly 1.5 times lower than that for the rounded keel edge. The right-angled keel edge data also show a slightly reduced resonant frequency which is consistent with the higher damping forces associated with this motion.

Figure 5 also displays results from potential-flow diffraction theory again computed with 624 panels on the barge surface. The measured rounded keel edge roll response agrees well with theory. However, close to roll resonance, the right-angled keel edge response shows a reduction which is of a similar magnitude to that obtained for the still water tests.

5. CONCLUSIONS

The results from tests described in this paper indicate that the effects of viscosity on flat-bottomed barge motions in gravity waves are only significant at roll resonance and for right-angled or sharp cornered keel edge profiles. A theory for the vortex shedding off these keel edges predicts the vortex movements that are observed in flow visualisation experiments. This research is being continued towards developing the vortex shedding model further to yield reliable predictions of the measured additional damping at roll resonance.

6. REFERENCES

1. Clements, R.R.: An inviscid model of two-dimensional vortex shedding. J. of Fluid Mechanics, 57, Pt. 2,(1973).

2. Lua, A.B.: An experimental study of the damping forces on flat-bottomed barges during roll and pitch motions. Department of Mechanical Engineering, University College London, M.Sc. thesis, October 1978.

3. Patel, M.H.: The influence of vortex shedding on the roll motions of a flat-bottomed barge. Paper presented at Euromech 119 Colloquium on Vortex Shedding from Bluff Bodies in Oscillating Flow - Imperial College of Science & Technology, London, 16-18 July 1979.

4. Petrides, P.C.: Investigation of flat-bottomed barge motions due to wave excitation. Department of Mechanical Engineering, University College London, M.Sc. thesis, October 1980.

5. Eatock Taylor, R. and Waite, J.B.: The dynamics of offshore structures evaluated by boundary integral techniques. International J. for Num. Methods in Engineering, 13, (1978).

ACKNOWLEDGEMENT

The authors wish to acknowledge the support of the Science Research Council towards this study under Grant No. GR/B/2181.2, carried out in the London Centre for Marine Technology, University College London.

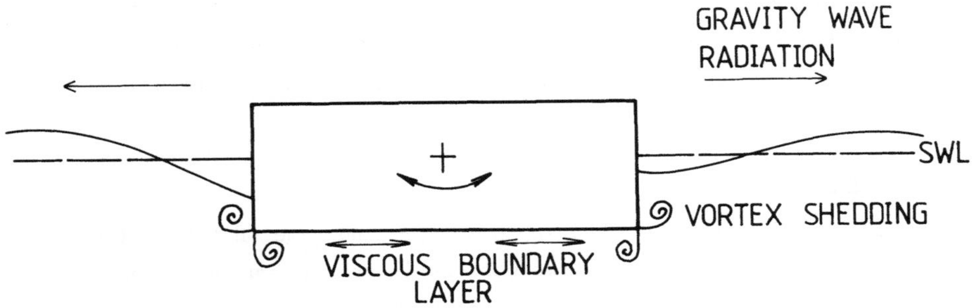

FIG 1: FLOWS ASSOCIATED WITH A BARGE IN ROLL MOTION.

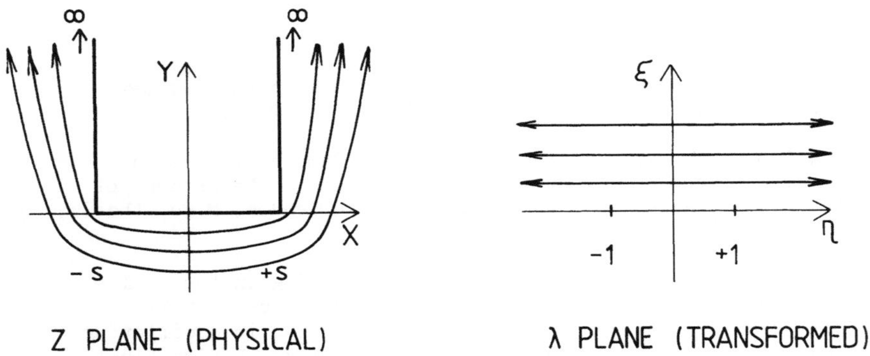

FIG 2: ARRANGEMENT OF AXES.

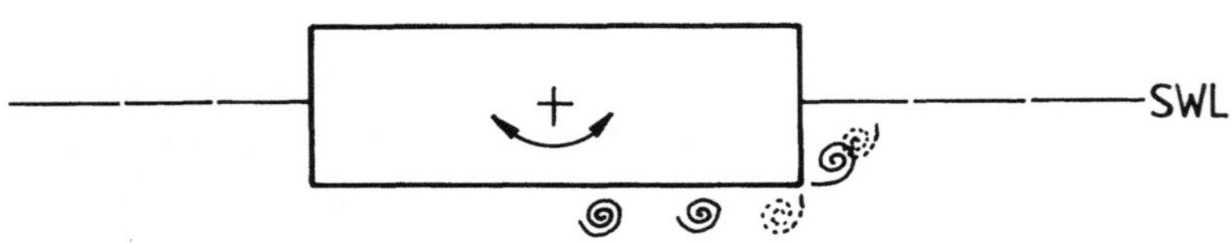

FIG 3: FLOW VISUALISATION RESULTS AT RESONANCE.

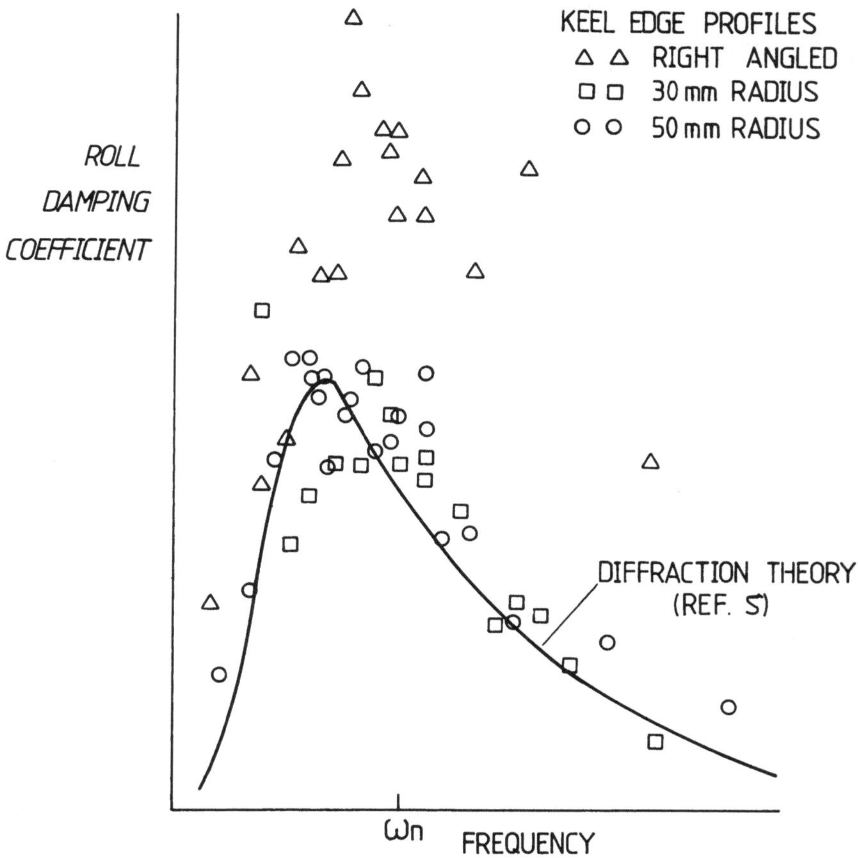

FIG 4: COMPARISON OF ROLL DAMPING COEFFICIENT WITH THEORY

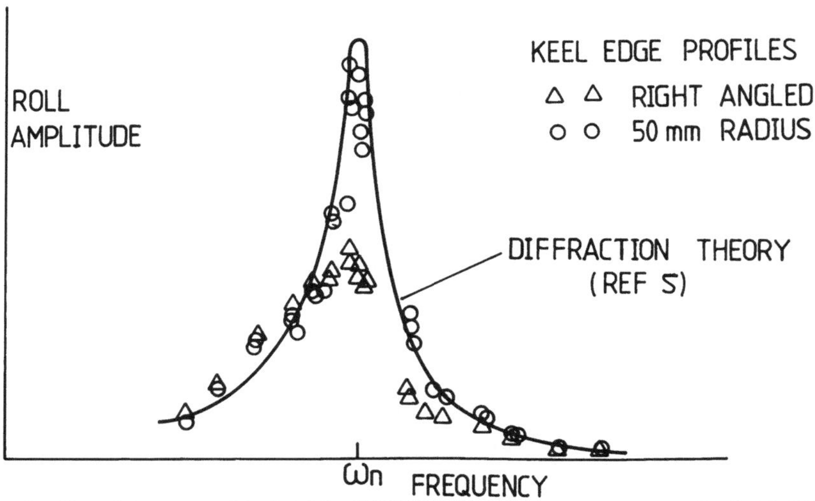

FIG 5: ROLL AMPLITUDE RESPONSE OF 1:36 SCALE BARGE

Kinematic Properties in a Cylinder of a Motored Reciprocating Engine

A. GERBER, J.P. MELINAND and G. CHARNAY

Ecole Centrale de Lyon
Laboratoire de Mécanique des Fluides, associé au CNRS
36, avenue Guy-de-Collongue
69130 Ecully (France)

ABSTRACT

The velocities measured in a combustion chamber are discussed. A forward scatter laser Doppler velocimeter system is used, a diameter of the cylinder being equiped of two windows. The output signal of a counter is digitally stored for delayed processing. Ensemble averages at each crankangle and time averages are computed. Large variations are observed all along the cycle and even without combustion. Such results obtained in a motored engine may be compared with those obtained with steady and isothermal conditions.

I. INTRODUCTION

This work is a part of a general study on the internal aerodynamic in combustion chamber of reciprocating engine. It is well known that fluid motions strongly influence the combustion and consequently act on the performance and the emissions of car or truck engines. For the experimental study a gradual approach is choosed. In a first part, the flow in a cylinder downstream a head has been investigated with steady and isothermal conditions (1). This arrangement provides a simulation of the intake flow and is used practically for the qualification of helical port of Diesel engine. The second part is a study in a single cylinder reciprocating engine but without combustion effects. The late results obtained whith these conditions are described in this note. In the last part, the overall problem in a fired single cylinder engine will be considered.

In recent years, results have been published of measurements in internal combustion engine. For the flow in the cylinder, with motion of the piston, the papers are equally related to motored engines (2) - (5) and to fired engines (6) - (9). Some works (10) - (12) present data gathered in both conditions. The hot wire anemometry has been initially used (6), but this use decreases for the benefit of laser technics (2), (3), (5), (7), (9), (10), (11) or visualizations methods (4), (8). In the most cases the engines are modified in order to obtain easy optical access.

In the following sections the experimental facilities are described first. Then some results are presented. Concluding comments on the variations and on the next measurements are given in the last section.

II. EXPERIMENTAL APPARATUS

II.1. Reciprocating engine.

The investigation is carried out on a RENAULT type 807 single cylinder engine (4 - stroke, bore x stroke = 77 x 84 mm, compression ratio = 8.6 to 1, 20 H.P. at 5000 rpm). The head is hemispheric with 43.45 cc internal volume including valves and spark plug. The intake port is direct. The flow measurements are made under motoring conditions, using a variable speed electric motor, with an operating range of 500 to 15000 rpm. Both reciprocating engine and electric motor are rigidly mounted on a large block, isolated from the floor by anti-vibrating elements.

II.2. Test section.

For LDA measurements inside the cylinder optical arrangements are necessary. In this engine, three quartz window (Ø 14 mm), are setted in one section, 18 millimeters under the headplane (fig. 1). Because of the localization of the test section, measurements are not possible just around the top-dead-center (± 25 %). But the main interest of the position of the windows is the possibility of measuring the axial component of the velocity, with cannot be done with only one window at the top of the head.

III. EXPERIMENTAL PROCEDURE (fig. 2)

III.1. Laser anemometry.

A two color DISA LDA optic system is used. The laser is a SPECTRA-PHYSICS Model 164.09 Argon laser. The power of which is 2W at 514,5 nm. The front lens focal length is 0.3 m and the forward scatter mode is used. Only one channel is equiped, so the measurement of two velocity components is obtained by rotating of the plane of the beams. The Bragg-cell frequency is 2 MHz. DOP particles are introduced at the intake port with a mean diameter of 1 μm.

III.2. Data processing.

The velocity is provided with a DISA counter. All the measurements were performed with a band pass filter in the range 64 kHz 4 MHz. The validated data obtained from the counter are digitally stored on the disk of a HP 2112 computer. The data are then processed through Fortran IV (averages defined in 4). The current information are satisfactory dense with about 200 data per each cycle.

IV. INSTANTANEOUS VELOCITIES : RESULTS AND ANALYSIS

With engine speeds f of 500, 1000 and 1500 rpm, the measurements have been made at one point in the test section, near the exhaust valve (fig. 1). The dispersion of the velocities can be seen in figure 3, where we have used mixed ensemble and time averages roughly defined as :

$$\bar{U}_\alpha = \frac{1}{\Delta\alpha} \frac{1}{n} \sum_{i=1}^{n} \sum_{\alpha}^{\alpha+\Delta\alpha} u_i(\alpha)$$

where
- u_i = instantaneous velocity of a particle crossing the measurement volume, in the cycle number i ;
- n = total number of cycles being processed ;
- α = crank angle.

In figure 3, the axial velocity is presented with $\Delta\alpha = 0.7°$, n = 1900 cycles for f = 500 rpm and n = 2750 cycles for f = 1500 rpm. With the same drawing the local intensities of the fluctuating velocity $u_i(\alpha) - \bar{U}(\alpha)$ are shown in figure 4. Therefore, there is more dispersion during exhaust stroke than during compression stroke. The random fluctuations are not the reason of this dispersion during this last period. On figure 5, for a single cycle the number of data in exhaust stroke is about 10 % of the number of data in compression stroke. Then, the averaging process is less complete in this last period.

The present problem is the discrimination between "true" turbulence and unsteadiness with discernable time scales. This last fluctuations are known as cycle to cycle variations and may be due to different conditions at the end of each cycle. A tentative examination is obtained with the curves of time mean velocities (α in the ranges 60-70° and 60-120°) versus of the number of successive cycles (fig. 6). In both cases no convergence appears with a dispersion of 2 m/s, values not small in comparison with the global intensity of about 3 m/s.

V. CONCLUSIONS

This investigation is developped on an engine which is very similar to those produce by manufacturers. The main features are the following :

- We confirm the ability of LDA technique to measure in a cylinder.
- The unsteady properties of the aerodynamic field are very important. In addition of inherent cyclic variations we observe large fluctuations which seem due to a cycle to cycle dispersion and a turbulent contribution. This fact appears even without combustion in motored engine.
- The rotating speed f act on the velocity distribution but the fluctuating field generated during the intake period is relatively small and constant up to the top dead center.

(1) L. CHABERT, J.G. JOUASSIN and G. CHARNAY, Caractéristiques turbulentes du champ aérodynamique permanent et isotherme en aval d'une culasse de moteur. 17ème colloque d'aérodynamique appliquée. Grenoble, 12-14 novembre 1980.

(2) P.O. WITZE, Application of laser velocimetry to a motored internal combustion engine. SANDIA Laboratories, July 11-13 1978, at Purdue University.

(3) F. BRANDL, I. REVERENRIC, W. CARTELLIERI et J.C. DENT, Turbulent air flow in the combustion bowl of a DI Diesel engine and its effect on engine performance. Congress and exposition Cobo Hall, Detroit, February 26 - March 2, 1979.

(4) I. WAKISAKA, Y. HAMAMOTO, S. OHIGASHI et M. MASHIMOTO, Measurements of air swirl and its turbulence characteristics in the cylinder of an internal combustion engine. IME Conference 9, Londres 12-14 juin 1979.

(5) J.B. COLE et M.D. SWORDS, Optical studies of the flow field in a motored EG Engine. IME Conference 9, Londres 12-14 juin 1979.

(6) J.C. DENT et N.S. SALAMA, The measurement of the turbulence characteristics in an internal combustion engine cylinder. Dept of Mechanical engineering - Loughborough University of Technology (United Kingdom). SAE - Automobile Engineering Meeting, Detroit, Michigan, October 13-17 1975.

(7) P.O. WITZE, Influence of air motion variation on the performance of a direct injection stratified charge engine. IME Conference 9, Londres, 25-26 novembre 1980.

(8) J.N. MATTAVI, E.G. GROFF et F.A. MATEKUNAS, Turbulence flame motion and combustion chamber geometry. Their interactions in a lean combustion Engine. IME Conference 9, Londres, 12-14 juin 1979.

(9) W.J. CORKILL, K.J. BULLOCK et G. WIGLEY, Flow and combustion measurements within a dual chamber stratified charge engine. IME Conference, Londres, 25-26 novembre 1980.

(10) T. ASANUMA et T. OBOKATA, Gas velocity measurements of a motored and firing engine by laser anemometry. Congress and exposition Cobo Hall, Detroit, February 26 - March 2, 1979.

(11) R.B. RASK, Laser Doppler Anemometer measurements in an internal combustion engine. Congress and exposition Cobo Hall, Detroit, February 26 - March 2, 1979.

(12) G.C. LUCAS, M.F. BRUNT et R. ANTON, The effect of squish on charge turbulence and flame propagation in an SI Engine. IME Conference 9, Londres, 12-14 juin 1979.

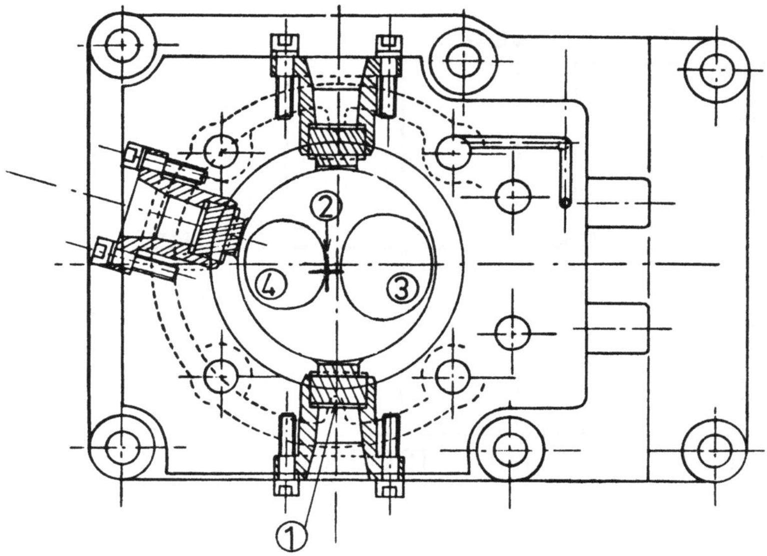

Fig. 1. Experimental set up. 1 : windows, 2 : measuring point, 3 : intake valve, 4 = exhaust valve.

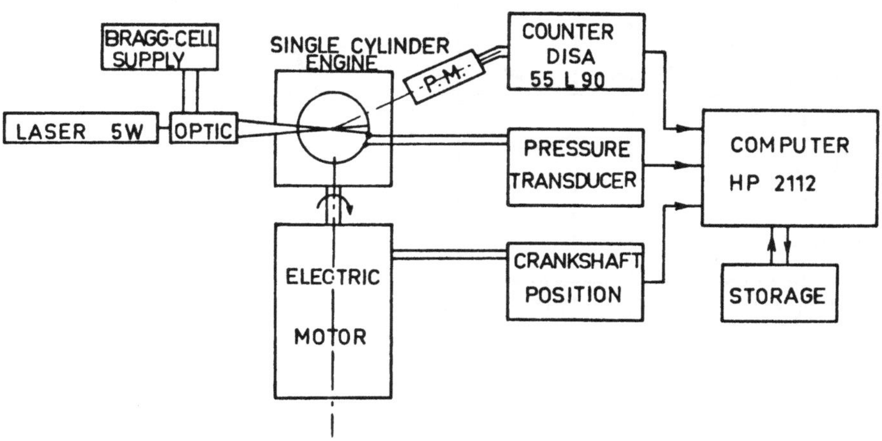

Fig. 2. Laser doppler anemometer and digital processing.

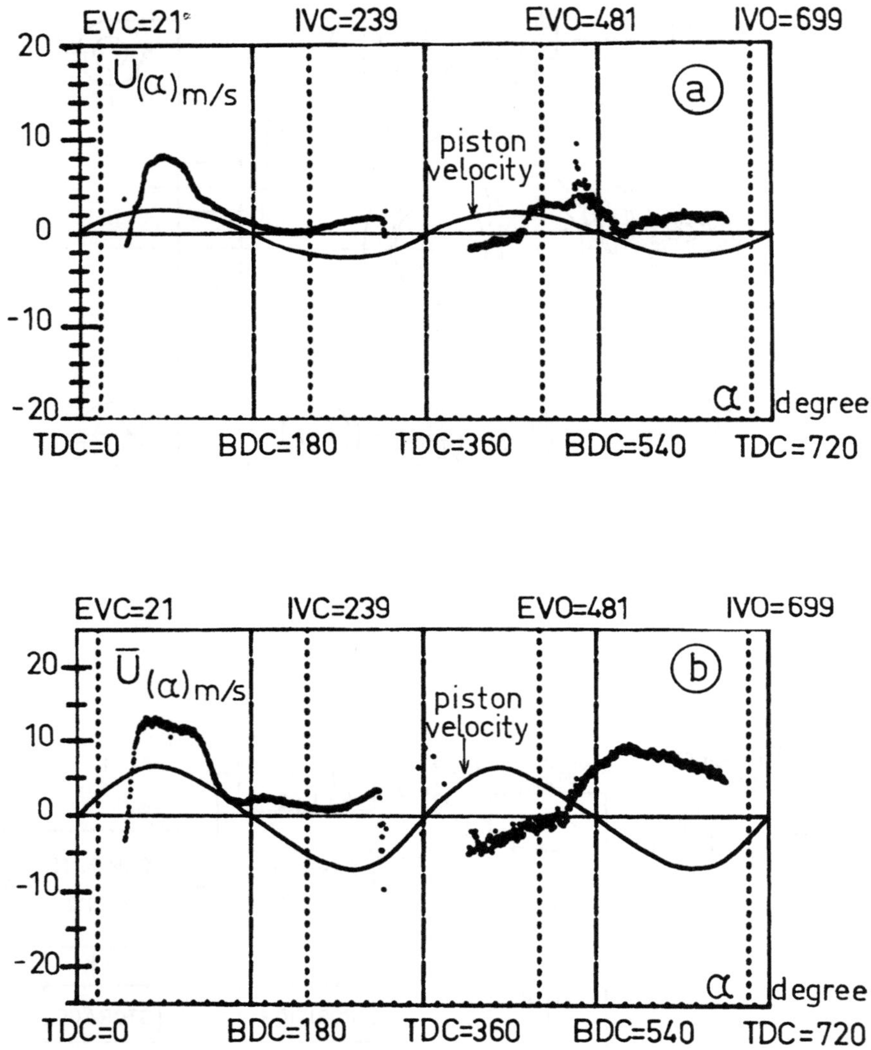

Fig. 3. Mixed ensemble and time averages of axial velocity. a : rotating speed f = 500 rpm ; b : f = 1500 rpm ; $\bar{U}$: positive from head to piston ; α : crank angle ; $\Delta\alpha$ = 0.7° ; n : 1900 and 2750. (E.V.C. : Exhaust valve closes, I.V.C. : Inlet valve closes, E.V.O. : Exhaust valve opens, I.V.O. : Inlet valve opens, B.D.C. : Below dead center, T.D.C. : Top dead center).

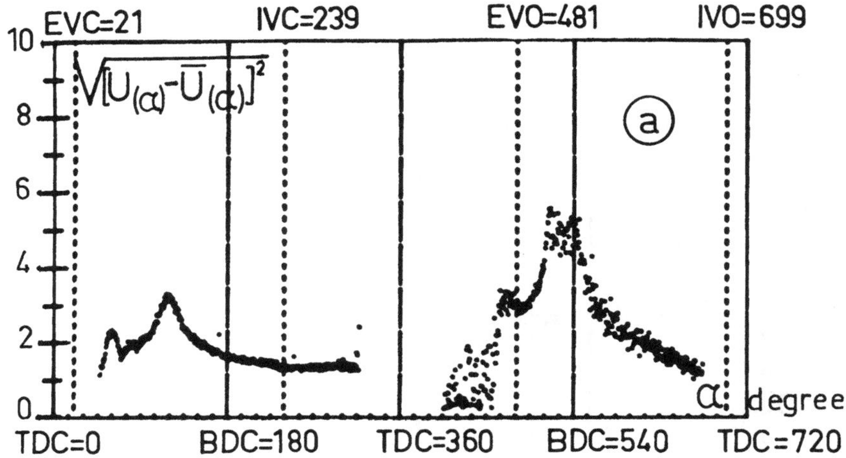

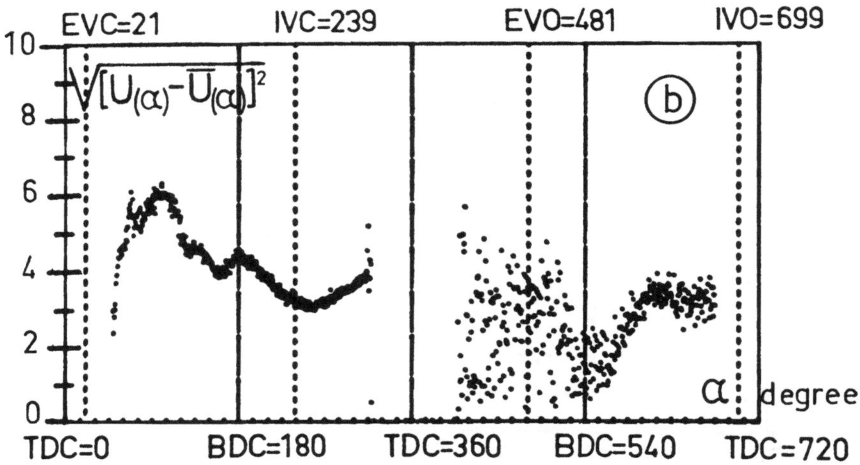

Fig. 4. R.M.S. values of the global fluctuating axial velocity.
a : f = 500 rpm ; b = f = 1500 rpm (see references fig. 3).

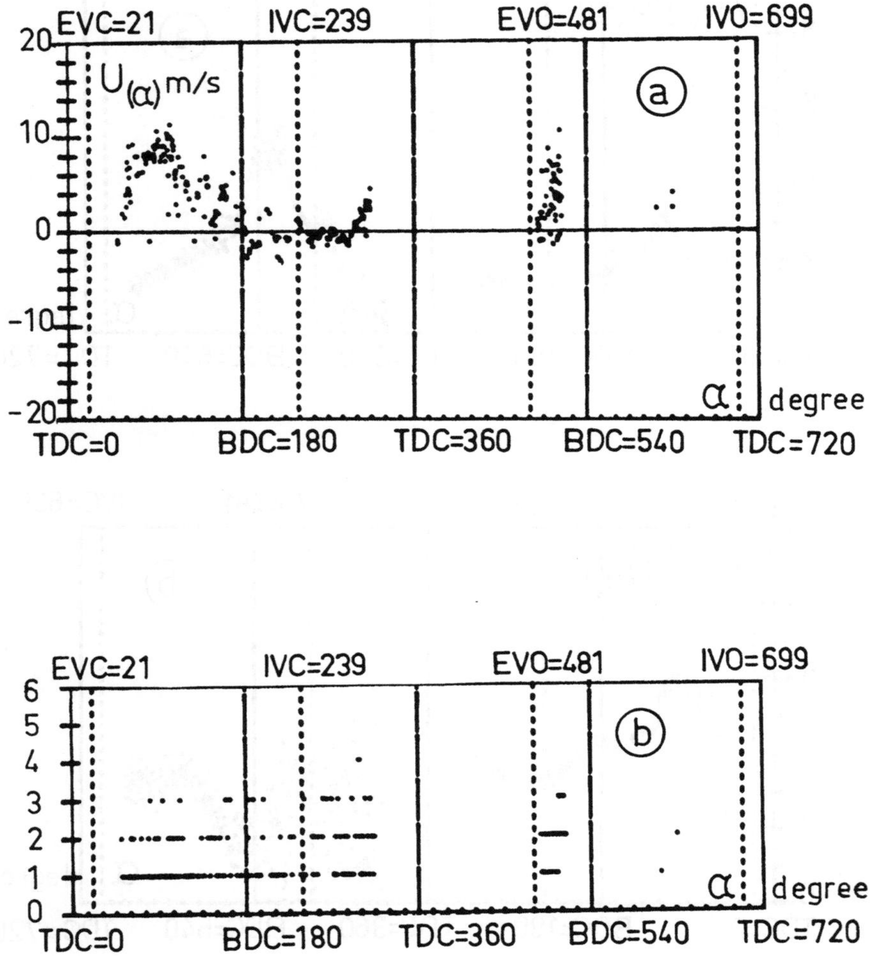

Fig. 5. Record of the axial velocity, for a single cycle.
a : velocity ; b : number of data for each interval $\Delta\alpha$ ($\Delta\alpha$ = 0.7 degree).

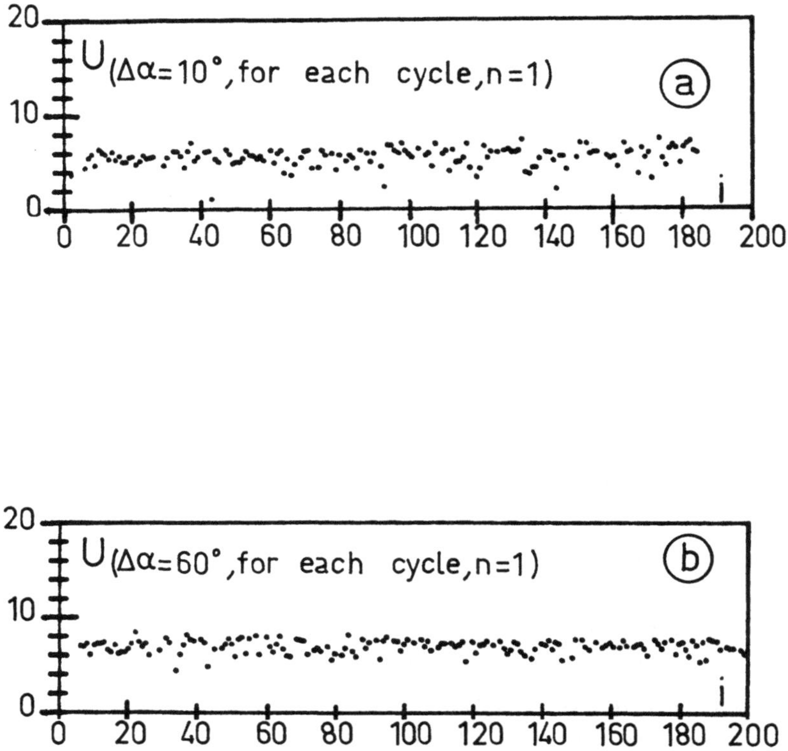

Fig. 6. Time averaged mean axial velocity for each successive cycle (number i). a : α in the range 60-70°; b : α in the range 60-120°.

An Oscillatory Approach to Turbulence

ENZO LEVI

Instituto de Ingeniería
Universidad Nacional Autónoma de México
México 20, D. F., México

Summary

Nature offers many examples of restrained fluid layers that an outer flow of velocity U forces to oscillate with a basic frequency $f = U/2\pi d$ approximately, d being the layer thickness. When the oscillations are convected by the current, waves of length $2\pi d$ are formed. Harmonics of the basic frequency and the corresponding fractional-length waves often arise. Taking into account the presence of those waves, an undulatory model for the turbulent flow is propounded. As an example of its application, a turbulent-energy spectrum is obtained through elementary wave-mechanics considerations.

A Universal Strouhal Law

Roshko (24) analysed the frequency of vortex shedding from cylindrical bodies taking into account the width d of the wake (i.e., the spacing of the free streamlines delimiting it) and the velocity U at the point in which these streamlines separate from the body, instead of the traditional parameters: body width and approach velocity. So the Strouhal number fd/U resulted to be independent of the body shape and flow Reynolds number; its average value was found to be

$$S = fd/U = 0.16 \qquad (1)$$

It was subsequently proved that the S value remains practically the same when the flow is constrained by a central splitter plate dividing the wake (24), by parallel walls confining the flow (23) or by forcing the cylinder to vibrate, in order to change artificially the shedding frequency (7).

Now, this value 0.16 for the Strouhal number is not peculiar to the wake vortices. In fact, it is not uncommon to find it, or a very near value, associated with other modes of fluid oscillations, as well as the value

$$\lambda = U/f = 6.2d \tag{2}$$

for the length λ of the travelling waves resulting from the convection of those oscillations by the main flow.

So for instance Crow and Champagne (4), observing the response of a round turbulent jet to a periodic surging imposed to its exit in the form of puffs emitted downstream, found that $\overline{f}d/U = 0.15$, f being the puffing frequency, d the nozzle radius and U the exit speed of the jet. Similarly, from Cervantes and Goldschmidt data (3), one infers that a plane jet flaps according to the formula $\overline{f}d/U = 0.154$, f being the flapping frequency, d the jet width and U the centerline mean velocity at a given section.

Measuring the frequency f of intermittent erect vortices that form upstream from a weir set across a rectangular water channel, U being the approach mean speed, Levi (16) found $\overline{f}d/U = 0.154$. Here d is the upstream water depth. For the frequency of orifice vortices upstream from a screen crossing the channel, he found $\overline{f}d/U = 0.176$, d being the screen submergency.

The length λ of wind waves produced with minor wind speeds U and fetches x appears to satisfy eq. 2, d representing the thickness of the wind laminar boundary layer. From Sen's laboratory measurements (26) one gets $\lambda/d = 6.21$ for $U = 5.12$ m/s, $x = 54$ cm, and $\lambda/d = 6.63$ for $U = 6.52$ m/s, $x = 49$ cm. From Sudolskiy's field measurements (27), one gets $\lambda/d = 5.89$, 6.99 and 6.22 for $U = 5$ m/s and fetches of 1, 2 and 5 km respectively.

Yalin (31) suggests that the length of dunes formed in a loose-bed river of depth d is on average equal to $2\pi d$. A similar result can be inferred from Thorpe's measurements of the increasing-with-time length of waves formed at a density interface between miscible fluids, provided that the thickness of the mixing layer is taken as d (28).

An Oscillatory Model

Birkhoff attempted to justify the value of the wake vortex-shedding Strouhal number through an analysis of the wake mechanism (2). However, the validity of eq. 1 for so many different flow modes suggests that we are in presence of a very general physical law, independent of the specific features of each single mode. The following simple reasoning (17) will lead us to corroborate this assumption and formulate the law.

Let us suppose that a restrained fluid layer of width d is forced to oscillate with the frequency f by the presence of a contiguous free flow of speed U, and that this frequency is the same that would correspond to an elementary oscillator of length d. The specific mechanical energy of the latter is

$$E = \frac{1}{2}(2\pi fd)^2 \qquad (3)$$

while the available kinetic energy is $U^2/2$. Equating both, one gets

$$S = fd/U = 1/2\pi \qquad (4)$$

that is, 0.159, which agrees with eq. 1. Oscillations governed by this law, if convected by the flow, will look to a stationary observer as undulatory perturbances of wavelength

$$\lambda = U/f = 2\pi d \qquad (5)$$

that agrees with eq. 2.

By the way, eq. 4 suggests the expedience of preferring the number $S' = 2\pi S$ to the usual Strouhal number S, in order that the value 1 should correspond to critical conditions, as it occurs for instance for Mach and Froude numbers.

Evidence of Strouhal-law Validity in Turbulent Flows

Boundary-layer transition. Three succesive stages characterize the transition from laminar to turbulent flow (13): at first a procession of longitudinal waves appears, then cross waves, and finally the resulting doubly-periodical waves shatter into "hairpin eddies" preluding to turbulence. Now, all these stages appear to obey the Strouhal law.

The correlation between the length of longitudinal waves and the boundary-layer thickness d can be deduced from an old Tollmien's result (29). In fact, he showed that, provided that the flow Reynolds number exceeds a certain critical value, the minimum wavelength of an oscillatory disturbance able to compromise the stability of a flat-plate laminar boundary layer is equal to

$(2\pi/0.36)d^*$, d^* being the displacement thickness. Now, this is about 6d, taking as usual $d = 2.9\ d^*$.

The transversal periodicity is usually visualized through the furrows grooved by the current into a fresh wall coating. Data from a relevant NACA technical note (10) give, on an average, a furrow spacing of 3.09 d, which agrees with eq. 5, because the furrows appear to be the result of an accumulation of paint at the nodes of standing transversal cross waves, and the node spacing is half the wavelength.

Klebanoff, Tidstrom and Sargent (11), measuring the frequency f of hairpin-eddy production obtained that $\bar{f}d^*/U = 0.13$, U being the free-flow velocity. Since in their case $d/d^* = 2.55$, it results that $\bar{f}d/U = 0.33 = 2\times0.165$. This is double the value given by eq. 4, suggesting the presence of a first harmonic.

<u>Wall layer</u>. Longitudinal and transversal waves of the same length λ_w appear to coexist also within the viscous sublayer, but they are much smaller than the transition waves, because they scale with the wall-layer thickness d_w.

Evidence of longitudinal waves can be found in a paper by Fage and Townend (6). When observing by ultramicroscope the motion of particles dragged by a turbulent current, they recorded regular oscillations of the particle paths inside a layer very near the wall, whose nondimensional thickness was about $y^+ = yu_\tau/\nu = 0.4$, u_τ being the friction velocity and ν the kinematic viscosity. Now, if λ_p is the particle-path wavelength, it should be to the local velocity u as λ_w is to the wave celerity c_w. Since from Fage data one infers that $\lambda_p^+ = 4.43$ and Morrison (18) finds that $c_w^+ = 8.2$, taking $y^+ = 0.2$ as a mean position for the observed path, we get $u^+ = 0.2$, and then $\lambda_w^+ = 182$. On the other hand, since $d_w^+ = 30$ (12), eq. 5 gives the theoretical value $\lambda_w^+ = 2\pi d_w^+ = 188$.

Coming now to the low-speed viscous-sublayer longitudinal striations, let us suppose that, as the transition ones, they correspond to nodes of transversal standing waves. Their spacing λ_w' should then be equal to $\lambda_w/2$, the theoretical value of $\lambda_w'^+$ being thus 94. In fact an experimental average for it is about 97 (22).

The other typical feature of wall layer is its bursting activity. Narahari Rao discovered that the burst frequency f scales with outer parameters, i.e., the boundary layer overall thickness d and the free-flow velocity U. His measurements (19) give for fd/U values between 0.14 and 0.33. More precise results are now available. For instance from the measurement of wall pressure fluctuations (that are closely related to bursting activity) by Schewe (25), one obtains $\overline{f}d/U = 0.172$.

<u>Fully developed turbulent flow</u>. Nychas, Hershey and Brodkey (21) pointed out the alternation of low-speed and high-speed fluid bodies in the region of fully-developed turbulent flow. Wallace, Brodkey and Eckelmann (30), working in a channel of d = 22 cm width, with a centerline velocity U = 21 cm/s, measured the time of passage T of a characteristic pattern of the fluctuation of the streamwise velocity component, that is likely to correspond to the passage of one of those bodies. A typical graph in their paper gives T = 3.3 s; therefore 2TU/d = 6.30, 2TU being the streamwise width of a low-speed-high-speed pair. Comparing with eq. 5, we get that 2TU = λ, λ being the length of a fundamental wave. This suggests that the speed alternation ensues from the passage of the wave, the low speed corresponding to the wave outward half-length, the high speed to the wallward half-length (see Fig. 1).

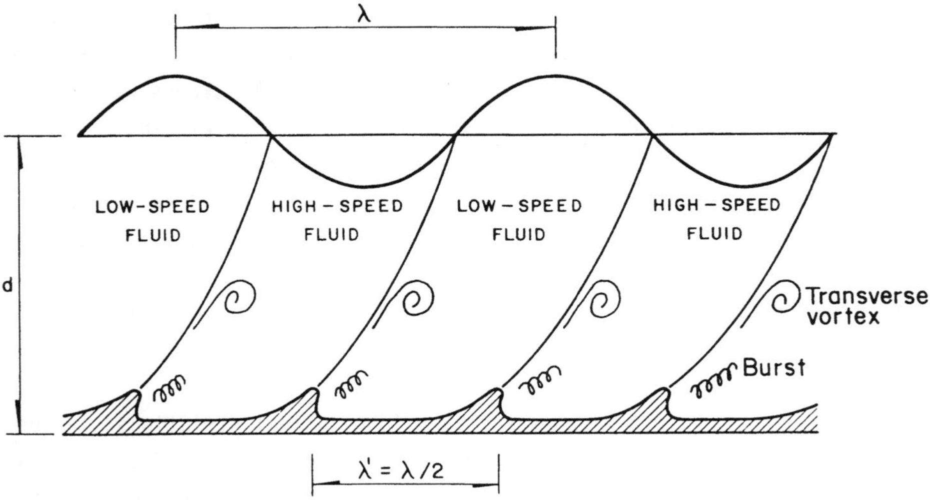

Fig. 1. Alternation of low-speed and high-speed fluid bodies and burst-inducing mechanism.

Turbulent structures display a near-periodicity. Badri Narayanan and Marvin (1), autocorrelating velocity and pressure fluctuations across the boundary layer at a wide range of Reynolds and Mach numbers, found out that $fd/U = 0.17 \pm 0.03$, f being the fluctuation frequency, d the boundary-layer thickness and U the free-flow velocity. From recent measurements by Hofbauer (9) one gets that $\overline{f}d/U = 0.152$.

Finally, let us assume (5) that the characteristic length l_0 of large eddies in a pipe flow be such that

$$l_0 = \tilde{u}/f \tag{6}$$

f being their frequency and $\tilde{u}$ the turbulent intensity at the pipe axis. If those eddies are envisaged as oscillators of length πl_0 and their energy is equated to the one given by eq. 3, one gets that $\pi l_0 = d$, that is,

$$l_0/d = 1/\pi = 0.32 \tag{7}$$

Therefore, taking into account eqs. 6, 7 and 4, one gets

$$\frac{\tilde{u}}{U} = \frac{fd}{U}\,\frac{l_0}{d} = \frac{1}{2\pi^2} = 0.050$$

U being the mean velocity at centerline. Experimental results by Laufer (14) give $\tilde{u}/U = 0.047$.

A New Turbulence Model

Let us accept that, as the foregoing results suggest, within a turbulent boundary layer of thickness d associated with an outer free flow of speed U, oscillations of frequency $f = U/2\pi d$ and wavelength $\lambda = 2\pi d$ occur, that manifest themselves in the alternation of low- and high-speed fluid bodies of width $\lambda/2$.

According to Nychas (21), in the shear layers between these bodies transverse vortices arise. They usually move outwards, and this motion seemingly rouses low-velocity tongues up from the viscous-sublayer streaks. As shown elsewhere (15), there are good reasons for assuming that the bursts are the wakes formed behind those tongues by the circumventing faster flow.

Bursts, possessing a velocity component normal to the wall inherited by the parent uprising tongue, leave the wall layer and spread into the region of fully developed turbulence, creating there structures endowed with vorticity.

Now, the travelling waves of length $\lambda = 2\pi d$ are not alone. They coexist with shorter waves of length $\lambda/2$, $\lambda/3$, ..., carrying the oscillations that correspond to the harmonics $f_2 = 2f$, $f_3 = 3f$, ... of the basic frequency $f_1 = f$. A progressive wave forces fluid particles to turn with the wave frequency, following oval orbits whose size diminishes as the wall is approached. It is thus reasonable to expect that, through this timing-and-shaping activity, the travelling waves control the coherent structures arised from ejected bursts, creating eddies of various frequencies (Fig. 2). Travelling waves should also control cascade processes, shaping into higher-frequency eddies the pieces into which a coherent structure would eventually disrupt.

On these premises, it seems reasonable to try to build an oscillatory theory of turbulence, that could use the analytic tools of wave mechanics. As an advance, we will solve the problem of obtaining a turbulent-energy spectrum by deterministic means (17).

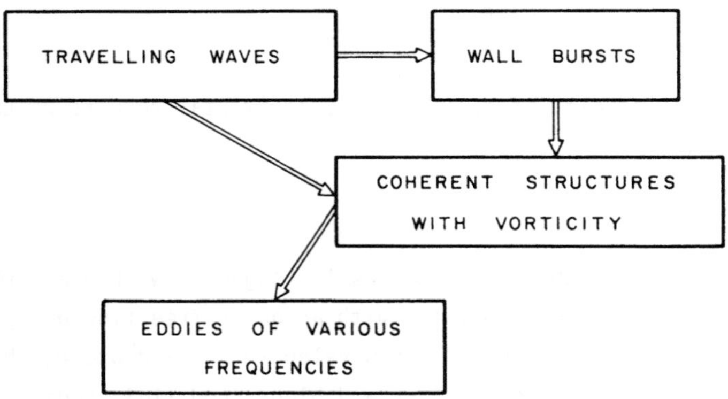

Fig. 2. How travelling waves should give rise to turbulent eddies.

The Turbulent Energy Spectrum

Let us admit that turbulent eddies of frequencies $f_n = nf (n = 1, 2, 3, ...)$ are able to receive or emit energy only through quanta ε_n. At a certain state of flow, d and U being given, one may expect by eqs. 3 and 4 that

$$\varepsilon_n = \alpha\, S_n^2 \tag{8}$$

α representing an energetic factor, function of the free-flow Reynolds number, and $S_n = nS = n/2\pi$. Now, let us observe that S_n represents also the ratio of the energy $(f_n dU)/2$ associated with the frequency f_n and the total kinetic energy $U^2/2$. In view of the considerable quantity of eddies that are present, this fact suggests that the probability of finding an eddy of frequency f_n endowed with a quantum of energy has to be proportional to e^{-S_n}, the probability of finding such an eddy endowed with two energy quanta has to be proportional to e^{-2S_n}, and so on. Therefore, the number of eddies with frequency f_n and k energy quanta (k = 1, 2, 3, ...) can be written as

$$N_k = c\, e^{-kS_n}$$

c being a numerical constant.

The total number of f_n-frequency eddies will then be

$$N = N_1 + N_2 + N_3 + \ldots = c(e^{-S_n} + e^{-2S_n} + \ldots) = c\, e^{-S_n}(1 - e^{-S_n})^{-1}$$

As a consequence

$$c = N\, e^{S_n}(1 - e^{-S_n}) \tag{9}$$

The total energy E_t corresponding to all the f_n-frequency eddies will be

$$\begin{aligned} E_t = N_1\varepsilon_n + N_2(2\varepsilon_n) + \ldots &= c\, e^{-S_n}\varepsilon_n(1 + 2e^{-S_n} + 3e^{-2S_n} + \ldots) \\ &= c\, e^{-S_n}\varepsilon_n(1 - e^{-S_n})^{-2} \end{aligned}$$

that is by eq. 9

$$E_t = N\varepsilon_n(e^{S_n} - 1)^{-1} \tag{10}$$

Introducing now eq. 8 into eq. 10 and dividing by N, the following expression results for the mean energy $E_m = E_t/N$ of the whole of f_n-frequency eddies:

$$E_m = \alpha S_n{}^2(e^{S_n} - 1)^{-1}$$

That is, since $S_n = n/2\pi$,

$$\frac{Em}{\alpha} = (\frac{n}{2\pi})^2(e^{n/2\pi} - 1)^{-1} \qquad (11)$$

Eq. 11 has been plotted in Fig. 3, showing $4\pi^2E_m/\alpha$ as a function of the frequency number n. The resulting curve agrees qualitatively with energy spectrum deduced on dimensional grounds (8).

To show its quantitative validity, three points have been marked on the n-axis, pointing out the typical values that, according to Davies (5), correspond, for medium Reynolds numbers, to (a) Prandtl eddies (i.e., those whose characteristic dimension is the Prandtl mixing length), (b) energy-containing eddies, and (c) energy-dissipating eddies. Their position has been ascertained according to the following considerations. Nikuradse (20), experimenting

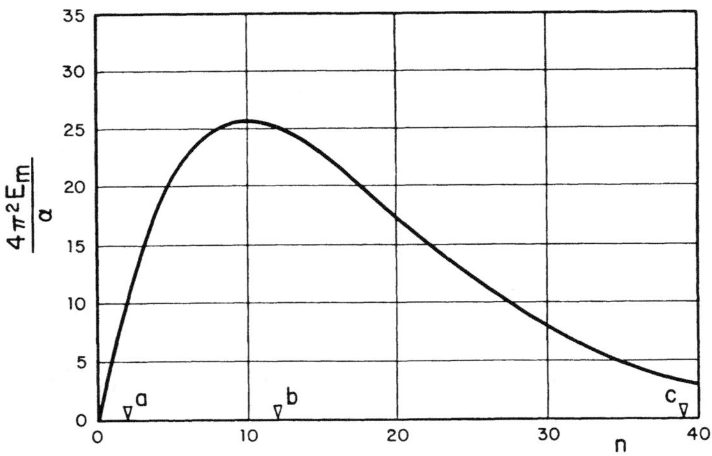

Fig. 3. Energy spectrum as a function of the frequency number n. Typical experimental values of n are shown for (a) Prandtl eddies, (b) energy-containing eddies and (c) energy-dissipating eddies.

with smooth circular pipes, was able to determine the mixing length l_m as a function of the distance from the pipe wall, for different Reynolds numbers. At values of 10^5 or more he found that, at the pipe axis, $lm/R = 0.16$, R being the pipe radius. Now if, as suggested before for axisymmetrical flows, we take d = R and compare with eq. 7, we find that $1_m = 1_o/2$, that is, that the Prandtl eddies correspond to n = 2. Having thus found the location of Prandtl eddies, a simple proportion applied to Davies values give n = 12 for energy-containing eddies and n = 39 for energy-dissipating eddies. These are the abscissas marked as a, b, c in Fig. 3. Their position with respect to the energy curve agrees with accepted beliefs (8).

References

1. Badri Narayanan, M.A.; Marvin, J.G.: On the period of the coherent structure in boundary layers at large Reynolds numbers. Workshop on Coherent Structure of Turbulent Boundary-Layers, Lehig University, Bethlehem, Pennsylvania (1978) 380-385.

2. Birkhoff, G.: Formation of vortex streets. J. Appl. Phys. 24 (1953) 98-103.

3. Cervantes de Gortari, J.: Goldschmidt, V.W.: The apparent flapping motior of a turbulent plane jet. Further experimental results. ASME Winter Annual Meeting, Chicago, Ill. (1980) 80-WA/FE-13.

4. Crow, S.C.; Champagne, F.H.: Orderly structure in jet turbulence. J. Fluid Mech. 48 (1971) 547-591.

5. Davies, J.T.: Turbulence phenomena. Academic Press 1972.

6. Fage, A.; Townend, H.C.H.: An examination of turbulent flow with an ultramicroscope. Proc. Roy. Soc. A135 (1932) 657-677.

7. Griffin, O.M.: A universal Strouhal number for "locking-on" of vortex shedding to the vibrations of bluff cylinders. J. Fluid Mech. 85 (1978) 591-606.

8. Hinze, O.: Turbulence. Mc Graw-Hill 1975.

9. Hofbauer, M.: Evidence for instability waves in the velocity field of a fully developed turbulent channel flow, AGARD Conference on Turbulent Boundary Layers, The Hague (1979) 271.

10. Hopkins, E.J.; Keating, S.J.; Bandettini, A.: Photographic evidence of streamwise arrays of vortices in boundary-layer flow. NACA Techn. Note D-328 (1960).

11. Klebanoff, F.S.; Tidstrom, K.D.; Sargent, LM.: The three-dimensional nature of boundary-layer instability. J. Fluid Mech. 12 (1962) 1-34.

12. Kline, S.J.; Reynolds, W.C.; Schraub, F.A.; Runstadler, P.W.: The structure of turbulent boundary layers. J. Fluid Mech. 30 (1967) 741-773.

13. Knapp, D.F.; Roache, P.J.: A combined visual and hot-wire investigation on boundary layer transition. AIAA J. 6(1968) 29-36.

14. Laufer, J.: The structure of turbulence in fully developed pipe flow. NACA Rep. 1174 (1954).

15. Levi, E.: Eddy production inside wall layers. J. Hydr. Res. 16 (1978) 107-122.

16. Levi, E.: Periodicidad de estructuras vorticosas. Proc. 9th. IAHR Latin-American Hydr. Congress, Mérida, Venezuela, 1 (1980) 143-151.

17. Levi, E.: Nuevas consideraciones sobre la periodicidad de estructuras vorticosas. Proc. 9th. IAHR Latin-American Hydr. Congress, Mérida, Venezuela, 2 (1980) 126-137.

18. Morrison, W.B.R.; Bullock, K.J.; Kronauer, R.E.: Experimental evidences of waves in the sublayer, J. Fluid Mech. 47 (1971) 639-656.

19. Narahari Rao, K.; Narashimha, R.; Badri Narayanan, M.A.: The bursting phenomenon in a turbulent boundary layer. J. Fluid Mech. 48 (1971) 339-352.

20. Nikuradse, J.: Gesetzmässigkeit der turbulenten Strömung in glatten Rohren. VDI-Forshungsheft 356 (1932).

21. Nychas, S.A.; Hershey, H.C.; Brodkey, R.S.: A visual study of turbulent shear flow. J. Fluid Mech. 61 (1973) 513-540.

22. Oldaker, D.K.; Tiederman, W.G.: Spatial structure of the viscous sublayer in drag-reducing channel flows. Phys. of Fluids 20 (1977) S133-S144.

23. Richter, A.; Naudascher, E.: Fluctuating forces on a rigid circular cylinder in confined flow. J. Fluid Mech. 78 (1976) 561-576.

24. Roshko, A.: On the drag and shedding frequency of two-dimensional bluff bodies. NACA Techn. Note 3169 (1954).

25. Schewe, G.: Untersuchung von Wanddruck-und Wanddruckgradientenshwankungen unter einer turbulenten Grenzschichtsrömung, Doctoral Thesis, University of Göttingen 1978.

26. Sen, M.: Interaction between scales in the problem of wave generation by wind, Doctoral Thesis, Massachussetts Institute of Technology 1974.

27. Sudolskiy, A.S.: Wind waves on Kayrak-Kumskiy Reservoir. Soviet Hydrology (1963) 366-388.

28. Thorpe, S.A.: Experiments on the instability of stratified shear flows: miscible fluids. J. Fluid Mech. 46 (1971) 299-319.

29. Tollmien, W.: The origin of turbulence. NACA Techn. Memo. 609 (1931).

30. Wallace, J.M.; Brodkey, R.S.; Eckelmann, H.: Pattern-recognized structures in bounded turbulent shear flows, J. Fluid Mech. 83 (1977) 673-693.

31. Yalin, M.S.: Mechanics of sediment transport, Pergamon Press 1977.

The Development of Vortices in a Mixing Layer

A. DYMENT

Université de Lille I 59655 Villeneuve d'Ascq Cedex France
and I.M.F.L., bd. Painlevé 59000 Lille France

Abstract

The theoretical scheme proposed in this paper brings some new materials to the knowledge of ordered unsteady phenomena occuring downstream of a separation in a two dimensional flow at high Reynolds number. This scheme explains the formation of large eddies and gives a simple description of the initial development of a mixing layer. Due to the complexity of the problem, only evolution laws can be formulated, but they give an admissible model which agrees fairly well with experiment.

Nomenclature

ν kinematic viscosity

L characteristic length of the body

U free stream velocity

$\mathcal{R}$ Reynolds number $\frac{UL}{\nu}$

q exponent defining the size of a Navier Stokes domain

N frequency

Γ circulation

S Strouhal number $\frac{NL}{U}$

x curvilinear abscissa of the center of a vortex

V propagation velocity of a vortex

t time

R vortex radius

b dimensionless quantity measuring the spreading of a vortex

$\xi = \frac{x}{L}$

$\tau = \frac{Ut}{L}$

$k = \frac{V}{U}$

$r = \frac{R}{L}$

D distance from the center of a vortex

w velocity induced by a vortex

Λ distance between two successive vortices

$\lambda = \frac{\Lambda}{L}$

α constant appearing in formula (8)

Δt time interval between two successive flashes

H height of the backward facing step

x_t abscissa of transition

Subcripts : 0 refers to shedding

n corresponds to quantities after the nth pairing

1. Introduction

It is well known that separated flows exhibit unsteady properties although boundary conditions do not depend on time. Experimentally, large eddies have been detected for a long time, but their formation has not yet been satisfactorily explained. Actually, even their evolution is badly known because the observation of rapidly varying phenomena is very difficult.

Our aim here is to solve, with the help of a remarkably simple theoretical scheme, the problem of the formation of large coherent eddies in a mixing layer, downstream of a two dimensional separation occuring in a laminar incompressible flow.

The instability of the shear layer which follows separation makes it appear as a vortex source. We show that the shed vortices must necessarily undergo successive pairings, the process of which is brought to light. The amalgamation phenomenon rapidly decreases downstream of the shedding point.

The experimental study of very quickly varying flows is difficult unless confined to mean statistical properties. In our experiments we use a technique of multiple sparks high speed visualization in which the time interval between two successive sparks can be adjusted. Results obtained in a mixing layer over a backward facing step show quite good an agreement with theoretical estimates.

2. Vortex shedding at separation

The streamline issuing from the separation point, at the juction of the flow coming from upstream and the recirculating flow, belongs to a shear layer. From a theoretical point of view, the simplest case corresponds to an infinite separating streamline. Numerous studies about the stability of such a flow have shown an amplification of the initial disturbances and a periodic rolling up [1], [2], [3]. These results reveal an essential character which lies in the evolution towards the formation of large eddies. In a real mixing layer this tendency is both temporal and spatial. It has been experimentally brought to light by Winant and Browand [4] and Brown and Roshko [5]. Later Dimotakis and Brown [6] have attempted to discuss the influence of the initial state of the boundary layer on the development of the mixing layer. Chandrusa et al [7] suggested that the existence of large structures is an exceptional phenomenon which does not appear when the initial flow is turbulent. However, Bernal et al [8] do not agree with this opinion : they observe large structures even when tridimensional disturbances are present. The same holds for Browand and Latigo [9] and for Dyment and Gryson [10].

It is usual to call the flow under consideration a turbulent mixing layer. This may appear a bit surprising as we are not concerned with a random phenomenon, showing chaotic motions in a wide range of frequencies. Our way to approach this problem is quite different [11][12]. We take into account the existence in a steady separated flow of small domains which surround separation points and where the flow is governed by the full Navier Stokes equations [13][14].

Let L be a characteristic length of the body, U the free stream velocity, ν the kinematic viscosity and R the Reynolds number $R = UL/\nu$. The order of magnitude of the size of the Navier Stokes domains and of the velocity therein are LR^{-q} and UR^{q-1}. The exponent q depends on the nature of separation : it is greater than 1/2 as Navier Stokes domains are embedded in the boundary layer.

Outside the Navier Stokes domain the shear layer is unstable because its representative Reynolds number is large. This instability gives brith to vortices. With respect to the size of the body vortex shedding can be regarded as happening at separation. In other words, this means that shedding only depends on the nature of separation which can obviously be the same for various bodies. Let N_o be the frequency of the emitted vortices,

$S_o = N_o L/U$ the corresponding Strouhal number and Γ_o their circulation. It has been shown that [11][12]

$$S_o \sim R^{2q-1}, \quad \Gamma_o \sim \nu . \qquad (1)$$

Thus, S_o is much larger than one and it is a growing function of R. On the other hand, the reduced circulation Γ_o/UL is of order R^{-1}.

The vortices shed at separation propagate downstream and induce an unsteady velocity field. Now, the flow has been supposed steady, so that our results are valid as a first approximation because the velocity induced by the vortices is small compared with U as will be shown in 3.

Likewise, previous results remain valid for an oscillatory body provided that its reduced frequency is very small compared to S_o, which is always true owing to the high values of S_o.

3. The pairing phenomenon

Let x be the curvilinear abscissa of the center of a vortex taken along its trajectory from the shedding source, and let V be its propagation velocity. We suppose that viscosity does not directly interfere in the translatory motion of vortices. However, we take into account the fact that during their propagation vortices spread out and that their intensity is less and less concentrated. According to the properties of viscous diffusion, their radius R is connected to time t by $b^2 \, d(R^2) = \nu dt$ where b is a dimensionless quantity which depends on the vortex position.

Let us introduce the dimensionless quantities defined by $\xi = x/L$, $\tau = Ut/L$, $k = V/U$, $r = R/L$. We can write $d\xi = k \, d\tau$, $R b^2 \, d(r^2) = d\tau$, i.e. $R \, k \, b^2 \, d(r^2) = d\xi$.

As convection effects are dominant V is much larger than $\frac{dR}{dt}$, whence

$$R \, k \, b^2 \, r \gg 1. \qquad (2)$$

The reduced velocity k is of order one because vortices are carried downstream with a velocity comparable with U, but b can be small. Obviously, the radius of a vortex cannot be defined accurately. For a motionless vortex the velocity induced at distance D is given by the classical formula $w = \frac{\Gamma_o}{2\pi D}\left[1-\exp\left(-\frac{D^2}{4\nu t}\right)\right]$. If the radius is considered to be the distance from the center where w is, say, 3 % of the maximum velocity, then b is of order 10^{-2}.

As $b^2\ R^2 \sim \nu t$ and $t \sim x/U$ we obtain

$$\mathcal{R}\ b^2\ r^2 \sim \xi\ . \tag{3}$$

The velocity w induced by a vortex is of order Γ_o/D, i.e. $UL/\mathcal{R}D$. According to (3) we may write $w/U \sim (\mathcal{R}\xi)^{-1/2} b\ R/D << 1$. This shows that, as it has been admitted in 2, the field induced by the vortices can be neglected.

Consider now two successive vortices and choose the origin of time at the moment when the second vortex is shed. At time t the radii R and R' of these vortices are respectively given by $R^2 = \int_{-1/N_o}^{t} \frac{\nu dt}{b^2}$ and $R'^2 = \int_{o}^{t} \frac{\nu dt}{b^2}$ and the distance between their centers is $\Lambda = \int_{t-1/N_o}^{t} \dot{V}\ dt$.

The three corresponding dimensionless quantities are r, r' and λ defined by

$$\mathcal{R}\ r^2 = \int_{-1/S_o}^{\tau} \frac{d\tau}{b^2}\ ,\quad \mathcal{R}\ r'^2 = \int_{o}^{\tau} \frac{d\tau}{b^2}\ ,\quad \lambda = \int_{\tau-1/S_o}^{\tau} k\ d\ \tau.$$

Suppose that at time t_1, when the center of the second vortex reaches the point of abscissa x_1, the two vortices under consideration come into contact (fig.1). We shall see hereafter that this cannot be avoided. Let us give the subcript 1 to any quantity corresponding to time t_1. We have $r_1 + r_1' = \lambda_1$. But, as $r_1^2 - r_1'^2 \simeq 1/\mathcal{R}S_o b_1^2$ and $\lambda_1 \simeq k_1/S_o$ we obtain $r_1 - r_1' \simeq 1/\mathcal{R}k_1 b_1^2$. It follows that $2r_1 \simeq k_1/S_o + 1/\mathcal{R}k_1 b_1^2$ and, according to (2), the last term can be neglected. Consequently, we obtain

$$\mathcal{R}\ k_1^2 \simeq 4\ S_o^2 \int_o^{\xi_1} \frac{d\xi}{kb^2}\ . \tag{4}$$

Considering only the orders of magnitude we have

$$S_o^2\ \xi_1 \sim \mathcal{R}\ k_1^3\ b_1^2\ . \tag{5}$$

Taking (1) into account we may write $\xi_1 \sim k_1^3\ b_1^2\ \mathcal{R}^{3-4q}$. Since $r_1\ S_o \sim k_1$ we also have $r_1 \sim k_1\ \mathcal{R}^{1-2q}$ and consequently $r_1 << \xi_1$.

As it has been anticipated, we see that overlapping of two successive vortices must necessarily occur. It indicates the beginning of the pairing of two neighbouring vortices. As we have noted, this pehnomenon is instigated by viscous diffusion, but it is performed under the effect of inertia forces which are dominant.

4. The successive pairings process.

The pairing detected above gives birth, beyond abscissa x_1, to vortices of frequency $N_1 = \frac{N_o}{2}$ and it is obvious that the pairing phenomenon will start again : a serie of successive pairings will then occur.

Let n be the subscript refering to all quantities after the nth paring. For simplicity we neglect the lengths necessary for all the pairings to be achieved and we suppose that the radius does not noticeably change during each coalescence. As a consequence, by virtue of the same arguments as previously, we obtain approximately

$$R\ k_n^2 = 4\ S_{n-1}^2 \int_o^{\xi_n} \frac{d\xi}{kb^2} .$$

With regard to orders of magnitude, as

$$S_n = 2^{-n}\ S_o, \tag{6}$$

we have

$$\xi_n \sim 4^{n-1}\ \xi_1 . \tag{7}$$

It can be seen that S_n decreases and x_n increases very rapidly. In other words, the pairing process considerably slows down as we progress downstream (fig.2). As a result, if the vortices can only be observed in a domain sufficiently far from separation, the measurements give a roughly constant frequency, much smaller than N_o.

It is convenient to represent the successive pairings by a continous curve (fig.2). From previous results it follows that this curve may be defined by

$$R\ k^2 = \alpha\ S^2 \int_o^{\xi} \frac{d\xi}{kb^2} \tag{8}$$

where the constant α can be taken equal to 4.

This relation only holds sufficiently far from separation. It may be written

$$4\ S^2\ \xi \sim R\ k^3\ b^2 . \tag{9}$$

Equivalent forms of formulae (8) and (9) are

$$V^2 = \alpha\ \nu\ N^2 \int_o^{x} \frac{dx}{Vb^2} \quad \text{and} \quad \frac{Nx}{V} \sim \left(\frac{Vx}{\nu}\right)^{1/2} .$$

The whole of previous results give a remarkably simple explanation of the complex formation of large eddies in a flow.

It is worth noting that, according to our formulae, similarity of coherent unsteady phenomena requires the invariability

Now it may be asked whether it is possible to conciliate the two points of view, the first considering that pairings results from a random process and the second which has been developped above. These positions seem to be poles apart. However, we must keep in mind that in our theoretical scheme the velocity field induced by the travelling vortices has been neglected. Now, at the issue of the nth pairing we have $\Gamma_n \sim 2^n \Gamma_o$. As a result, the corresponding velocity w is such that $\frac{w}{U} \sim \frac{2^n \Gamma_o}{DU} \sim \frac{2^n L}{\mathcal{R}D}$ and, according to (3), $\frac{w}{U} \sim 2^n (R\xi)^{-1/2}\, b\, \frac{R}{D}$. Taking (6) and (9) into account we get $\frac{w}{U} \sim \frac{S_o}{\mathcal{R}} \frac{R}{D}$. If (1) is admitted we may write $\frac{w}{U} \sim \mathcal{R}^{2(q-1)} \frac{R}{D}$. Therefore, this quantity always remains small, even at distances of order R. Thus, it is correct to neglect w with regards to V.

Nevertheless, although the induced velocity is small it can produce a disturbing effect which may no longer be disregarded. Finally, viscous diffusion is certainly dominant at the beginning of the mixing layer, but farther on, where vortices are stronger and less concentrated, the instability resulting from interaction between vortices must be taken into account : the phenomenon thus gains some pseudo randomness that may partly hide its determinist character.

To sum up, apart from the Navier Stokes domain and the region of formation of vortices which both are almost reduced to a point, two domains can be distinguished : a first one with well ordered pairings, then a second one where some instability gradually takes place through the interaction between vortices. Experiments will corroborate this point of view.

5. Experiments

Following vortices in a moving fluid is a very difficult task and it can only be done by visualization. But, visualization performed with foreign particles often gives a distorbed picture of the phenomenon under observation as, because of inertia effects, the tracers cannot correctly respond to velocity gradients or to high frequencies.

Optical visualization methods do not show this drawback, but they require sufficiently high speeds in order to be able to dectect density variations.

To visualize a phenomenon at a given moment it is necessary

that the observation duration be much smaller than a characteristic time scale. To restore the development with time of a phenomenon it is necessary to dispose of a serie of such pictures separated by known time intervals. This aim can be achieved by using a multiple spark apparatus, called Cranz-Schardin system [15], and by making the time interval between two flashes adjustable. The main drawback of this method lies in the parallax errors, as, if the light sources are numerous, it is not possible to concentrate them in a small volume. These errors are minimized by putting the flash sources and the camera far from the test section and they are taken corrected for during the analysis of the pictures.

Our device is made up of 24 spark sources and 24 photographic lenses. The time interval Δt between two successive flashes can be continuously adjusted from 1 second to 10^{-7} s [10][12].

The measurement of the time between the passing by a given abscissa of two successive vortices gives the frequency N. An interpolation is almost always necessary. Furthermore, as the localization is little precise because they are not point vortices, the measurement only provides an order of magnitude. The same thing can be said about the propagation velocity V and the radius R, both obtained from the displacement of a given vortex between two successive pictures.

Our experiments have been made over a backward forcing step of height H = 25 mm [16]. In all the quantitative results given later on the reference length used in any dimensionless parameter will always be H.

Figure 3 shows an example of visualization with a Mach at separation equal to .72 : the three successive pictures presented are separated by an interval $\Delta t = 5 \, 10^{-6}$ s. It has been verified that the mean pressure is almost uniform in the mixing layer : therefore k and b vary little and relation (8) becomes

$$S^2 \xi = \alpha \, k^3 \, b^2 \, \mathbf{R} = cst \, .$$

The results plotted on figure 4 show that this formula is satisfied. Rough estimates of k and b leads to $k \simeq .5$ and $b \simeq 1.4 \, 10^{-2}$, so that the constant is close to 1.5 whereas the experimental curve gives .7. This apparent discrepancy has indeed no importance when taking into account first the limitative assumptions made in our theory and the fact that (8) is valid only in incompressible flow, second the inaccurary in our estimates of k and especially of b.

All our experiments reveal that beyond an abscissa x_t vortices gather in clusters rather than coalesce. For $x > x_t$ the mixing layer neatly seems less organized and its boundaries become fuzzy.

Ultimately, the analysed phenomenon can be outlined as follows : downstream of the domain where well ordered successive pairings occur there is a domain which does not depend on initial conditions and which exhibits random properties. In the first domain pairings actually are amalgamations whereas in the second vortices gather in clusters without merging.

An accurate estimation of the abscissa x_t where the ordered unsteady regime ends is very uneasy because transition gradually occurs. In our experiments x_t lies between $\frac{H}{2}$ and H : its approximately corresponds to the beginning of the plateau on the curve giving S versus ξ .

The previous observations agree, at least qualitatively, with the theoretical precisions given formerly.

References

[1] Betchov, Criminale Stability of Parallel Flows. Academic Press, 1967.

[2] Gaster Progr. Aero. Sces, 6 , 251.

[3] Stuart J. Fl. Mech, 29, 417.

[4] Winant, Browand J. Fl. Mech, 63, 237.

[5] Brown, Roshko J. Fl. Mech, 64, 775.

[6] Dimotakis, Brown J. Fl. Mech, 78, 535.

[7] Chandrusa et al J. Fl. Mech, 85, 693.

[8] Bernal et al 2nd Symp. Turb. Shear Flows, London 1979.

[9] Browand, Latigo Ph. Fluids, 22, 1011.

[10] Dyment, Gryson Colloque AAAF, Marseilles 1978.

[11] Dyment Note C.R.Ac.Sces, 290, B, 47.

[12] Dyment, Gryson AGARD CP n° 227.

[13] François Publ. ONERA n° 128, 1969.

[14] Stewartson Adv. Appl. Mech, 14, 145.

[15] Merzkirch Flow Visualization, Academic Press, 1974.

[16] Dyment, Gryson, Ducruet, Flodrops. IMFL Report n° 80-43, Lille 1980.

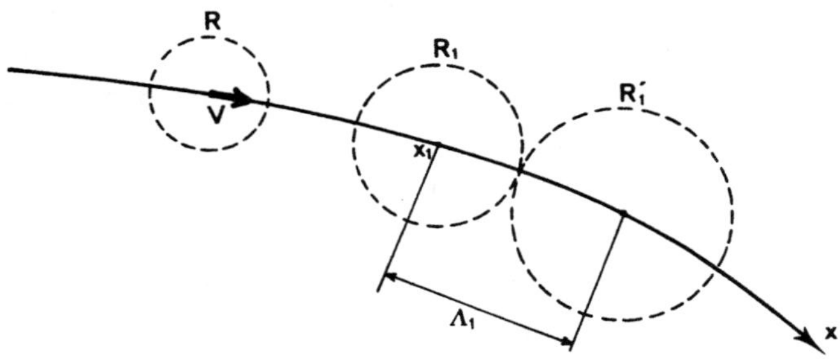

Fig.1

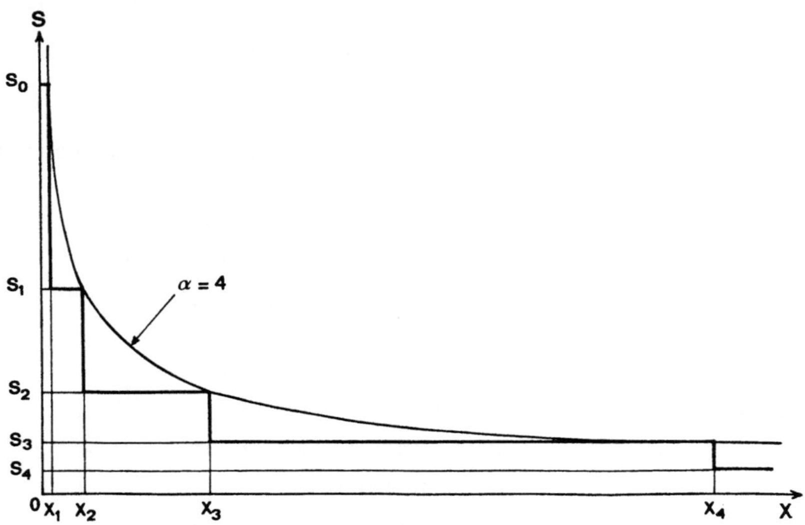

Fig. 2

$\Delta t = 5\ 10^{-6}$ s

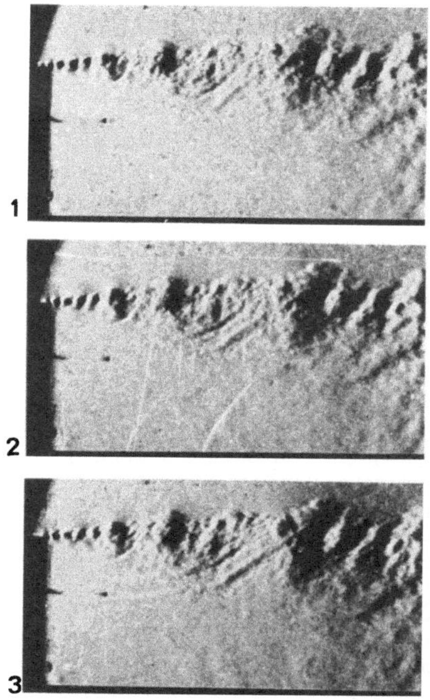

Fig. 3

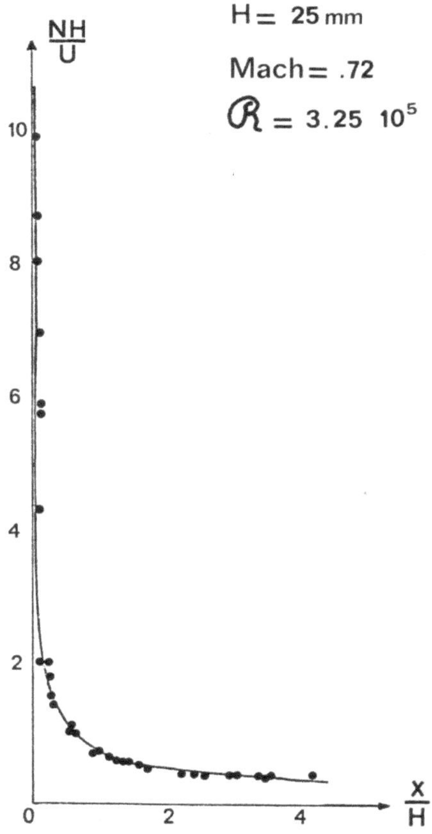

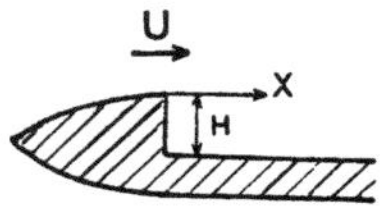

Fig. 4

Some Characteristics of Pulsating or Flapping Jets

G. BINDER - M. FAVRE-MARINET

Institut de Mécanique de Grenoble, B.P. 53X, 38041. GRENOBLE-CEDEX (France)

Summary

Axisymmetric jets subjected to large amplitude pulsations and plane jets forced to flap about a mean direction have been investigated experimentally. Both types of forcing increase the jet entrainment in the initial region but while the pulsating jet relaxes into the same axymptotic regime as the unforced flow the expansion of the plane flapping jet is affected far downstream. This behaviour may be accounted for by the induced pressure gradient. The periodic motion is at first amplified and then decays in both cases. The turbulent intensity in the initial region grows more rapidly than in the corresponding unforced jets and overshoots the asymptotic level. Some properties of the periodicmotions are analysed and compared with predictions of stability calculations.

Many practical applications of jets, as for instance diffusion flames or ejectors, depend crucially upon their mixing and entrainment rate. One naturally has attempted to enhance or in general to control these properties in various ways one of them being the forcing of perturbations. This aspect links this basically practical problem to the more fundamental research on coherent structures in turbulent shear flows. Such structures where they exist are the active agents of mixing. Once they are identified and their properties are known one may hope to be able to act upon them and theirby influence the mixing rate. Conversely, if some forced perturbations influence the mean flow it may be expected that they trigger such coherent structures.

The present investigation was primarily motivated by the improvement of thrust augmenting ejectors. In order to produce substantial modifications in the entrainment rather large perturbations were applied to the jets. In spite of this, some details of the flow ressemble the coherent structures resulting from the natural instabilities of the jets.

Two flow families were investigated, pulsating axisymmetric and flapping plane jets, the former being generated by varicose perturbations on the jet column and the latter ones by sinuous perturbations which are only relevant to the two-dimensional geometry..

EXPERIMENTAL SET-UP

The air jet (diameter $d_o = 2.5$ cm) was pulsed by a butterfly valve placed at the exit of the caisson, followed by a pipe about 20 diameters long and termined by the nozzle (contraction ratio : 2.6). The turbulence on jet axis in the exit plane was less than 1%. The valve was driven by a variable speed motor which imposed the frequency of pulsation. The amplitude was changed by using valve plates of different diameters. Independent adjustment of the amplitude and of the Strouhal number was not easy, because the jet discharge and the frequency also influence the amplitude. The main characteristics of the facility were(for more details see (1)): jet exit velocity $7 \leqslant U_o \leqslant 30$ m/s, $10^4 \leqslant \frac{u_o d_o}{\nu} \leqslant 4.10^4$

- frequency of pulsation $0 \leqslant f \leqslant 200$ hz
- Strouhal number $0 \leqslant S \leqslant 0.8$
- amplitude of pulsation $0 \leqslant \alpha_o \leqslant 40\%$

The flapping motion (2) of the plane jet was generated by latteral blowing of two narrow control jets placed on each side of the rounded nozzle edge (radius of curvature = d_o) as commonly realized in fluidics. Let $\bar{u}_o$ be the modulus and $\theta(t)$ the angle with mean direction of the jet exit velocity. Then the instantaneous velocity components are :

$$\langle u(t)\rangle = \bar{u}_o \cos \tilde{\theta}(t) \qquad \text{and} \qquad \tilde{\theta}(t) = \Theta F(\omega t)$$
$$\langle v(t)\rangle = \bar{u}_o \sin \tilde{\theta}(t)$$

where $F(\omega t)$ is a periodic function of time. The flapping angle Θ_1 is defined as the amplitude of the fundamental mode of ΘF. Different values of Θ_1 were obtained by changing the generating pressure of the control jets (ratio with respect to main jet generating pressure : C_p) and the flapping frequency was imposed by modulating their flow such that their phases were in opposition (for details see (2)). The main characteristics of the facility were :

- main jet slot width $d_o = 10$ mm aspect ratio : 150
- mean exit velocity $\bar{u}_o = 16$ m/s
- flapping angle $0 \leqslant \Theta_1 \leqslant 8°$ frequency $0 \leqslant f \leqslant 100$ hz
- Strouhal number $0 \leqslant fd_o/u_o \leqslant 0.062$

two-dimensionality : standard deviation of $\bar{u}_o$ less than 1% , of Θ_1 less than 0.2° , of phase of Θ_1 less than 2°.

There are end plates at $x = 0$ in both facilities and also at $z = \pm 75\, d_o$ in the plane jet.

Measurements were performed with single or X-wire hot-wire anemometers. Data acquisition was performed by phase-locked ensemble averages on a multi-channel analyser and processed on a computer.

The triple decomposition is applied as usual now in periodic turbulent flows. Thus for any quantity q : $q = \bar{q} + \tilde{q} + q'$ where $\bar{q}$ is the mean value, $\tilde{q}$ the periodic fluctuation and q' the turbulent fluctuation. q and q' are separated by measuring $\langle q \rangle$ and $\langle q^2 \rangle$.

PULSATING JET RESULTS

i) The asymptotic regime of the pulsating jet is the same as that of the steady jet but it is reached more rapidly. The mean velocity profiles are already similar and identical to the profile of (3) beyond $7d_o$. The mean velocity decay and the growth of the half-width are parallel to those of the steady jet measured in the same facility beyond $15d_o$ (fig. 1, 2)

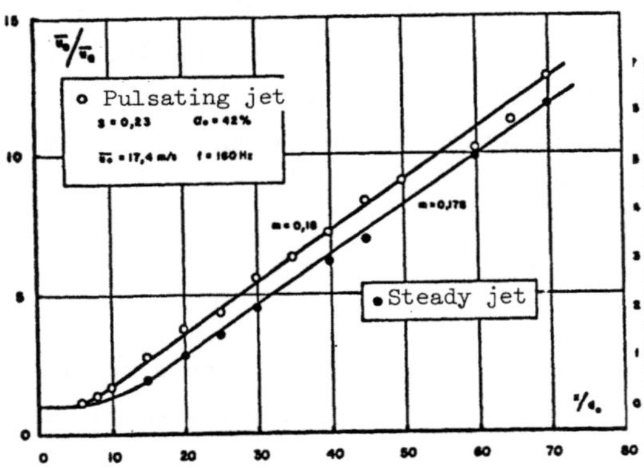

Fig. 1 - Velocity decay of pulsating unsteady jet

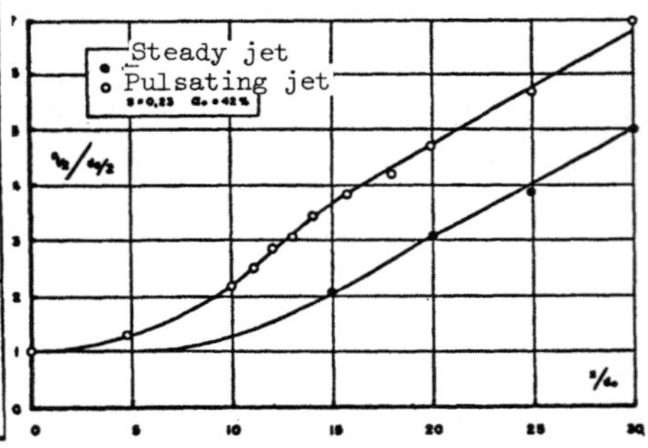

Fig. 2 - Jet spread

This is in agreement with the observations in self-excited jets by means of the whistler nozzle (4) except that in the present case of large amplitude pulsations the asymptotic regime of the mean flow is reached somewhat later ($15d_o$ instead of $10d_o$). The forcing produces finally an upstream shift of the virtual origine.

The same conclusion applies to the Reynolds stresses (Fig. 3a)

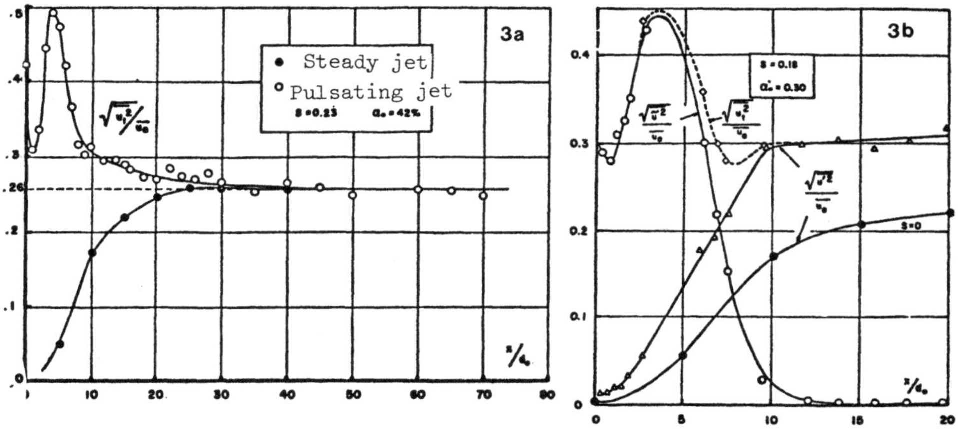

Fig. 3 - Evolution of periodic fluctuation and turbulent intensity

The shear stress $\overline{u'v'}/u_a^2$ follows the equilibrium profile already at $20d_o$ when the jet is pulsed but lies clearly underneath it in the steady jet. In (3) this profile was only obtained at $60d_o$.

ii) The initial development of the mean flow is affected by the pulsations which produce a faster velocity decay and rate of spread (fig. 1,2). The potential core is shortened to about $1d_o$. At 15 diameters downstream the pulsed jet may be twice as large as the steady jet. Measurements in the Strouhal number range 0-0.4 and with initial amplitudes of 20 and 30% show that the effect increases with S and saturates for $S \gtrsim 0.3$ and increases with amplitude.

The intensity of the periodic fluctuation $\tilde{u}$ on the axis (fig. 3b) is at first amplified despite the high initial level and then decays very rapidly. Autocorrelations show that beyond $10d_o$ the fluctuating signal is purely random ; there are also no more periodic nor quasi-periodic structures in the flow. The maximum amplification depends very strongly on the initial forcing level (1) showing a tendency towards saturation.

The initial growth of the turbulent intensity is faster than in the unforced jet (fig. 3b) and overshoots the asymptotic level. This strongly suggests an energy transfer from the periodic to the turbulent motion. The decrease in the total longitudinal fluctuation intensity $\overline{u_t^2} = \overline{\tilde{u}^2} + \overline{u'^2}$ between 3 and $8d_o$ is not only due to a transfer to other velocity components but also to a transfer to the mean motion by the work of the oscillating pressure $-\frac{\partial}{\partial x_j}(\overline{\tilde{u}_j \tilde{p}})$. These rather complex interactions are in qualitative agreement with the theoretical results of LIU (5) who have performed stability calculations in taking into account the "fine-grained turbulence" for which they use a BRADSHAW-type closure.

iii) The periodic structure of the flow field is illustrated by figure 4 which shows instantaneous profiles of the velocity on the axis at successive times or phase angles ϕ_f .

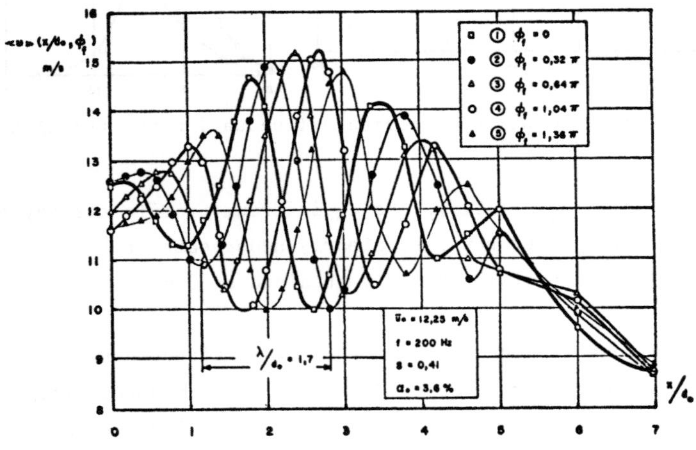

Fig. 4 - Instantaneous axial $<u>$ - profiles

This clearly shows the propagation, amplification and decay of the forced structures. Right after the nozzle exit plane the phase speed is larger than $\overline{u}_o$ as predicted by spacial stability theory and a little further it is about $0.7\,\overline{u}_o$ for various values of S and α_o . The wave-length is in good agreement with the calculations which takes into account the spreading of the jet (6).

FLAPPING JET RESULTS

i) The asymptotic characteristics of the mean flow as measured between 50 and 100 d_o are NOT the same for the flapping as for the steady jet in striking contrast to the pulsating axisymmetric jet. This is clearly illustrated by the axial velocity decay of Fig. 5 which also shows the influence of the flapping angle.

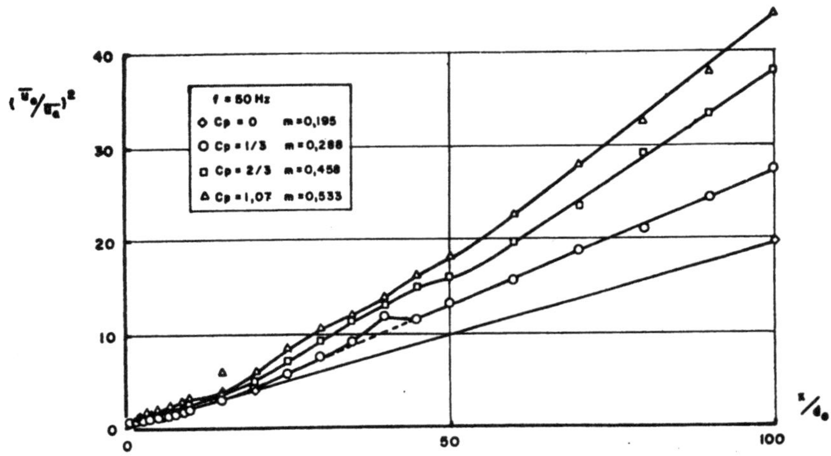

Fig. 5 - Axial velocity decay

The velocity profiles are closely similar beyond 50 d_o . The final rate of spread (fig. 6) clearly depends upon the frequency and amplitude of the angular perturbation ($y_{1/2}$ = half width) ; it can be almost twice as large as that of the unforced jet. Forced flapping offers also a good means of control of the jet spread. It is quite remarkable that there is little influence at low frequency (10hz, S = 0,006) because one is at first lead to think that these jets are quasi-steady and should, therefore, have a rate of spread roughly equal to the rate of the steady jet plus Θ_1 . The results show that this is obviously not so. A persistent influence on the spread of plane jets although not of the same magnitude, has also been found by HUSSAIN (7) who obtained different rates depending upon the state of the boundary layer at the nozzle exit. The changes produced by the flapping are so large that they drown the effect of the exit boundary layer. Moreover, the comparison is made here with the steady jet produced by the same facility.

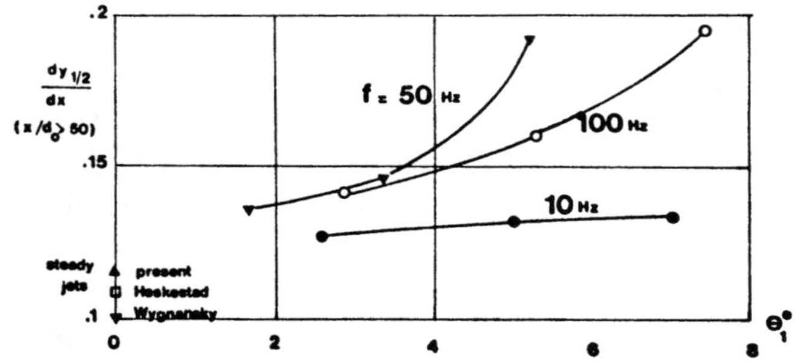

Fig. 6 - Final rate of spread of flapping jet

The longitudinal turbulent intensity (fig. 7) is, however, little different from that in the unforced case and so is the shear stress.

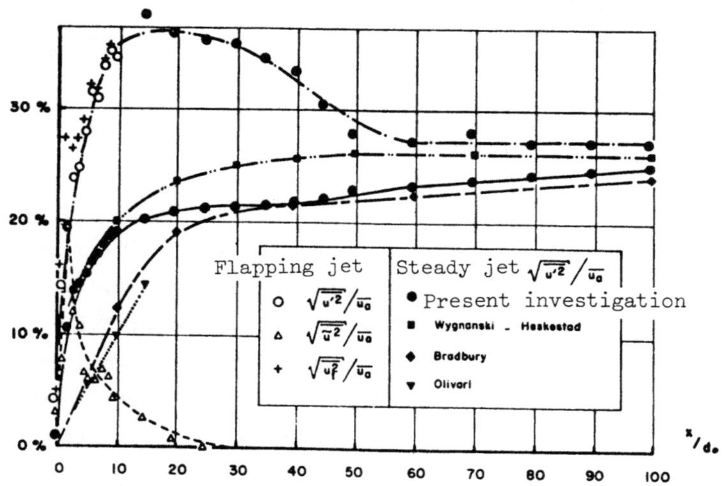

Fig. 7 - Evolution of periodic fluctuation and turbulent intensity in flapping unsteady jets.

Now in the absence of a pressure gradient $dy_{1/2}/dx \propto \overline{u'v'}/\overline{u}_a^2$. The faster expansion of the flapping jets must therefore be produced by the pressure gradient induced by the entrainment of the outer flow. This is qualitatively confirmed by pressure measurements on the jet axis and by the decrease of the momentum flux. Such an explanation has also been proposed by KOTSOVINOS (8) to account for the different rates of spread of steady jets observed by various investigatiors.

ii) The initial development of the flapping jets is completely dominated by the forced perturbation. The mean velocity develops at first a profile with a double hump which is due to the oscillation of the instantaneous velocity profile. The dip on the axis is then progressively filled by the developping turbulence.

The periodic fluctuation $\tilde{u}$ is at first rapidly amplified then decays and has vanished toward 20 d_o (fig. 7), a behaviour very similar to that of the pulsating jet. The turbulent intensity $\sqrt{\overline{u'^2}}$ increases also faster than in unforced jet but the overshoot is much larger than in the pulsating case.

iii) The periodic structure has been investigated (9) in the jet forced at S = 0,062 (100hz) and Θ_{10} = 7.4° . The amplification and decay of the average flapping angle based on the ratio of the lateral to the longitudinal momentum flux is shown on fig. 8.

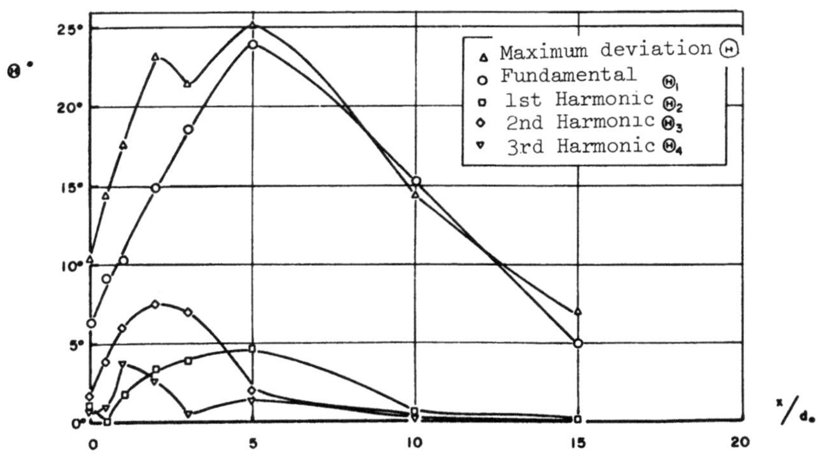

Fig. 8 - Amplification and decay of the flapping angle

It is observed that this angle reaches 25 degrees also a very large value. The higher modes evolve faster than the fundamental ; in fact, plotted non dimensionally in terms of the initial amplitude and the wave-length the evolutions are very similar. It also shows that the whole process takes place over only two wave-lengths. The amplification rate is much larger than the one predicted by linear stability theory.

The phase speeds, on the contrary, are in good agreement with the theory. For the sinuous mode this speed is smaller than $\bar{u}_o/2$ and tends to zero with the wave number. The wave-length does therefore not increase indefinitely as the Strouhal number goes to zero. This explains why the flapping jet at low frequency does not become quasi-steady.

Experimentally determined streaklines originating from the center of the jet at $x = 0$ are shown on fig. 9.

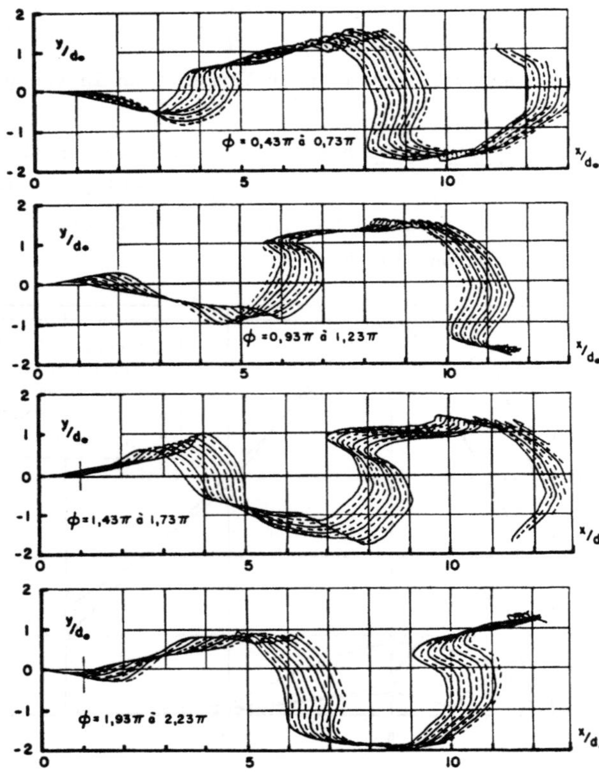

Fig. 9 - Streakline patterns

Besides, the amplification of the perturbation, they show the deep crevasses which are formed and which engulf the outer fluid. This explains the entrainment efficiency of the flapping jet in the initial region.

ACKNOWLEDGEMENT

Financial support of the Direction des Recherches, Etudes et Techniques from the Ministère des Armées is greatfully acknowledged.

REFERENCES

(1) FAVRE-MARINET M., BINDER G., Structure des jets pulsants, J. de Mécanique vol. 18, n° 2, 355-394, 1979.

(2) FAVRE-MARINET M, BINDER G., TE VENG HAC, Generation of Oscillating Jets, J. Fluids Eng., to appear.

(3) WYGNANSKY I, FIEDLER H., Some measurements in the self-preserving jet, J. Fluids Mech., vol. 38, 3, 577-612, 1969.

(4) HASAN M.A.Z. , HUSSAIN, The self-excited axisymmetric jet, submitted to the J. Fluid Mech.,

(5) LIU J.T.C., MANKBADI R., A study of the interactions between Large-scale coherent structures and fine-grained turbulence in a round jet, Phil. Tran. Roy. Soc., London, Serie A, to appear.

(6) CRIGHTON D.G., GASTER M., Stability of slowly diverging jet flow, J. Fluid Mech., vol. 77, 2, 397-413, 1976.

(7) HUSSAIN A.K.M.F., RAY CLARK A., Upstream influence on the near field of turbulent jets, Physics Fluids, vol. 20, 9, 1416 1426, 1977.

(8) KOTSOVINOS N.E., A note on the conservation of axial momentum of a turbulent jet, J. Fluid Mech., vol. 87, 1, 55-63, 1978.

(9) SOUTIF M., Diffusion et structure périodique des jets battants, Thèse de Doctorat de Spécialité, Université de Grenoble, 1977.

Diffusion of Heat as a Passive Contaminant in a Slightly Pulsating jet

H. LEMONNIER - M. FAVRE-MARINET - G. BINDER

Institut de Mécanique de Grenoble, B.P. 53X, 38041. GRENOBLE-CEDEX

Summary

Results on the diffusion of a passive contaminant (heat) by a round jet subjected to small amplitude periodic forcing at the frequency of the preferred mode are presented. The instantaneous isotherm pattern show the growth of coherent structures in the mixing layer from the nozzle to a distance $3d_o$ downstream. The temperature in the core of these structures is approximately uniform whereas on both sides of the mixing layer a sharp temperature gradient gives rise to intense θ - fluctuations. The structures begin to lose their coherence beyond $3d_o$ and are partially blurred by the turbulence which is developing. It is nevertheless observed that the structures convected downstream have a strong influence on the temperature on the jet axis. The isotherms are slowly stretched and then rapidly compressed during a cycle. Hot fluid is ejected outwards from the axis on the leading edge of the structures and cold fluid is trapped inwards on the trailing edge.

The presence of coherent structures in the first diameters of axisymmetric jets has been demonstrated by many experiments. These structures exist whether the initial boundary layer is laminar or turbulent even if the Reynolds number is as high as 4×10^5 [1] [2] [3] [4] [5] . When the exit boundary layer is laminar, axisymmetric vortex rings appear in the shear layer, interact by coalescing and give rise to three-dimensional structures at the end of the potential core [2] . Pairing of large scale structures may be an important feature of the jet development even when the exit boundary layer is turbulent [4] . The coherent structures play an important role in the mixing between the jet and the ambient fluid, giving rise to large fluctuations of the Reynolds stresses for example. They have been studied either by conditional sampling or by phase-averaging when the jet is forced at a given frequency by pure-tone excitation.

In order to study the mixing process in detail, a jet is slightly heated (12° above ambient) and coherent structures are triggered by forcing small amplitude ($<2\%$) pulsations on the flow with a loudspeaker in the settling chamber at the frequency f of natural oscillation of the jet at $4d_o$ (d_o = nozzle diameter = 80 mm, U = 12m/s mean exit jet velocity). It is found that $f = 67$ Hz and the corresponding Strouhal number is $S = fd_o/U = .45$. For these conditions the value of the Reynolds number Re_{do} is 6×10^4. A trip ring is placed in the nozzle and the mixing layer of the jet is initially turbulent.

Experimental procedure

Temperatures in the jet are measured with a cold-wire (diameter : 1μ) anemometer. The output is recorded on magnetic tape and is then ensemble-averaged on a computer (Norsk Data 10). The ensemble-averaging is phase-locked on the signal produced by the loudspeaker itself and gives the temperature $\langle\theta\rangle (x,r,\phi)$ (x : distance from the nozzle, r : from the axis ; ϕ : phase $0 \leqslant \phi \leqslant 2\pi$) associated with the large-scale motion. (θ is normalized by the temperature difference between the jet at the nozzle exit and the ambient fluid ; $0 \leqslant \theta \leqslant 1$). By ensemble-averaging the square of the temperature, one obtains the phase-averaged r.m.s. $\langle\theta'^2\rangle = \langle\theta^2\rangle - \langle\theta\rangle^2$ Typically $\langle\theta\rangle$ and $\langle\theta'^2\rangle$ are determined at 100 points equally spaced over the cycle and the phase-average is performed over 800 cycles ; the results are stored on the computer.

The measurements were performed in a horizontal half-plane between the distances $25d_o$ and $6.5d_o$ from the nozzle in the region where the jet develops and where periodic and turbulent temperature fluctuations are simultaneously present.

The points of measurements are located on a grid with a mesh $\Delta x/d_o = .25$ $\Delta r/d_o = .0625$ for $0.5 \leqslant x/d_o \leqslant 2.75$. For $3 \leqslant x/d_o \leqslant 6.5$, the mesh-size is doubled.

Results

1) Development of the shear layer

In the first 3 diameters the low-amplitude pulsation forced on the jet produce temperature perturbations which grow in the shear layer whereas the potential core remains perfectly isothermal. The r.m.s. of the periodic temperature fluctuations $\sigma_{\tilde{\theta}}$ is negligible in the first half diameter then increases very rapidly from $.75d_o$ on.

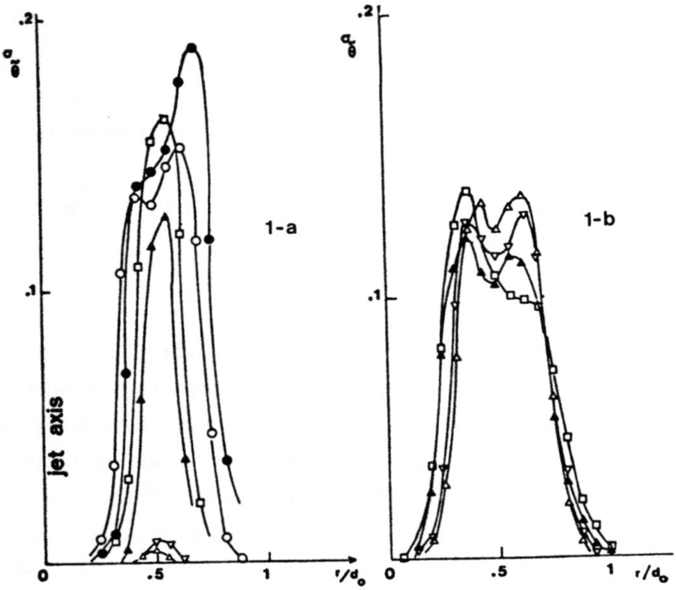

Fig. 1. Profiles of periodic temperature fluctuations r.m.s.
1.a x/d_o = △ .25 ▽ .5 ▲ .75 □ 1 ● 1.25 ○ 1.5
1.b △ 1.75 ▽ 2 ▲ 2.25 □ 2.5

From there on $\sigma_{\tilde{\theta}}$ is high (.15) and varies little with x. The maximum ($\sigma_{\tilde{\theta}}$ = .19) is reached at $x/d_o = 1.25$, $r/d_o = .68$. The $\sigma_{\tilde{\theta}}$-profiles spread progressively on both sides of the line $r/d_o = 0.5$. The width L associated with the periodic temperature fluctuations may be defined with the points where $\sigma_{\tilde{\theta}}$ is one half of the maximum for a given profile

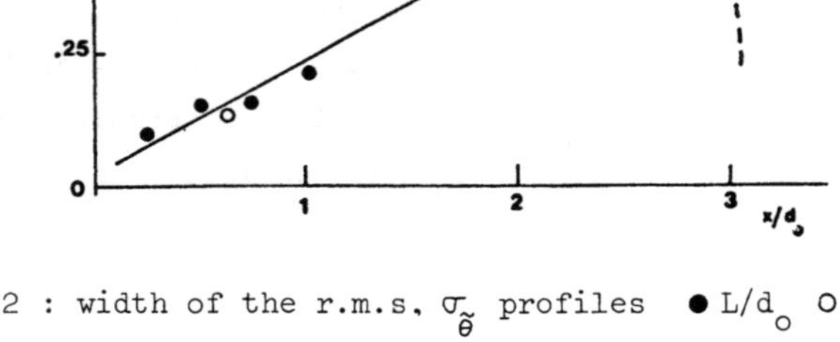

fig. 2 : width of the r.m.s. $\sigma_{\tilde{\theta}}$ profiles ● L/d_o ○ mixing layer width

The growth of L is approximately linear and close to the growth of the mixing layer defined by the distance of the isotherms $\bar{\theta}= .1$ and $\bar{\theta}= .9$. Beyond $2.75d_o$, L decreases because the jet-core itself is affected by the perturbation and the mixing layers interact.

The isotherm pattern of $<\theta>$ at a given phase angle (fig. 3) clearly shows the formation of structures whose transverse dimension is equal to the width of the mixing layer and which propagate downstream. This result resembles very closely to certain visualizations of Hussain and Clark [Ref. 5, p. 275] . On fig. 3, the lateral scale is expanded : ratio 2.

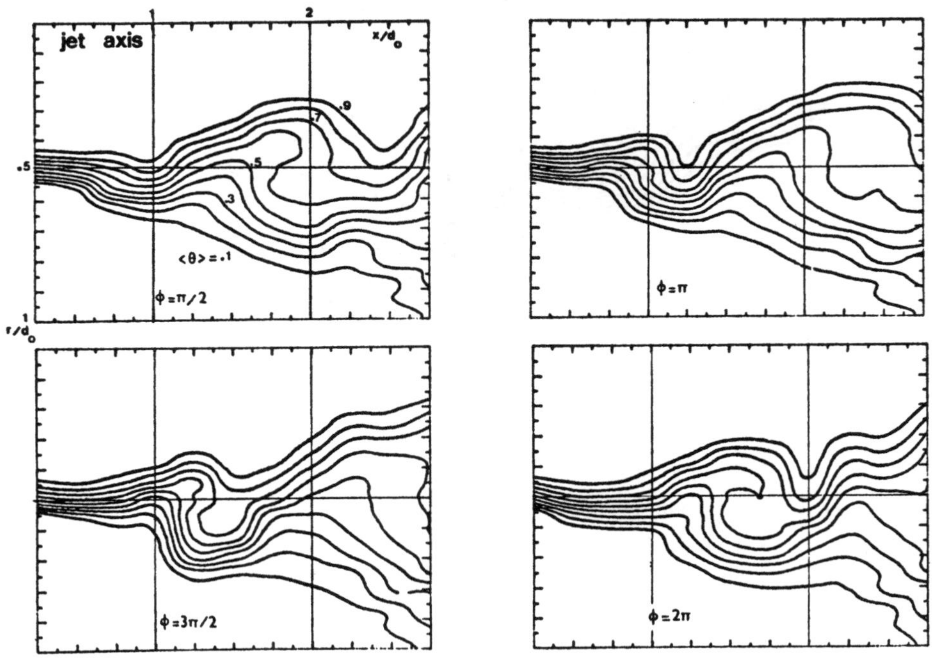

Fig.3 - Pattern of isotherms at a given phase angle

One observes at first a slight displacement of the isotherms to the outside for $x/d_o = 1$ $\phi = \pi/2$. A quarter of a period later ($\phi = \pi$). This varicose has been strongly amplified while it propagates downstream whereas an undulation in the opposite direction appears upstream. At $\Phi = 3\pi/2$, this bulge has grown, the distance between the isotherms $\langle\theta\rangle = .7$ and $\langle\theta\rangle = .8$ has increased resulting in a pocket of isothermal fluid. This pocket becomes wider as it travels downstream and its temperature stabilizes near .6. The analysis of the turbulent temperature fluctuations $\langle\sigma_{\theta'}\rangle$ shows that $\langle\sigma_{\theta'}\rangle$ is minimum in the pocket and maximum around it.

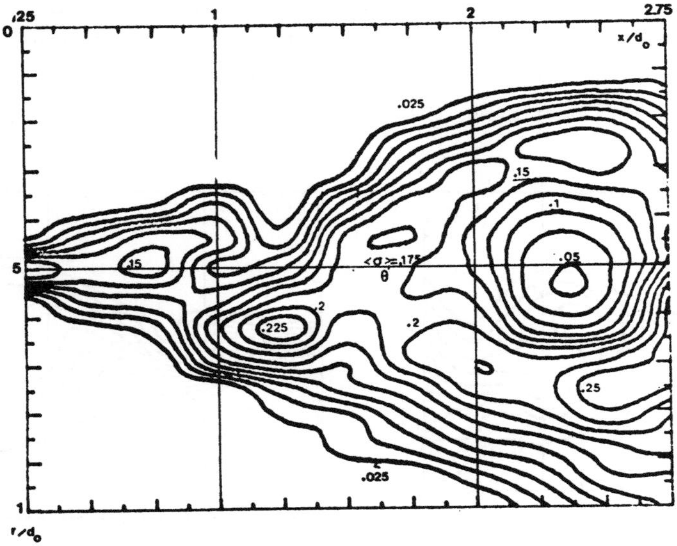

Fig. 4. Pattern of r.m.s. $\langle\sigma_{\theta'}\rangle$ for $\Phi = \pi$

$\langle\sigma_{\theta'}\rangle = \sqrt{\langle\theta'^2\rangle}$

A similar result is shown by the numerical experiment of the stratified shear layers by Patnaik, Sherman and Corcos [6]. The pattern of isopycnics obtained depends only slightly upon the Reynolds number and the Richardson number if this latter one is sufficiently small. The final state of the calculated isopycnics resembles to the pattern of the temperature at a distance $x/d_o = 1.5$. However, turbulent diffusion present in the physical experiments but not in the numerical experiments blurrs the isothermal-lines in the core of the structure and widens the braids with respect to the calculated pattern. Moreover the spacial resolution is not the same in the two cases, a finer resolution being much more easily achieved on the computer than in the laboratory.

The comparison can only be qualitative because in the jet the influence of the second length scale d_o on the evolution of the perturbation is rapidly felt and moreover, it is a spacial evolution in the experimental case

instead of a temporal evolution in the numerical experiment. The experimental wavelength varies little with x , yielding a coefficient:
$\alpha = 2\pi\delta/\lambda = 1.3$ which is much larger than the value (.43) used in the numerical calculation which corresponds to the most amplified mode.

2) Transition region of the jet

At 3 diameters from the nozzle, the mixing layers merge on the jet axis, $\sigma_{\tilde{\theta}}$ decreases on the edge of the jet and begins to grow on the axis.

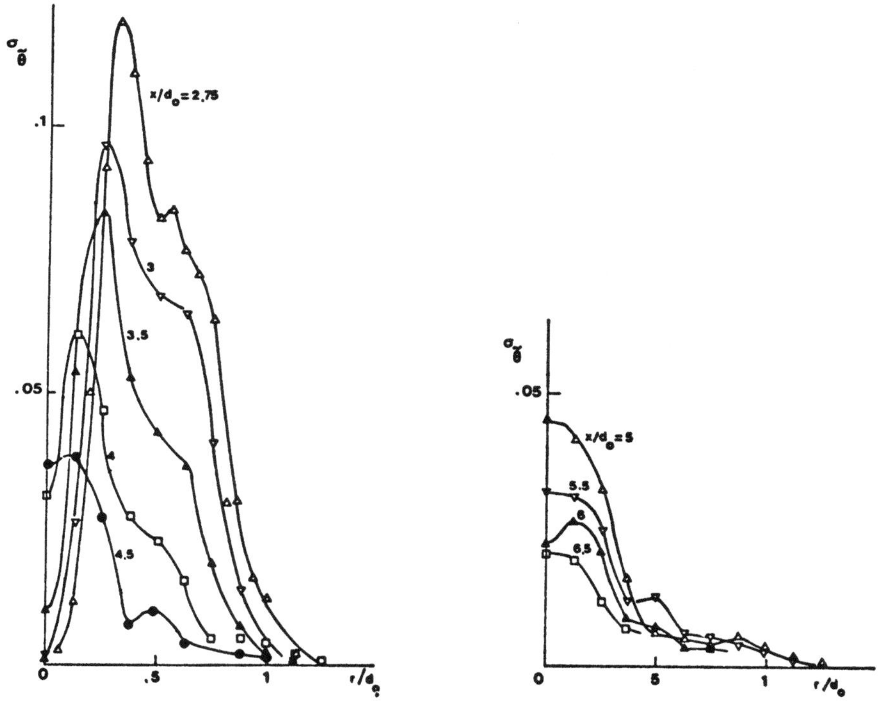

fig. 5. Profiles of periodic temperature fluctuations r.m.s.
5.a. x/d_o = △ 2.75 ▽ 3 ▲ 3.5 □ 4 ● 4.5
5.b. △ 5 ▽ 5.5 ▲ 6 □ 6.5

Beyond $4.5d_o$ the periodic temperature oscillations are practically concentrated in the jet core ; $\sigma_{\tilde{\theta}}$ reaches a maximum towards $5d_o$ ($\sigma_{\tilde{\theta}} \neq .05$). Beyond $7d_o$ the periodic temperature fluctuations are hardly detectable. Consequently, between 3 and $7d_o$ the coherent structures on the edge of the jet are less sharply defined on the isotherm pattern than upstream. On the contrary the periodic temperature perturbation shows up as large oscillations of the isotherm in the core of the jet.

Ten successive positions of the same isotherm ($\langle\Theta\rangle = 0.9$) at equally spaced times covering the whole period are shown on fig. 6. It is seen that the convected downstream coherent structures have a very strong influence on the shape and the position of this isotherm. It shows first a very large blob (instants 9-10) that propagates downstream, then a sharp lengthening (instants 5-6) and a very rapid shortening (instant 7). The same motion occurs for the isotherms $\langle\Theta\rangle = 0.8$ and 0.7 but at times shifted by T/3 between 2 isotherms. (T being the period, T = 1/f).

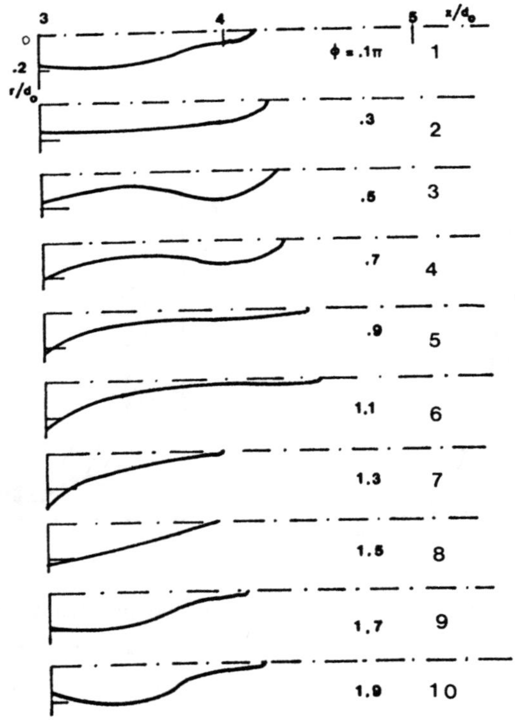

Fig. 6 - Successive positions of the isotherm $\langle\Theta\rangle = 0.9$

Fig. 7 represents the position of the intersection with the jet-axis of the isotherms $\langle\Theta\rangle = .7$, $.8$, $.9$ during one cycle. The forward translation velocity of the isotherm $a/U = 0.3$ is much smaller than the phase-speed of the velocity waves $c/U = .73$

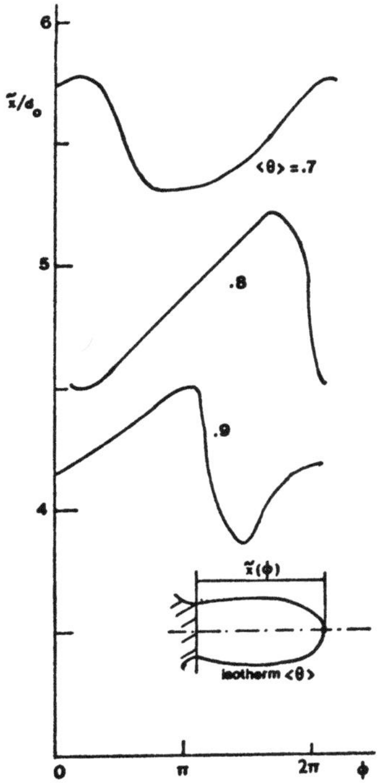

Fig. 7. Position of the isotherms on the jet axis vs the phase angle

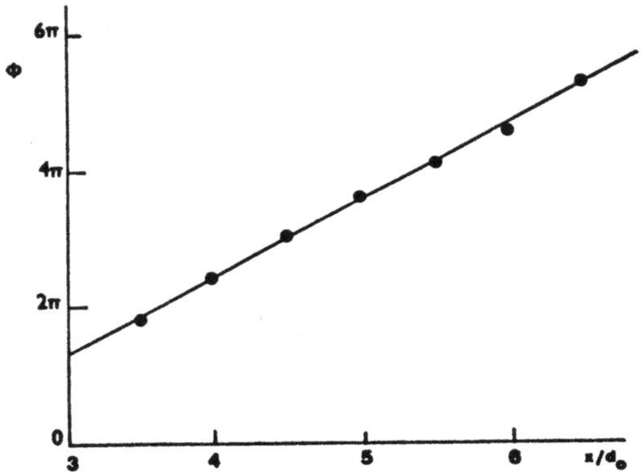

Fig. 8. Phase shift along the x-axis

The isotherm pattern near the axis is not purely transported by the coherent structures propagating downstream but is shifted with respect to these by the effect of turbulent diffusion. It should be stressed that $\langle\Theta\rangle$ contains the mean value $\bar{\Theta}$ and the periodic oscillation. The phase of the purely oscillating perturbation on the jet axis varies linearly with x (fig. 8) and yields a phase-speed which is nearly the same as that of the velocity oscillation ($c_\theta/U = .78$ $c_u/U = .73$) which corresponds to a wavelength $\lambda/d_o = 1.77$.

The trajectories in a frame moving with the speed $C_u/U = .73$ has been calculated with the measured values of the velocities $\langle u \rangle$ $\langle v \rangle$ for $x/d_o = 4$ and with the Taylor hypothesis based on this wave speed C_u

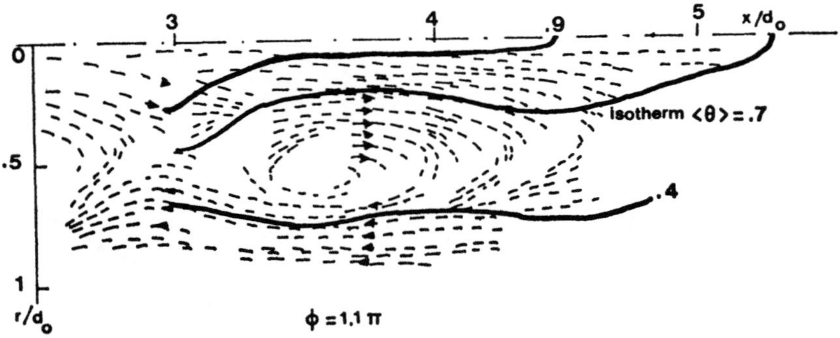

Fig. 9. Pattern of trajectories and isotherms in moving axes

The well known cat's-eye pattern appear in this moving frame. It shows ejection of hot fluid towards the outside of the jet at the leading front and suction of cold fluid towards the axis at the rear .

Contrary to what happens on the axis the isotherms $\langle\theta\rangle$ .7 , .4 are not shifted with respect to this structure. The stretching and rapid contraction of the isotherm .9 near the axis corresponds to the passage of the vortex core at $x/d_o = 3.5$.

References

1. Crow, S. , Champagne, F.M. : Orderly structure in jet turbulence, J. Fluid Mech., 48, 547-591, 1971.

2. Yule, A.J. : Large-scale structure in the mixing layer of a round jet, J. Fluid Mech., 89, 413-432, 1978.

3. Favre-Marinet, M., BINDER, G. : Structure des jets pulsants, J. de Mécanique, vol. 18, n° 2, 355-394, 1979.

4. Zaman K.B.M.Q., Hussain A.K.M.F. : Vortex pairing in a circular jet under controlled excitation, J. Fluid Mech., vol. 101, part 3, 449-544, 1980

5. Hussain A.K.M.F. and Clark A.R. : On the coherent structure of the axi-symmetric mixing layer : a flow-visualisation study, J. Fluid Mech., vol. 104, 263-294, 1981.

6. Patnaik P.C., Sherman F.S. and Corcos G.M. : A numerical simulation of Kelvin-Helmholtz waves of finite amplitude, J. Fluid Mech., vol. 73, part 2, 215-240, 1976.

The Preferred-Mode Coherent Structure in the Near Field of an Axisymmetric Jet With and Without Excitation

A. K. M. F. Hussain and K. B. M. Q. Zaman

Department of Mechanical Engineering
University of Houston, Texas 77004

Abstract

The 'preferred-mode' coherent structure in the near field of an axisymmetric jet is educed for different Reynolds numbers and for the two limiting asymptotic states of the exit boundary layer: laminar and fully-developed turbulent. Distributions of phase-average longitudinal and lateral velocities, coherent vorticity, coherent and incoherent turbulence Reynolds stresses, and coherent strain rate and production are obtained via phase-locked hot-wire measurements over the spatial extent of the coherent structure, for controlled excitation at the jet Strouhal number of 0.3. These properties are found to be independent of the initial condition but show weak dependence on the Reynolds number up to about 10^5. The naturally-occurring coherent structure in the unexcited jet, educed via an optimized conditional sampling scheme, shows no clear dependence on the jet Reynolds number up to 8×10^5.

I. Introduction

Recent years have seen vigorous studies of the large-scale coherent structures in turbulent shear flows. These studies have been spurred by the expectation that the structures dominate transport processes and noise production mechanisms in turbulent flows. Large-scale coherent structures are being studied by us with three goals in mind: (i) understand the flow physics, (ii) obtain detailed structure properties so that a viable theory can be developed by directly incorporating these structures, and (iii) turbulence management: i.e., manipulate and control the structures and thus presumably modify/control turbulent transport and noise production.

For a definition of the large-scale coherent structure and discussion of analytical and experimental considerations involved in the investigations of these structures, see Ref. [1], which has also briefly summarized a few of our results.

In the case of the near field of the axisymmetric jet, the flow is unique in that there is a single preferred-mode. The structure passage frequency, as measured by the $\tilde{u}$-spectrum in the potential core of the axisymmetric jet, falls in a narrow range near the end of the potential core corresponding to the jet Strouhal number $St_D (\equiv f_p D/U_e)$ of 0.3-0.5. Here, f_p is the excitation frequency, D is the jet diameter and U_e is the exit velocity. No prior study has attempted to educe the coherent structure associated with this preferred-mode, nor determine the dependence of the structure properties on the jet Reynolds number and the initial condition. The attractiveness of this study lies in the prospect that the physics of the jet near field could be essentially represented by the preferred-mode coherent structure. If we could show that the details of this structure were insensitive to the Reynolds number and the initial condition, then the jet near field, say between 1 to 3 diameters, could be viewed as essentially solved.

The large-scale coherent structures in a turbulent shear flow have large dispersions in their characteristic measures like shape, size, orientation, strength and convection velocity. Eduction of the structures is complicated by these dispersions and by the random formation and breakdown of these structures.

In an attempt to reduce the smearing due to the otherwise unavoidable dispersion in the characteristic measures of the structure, the structure has been enhanced via controlled axisymmetric excitation and the periodic occurrence of the structure in this condition has been taken advantage of to educe the structure through simple, periodic phase-averaging technique. We have contended that the excitation does not induce an artificial structure but paces the formation of the natural structure at regular intervals, thus facilitating the eduction of the structure with good accuracy and in such details as are unlikely to be possible in the absence of the excitation [2]. The excitation is used to merely prevent any other disturbances that are present naturally [1] from triggering the formation of these structures. Efforts to educe natural structures in the absence of the excitation are also briefly discussed in the concluding section.

The experiments have been carried out in two axisymmetric jet flow facilities. The lower Reynolds number studies have been performed in a 7.62cm jet described by Zaman & Hussain [3] and the larger Re_D studies in a 27cm jet described by Husain & Hussain [4]. The controlled excitation has been induced with the help of cavity resonance induced by a loudspeaker attached to the upstream settling chamber of the two-chamber facility. Unless otherwise stated, the excitation amplitude u_e'/U_e was 2% of the exit velocity. For details of experimental techniques, see [2,5]. The Strouhal number of excitation St_D was 0.3 in all cases. In order to reduce the effect of jitter, the eduction has been triggered by a detector probe located on the jet centerline, near the measurement location [2].

II. Results

Both visualization and the hot-wire measurements show that the preferred-mode excitation produces no pairing for either laminar or turbulent initial condition as the structure is initiated at the 'terminal Strouhal number.' On a time-average basis, the excitation produces a relative widening of the axisymmetric mixing layer and a shortening of the potential core. These effects are significant but considerably lesser than the effect of excitation at $St_D \approx 0.85$ which induces stable pairing, in which case rapid changes in the width of the shear layer as well as a large region of 'negative production' are depicted even in time-average measurements [3,5].

For the 7.6cm (tripped) jet at $Re_D = 55,000$ and for excitation at $St_D = 0.3$, contours of constant vorticity are shown in Fig. 1. For these data, the reference probe providing the triggering signal was located at $x/D = 4.5$ on the centerline. It is clear that even though the structure is induced by excitation at the preferred mode, the contour periodicity is clearly lost by $x/D \simeq 8$. Both flow visualization and phase-locked azimuthal correlation data [3,4] show that the structures divide into azimuthally lobed substructures starting at $x/D \simeq 3$. Thus, it is most likely that structure breakdown is complete by $x/D \simeq 8$.

It should be clear from Fig. 1 that the phase-average longitudinal velocity on the centerline will be maximum at the points where the vorticity contours have inward bulges. This has been checked by comparing the contours of u_p with the vorticity contours. On the basis of the u_p signal or the vorticity contours shown in Fig. 1, the structure spacing measured on the centerline is about 2.4D. This value agrees with that measured by Crow & Champagne, but is noticeably different from the value of 1.75D obtained in the shear layer; see the locations of the vortex centers marked by + in Fig. 1. The center-to-center structure spacing value of 1.75D has been checked by phase-locked flow visualization also. The wavelength on the jet centerline, on the other hand, has been directly calculated from the phase variation $\phi(x)$ of the $\tilde{u}(t)$ signal, measured by a lock-in-amplifier (Fig. 2). The linear variation of phase $\phi(x)$ gives a wavelength of about 2.25D. From the vorticity contours, one can see that the wavelength should gradually decrease from the jet centerline to the middle of the shear layer, as denoted by the two dotted lines. Since excitation at the preferred mode produces no pairing and the frequency remains unchanged everywhere, the wavelength variation corresponds to a convection velocity of $0.72U_e$ on the jet centerline and $0.53U_e$ in the middle of the shear layer. This variation in the convection velocity agrees with that measured for the most dominant structure of an unexcited axisymmetric mixing layer via double-Fourier transformation on the space-time correlation of $\tilde{u}(t)$ [6].

IIa. Effects of Reynolds Number and Initial Condition

The possible dependence of the preferred-mode coherent structure on the Reynolds number and the initial condition has been investigated. In the following, contours of the coherent structure properties will be discussed for the jet Reynolds numbers of 25000, 55000, and 110000, all excited at the exit perturbation amplitude of 2%. The exit boundary layer is fully-developed turbulent as denoted by the profiles of the mean velocity and longitudinal fluctuation intensity and u-spectra. These data were also obtained with a laminar exit boundary layer at the Reynolds number of 55000. Thus, data are presented in (a) and (b) for two initial conditions (laminar and fully turbulent) at the fixed Reynolds number of 55000. In each figure, (a), (c), (d) correspond to the data for the exit Reynolds numbers of 55000, 25000, and 110000, respectively, all for the case of the same initial condition (i.e., fully turbulent). The corresponding excitation frequencies f_p are 44Hz, 178Hz, and 87Hz, respectively.

Fig. 3 shows the contours of phase-average vorticity in the radial plane; the vorticity has been nondimensionalized by f_p. Note that in spite of the large differences among the jet exit velocities (two times), diameters (three times) and the frequencies (four times), the nondimensional peak vorticity is the same for the four cases. While there is no variation with the initial condition, the structure shape is elongated at a lower Reynolds number and becomes more rounded at a higher Reynolds number. Note that the vertical scales are magnified with respect to the streamwise coordinates. Contours of the coherent Reynolds stress $\langle u_p v_p \rangle$ are shown in Fig. 4. The coherent Reynolds stress contours are essentially identical for both initial conditions. However, there seems to be a noticeable and systematic dependence on the Reynolds number Re_D. With increasing Re_D, both positive and negative peaks increase; consequently, the difference between the two peaks also increases with Re_D. The extent of variation between the alternating positive and negative regions of $\langle u_p v_p \rangle$ would not be at all apparent from the time-average Reynolds stress. Controlled excitation at certain frequencies can change the relative sizes and strengths of these two regions and thus induce time-average Reynolds stress of the same or opposite sign as that of the mean strain rate. Controlled excitation has been shown to produce a large region of 'negative production' in mixing layers [3].

The progressively increasing peak values of the coherent Reynolds stress with Re_D raises the question that the time-average Reynolds stress of the jets would also increase with Re_D. The corresponding contours of constant values of $\overline{uv}/U_e^2$ are shown in Fig. 5. There is a suggestion that the time-average $\overline{uv}$ also slightly increases with Re_D. Note the contour dependence on Re_D.

In order to further understand the dependence on Re_D and the initial condition, contours of coherent strain rate $\langle S\rangle = \partial\langle v\rangle/\partial x + \partial\langle u\rangle/\partial y$ and coherent shear production $\langle P\rangle = -\langle u_r v_r\rangle\, [\partial\langle v\rangle/\partial x + \partial\langle u\rangle/\partial y]$ are shown in Figs. 6 and 7, respectively. Note that both $\langle S\rangle$ and $\langle P\rangle$ show dependence on both Re_D and the initial condition. For comparison, the contours of time-average strain rate $\bar{S} = \partial V/\partial x + \partial U/\partial y$ and production $\bar{P} = -\overline{uv}\, [\partial V/\partial x + \partial U/\partial y]$ are shown in Figs. 8 and 9, respectively.

Note that even the time-average production is dependent on Re_D. It would appear that the structures are 'stronger' at higher Re_D, as is also evident from the $\langle u_p v_p\rangle$ contours. The spatial localization of even time-average measures like $\overline{uv}$ (Fig. 5), and especially $\bar{P}$ (Fig. 9), suggests that the structures are more effective in production at certain orientations which happen to occur at specific locations (for example, see Fig. 7). Such localization is not expected in an unforced jet as these 'productive' orientations will occur at random locations.

Contours of phase-average measures like $\langle S\rangle$ and $\langle P\rangle$ show even stronger dependence on Re_D. The contours of $\langle P\rangle$ are quite similar to those of incoherent turbulence Reynolds stress $\langle u_r v_r\rangle$ (see Ref. [2]). Therefore, contours of $\langle u_r v_r\rangle$ may serve as a good indicator of spatial distribution of $\langle P\rangle$; similar is the correspondence between $\overline{uv}$ and $\bar{P}$ contours. A generalization of these observations, however, must await further experimentation.

III. Discussion and Concluding Remarks

The shear-induced tilting of the structure towards the jet centerline has the equivalent effect of producing a radial variation of the measured wavelength (i.e., structure spacing). This is the reason why convection velocities measured by different investigators varied depending on the radial location of the measurement. The detailed contours of structure properties are found to be independent of the initial condition but show small but systematic dependence on the Re_D up to about 10^5. The structures which are elongated in the streamwise direction at low Re_D become progressively more rounded with increasing Re_D and thus become more efficient in transverse momentum transport. Consequently, both positive and negative peaks of coherent Reynolds stress increase with increasing Re_D.

Can the observed mild dependence of the structure shapes and properties on the Reynolds number continue indefinitely for progressively increasing Re_D? If the answer were affirmative, this would suggest, of course, a violation of the Reynolds number similarity or asymptotic invariance. Even though conclusive evidence confirming this invariance is still lacking, the overwhelming belief among the researchers in this invariance would imply that this dependence cannot continue indefinitely for progressively larger Re_D. Another question that arises is: Are the induced structures the same as the natural structures? Even though we have suggested that the excitation merely paces the formation of natural structures and does not produce artificial structures, many peers have remained unconvinced.

We have addressed both of these questions by educing the naturally-occurring structures through conditional sampling techniques, which have been described elsewhere [7]. It should be emphasized that the naturally-occurring structures not only have large dispersion but, contrary to the suggestions by Lau & Fisher [8] and Lau [9], the large-scale preferred-mode structures are interspersed between substructures orginating via tearing and fractional pairing [6]. Our measurements showing poor correlation between the positive spikes in $\tilde{u}(t)$ signal on the low-speed side of the mixing layer and negative spikes in $\tilde{u}(t)$ on the high-speed side further support this observation. It should also be noted that lack of correlation between the low-speed side positive peaks and high-speed side negative peaks must be unique to the axisymmetric mixing layer. Browand & Wiedman [10] used a joint triggering criterion based on hot-wire signals on both the high-speed and low-speed sides, which would suggest that there is a good correlation of the two peaks across the plane mixing layer.

After considering a number of alternatives, it was found that eduction of the structure is the most successful when triggered on the positive peaks of the $\tilde{u}(t)$ signal obtained from the high-speed side. To validate the conditional sampling technique, contours of the structure vorticity were educed first for 0.1% excitation at the preferred-mode. The same structure was then educed without the excitation by using the conditional sampling technique. For the latter case, the educed structure is slightly more smeared, as to be expected. By setting a higher threshold level and a finer threshold window, however, the smearing could be reduced but only at the sacrifice of experimental time, as fewer realizations are accepted with a more refined sampling criterion. The structures educed by employing the same conditional sampling criterion for excited and unexcited jets were considered identical within the experimental uncertainty. These data confirmed our earlier suggestion that the small-amplitude excitation does not induce artificial structures, but only controls the phases of formation of natural structures. The details of the considerations in the conditional sampling measurement techniques are described in [7] and further details of the large-scale structures at different Re_D and IC are currently being investigated.

The same conditional sampling criteria were used in different unforced jets up to $Re_D = 8 \times 10^5$ and the educed structures appeared identical within the experimental uncertainty. The structure properties for $10^5 < Re_D < 8 \times 10^5$ were also found to be independent of the initial condition. Thus, even though the structure contours indicated some dependence on Re_D up to 10^5, no Re_D dependence was found above this value. This is consistent with the speculation of Crighton [11] and the self-excited jet data of Hasan & Hussain [12]. Therefore, high Re_D jet flow characteristics should not be inferred from jets at Re_D significantly below 10^5.

Since these details pertain to the 'preferred-mode', which is unique for the axisymmetric jet, the present phase-average data provide the instantaneous organized motion for the near field of any jet.

The preferred mode structure is attained in an initially laminar jet through successive pairings of the initially rolled-up structures. When initially fully turbulent, the shear layer may also roll up and even undergo pairing. However, if the Reynolds number is high enough, all pairing activity is complete before $x/D \simeq 1$. This observation should convince one that pairing is not a principal source of jet noise in spite of persistent claims to the contrary [13,14]. That is, even though pairing, when induced near the end of the potential core, can produce large amounts of noise in forced jets, pairing seldom occurs in the noise-producing regions of practical jets. These reasonings led us to claim [2] that it is not pairing, but the breakdown process of the initial toroidal vortical structure into azimuthally spaced substructures (and perhaps their interactions), which produces a dominant part of the jet noise.

The financial support of NASA Langley Research Center under grant NSG-1475 and the National Science Foundation under grant ENG-7822110 is gratefully acknowledge.

References

1. Hussain, A. K. M. F., 1980 Lecture Notes in Physics 136, 252-291.
2. Hussain, A. K. M. F. & Zaman, K. B. M. Q., 1981 J. Fluid Mech. (to

appear).

3. Zaman, K. B. M. Q. & Hussain, A. K. M. F., 1980 J. Fluid Mech. 101, 449-491.
4. Husain, Z. D. & Hussain, A. K. M. F., A.I.A.A. J. 17, 48-55.
5. Hussain, A. K. M. F. & Zaman, K. B. M. Q., 1980 J. Fluid Mech., 101, 493-544.
6. Hussain, A. K. M. F. & Clark, A. R., 1981 J. Fluid Mech. 104, 263-294.
7. Zaman, K. B. M. Q. & Hussain, A. K. M. F., 1981 (submitted to J. Fluid Mech.).
8. Lau, J. C. & Fisher, M. J., 1975 J. Fluid Mech. 67, 299-337.
9. Lau, J. C., 1979 Proc. Roy. Soc. A 368, 547-571.
10. Browand, F. K. & Wiedman, P. D., 1976 J. Fluid Mech. 76, 127-144.
11. Crighton, D. G., 1980 Lecture Notes in Physics, 136, 341-362.
12. Hasan, M. A. Z. & Hussain, A. K. M. F., 1981 J. Fluid Mech. (to appear).
13. Laufer, J., 1974 in Omaggio A. Carlo Ferrari, 451-464.
14. Kibens, V., 1979 in Mechanics of Sound Generation in Flows (ed. E.-A. Müller) Springer-Verlag, 174-180.

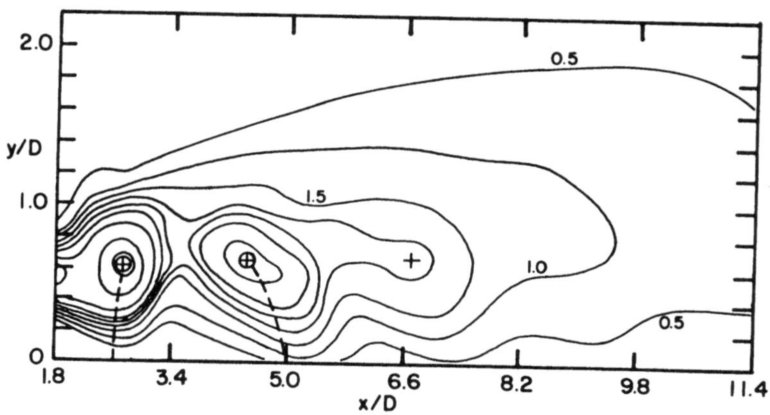

Fig. 1 Contours of Ω_z/f_p. Measurements in the 7.6cm (tripped) jet at Re_D = 55000; St_D = 0.3.

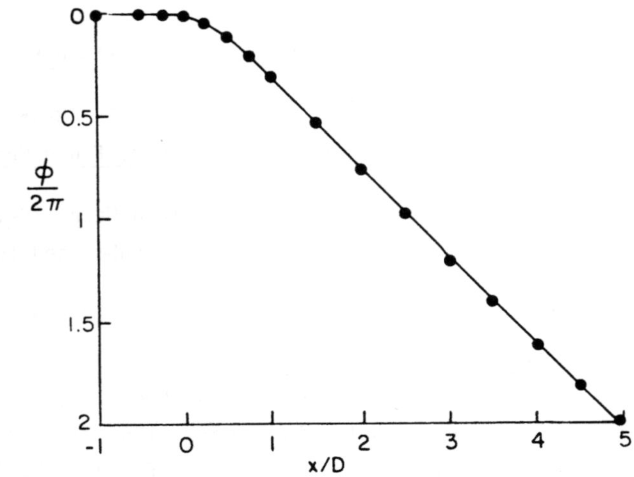

Fig. 2 $\phi(x)$ on the jet centerline for the flow in Fig. 1.

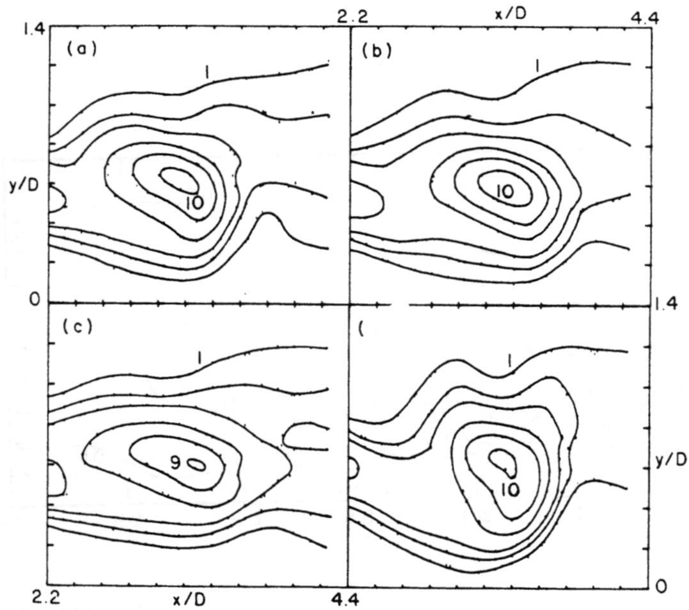

Fig. 3 Ω_z/f_p contours for excitation at $St_D = 0.3$. (a) 7.6cm (tripped) jet at $Re_D = 55{,}000$; (b) 7.6cm (untripped) jet at $Re_D = 55{,}000$; (c) 2.54cm (tripped) jet at $Re_D = 25{,}000$; (d) 7.6cm (tripped) jet at $Re_D = 110{,}000$. Unmarked contour levels are in the sequence 7,5,3, and 2.

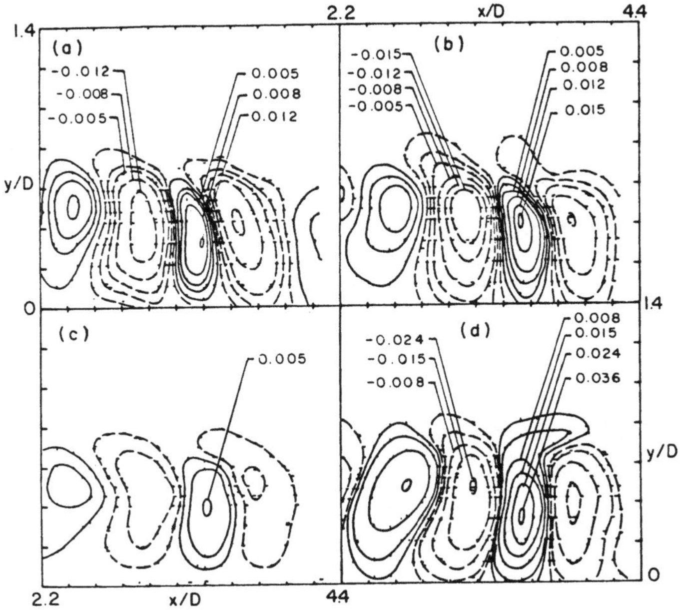

Fig. 4 $<u_p v_p> / U_e^2$ contours corresponding to the four cases in Fig. 3. The unmarked outer contour levels are ± 0.003 and ± 0.001

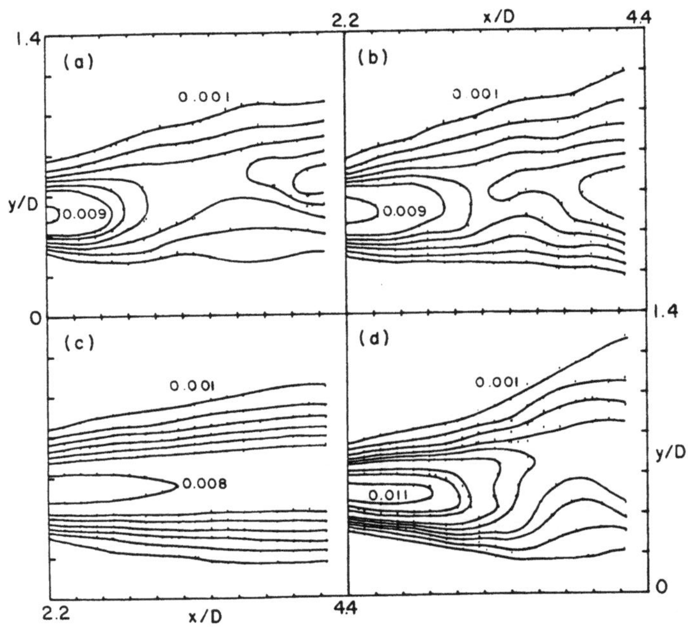

Fig. 5 Reynolds stress $\overline{uv}/U_e^2$ contours for the four cases in Fig. 3. Unmarked contour levels are in the sequence 0.006,0.005, 0.004,0.003 and 0.002.

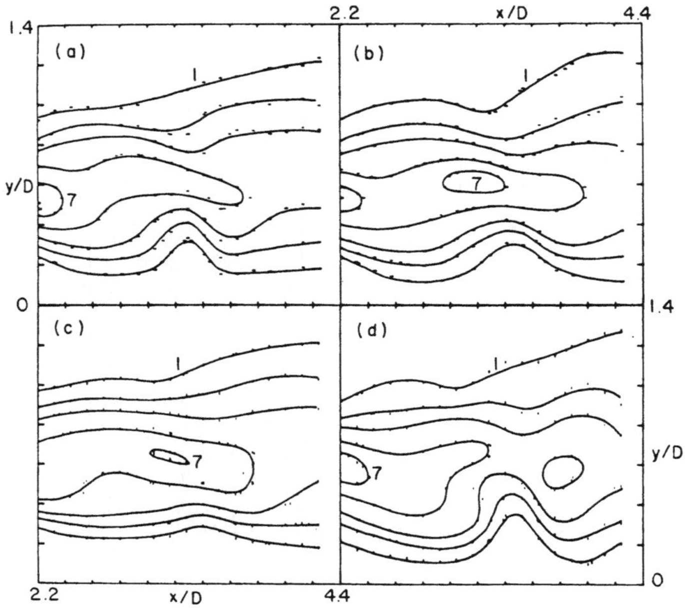

Fig. 6 The phase-average strain rate $<S>/f_p$ for the four cases in Fig. 3. Unmarked contour levels are in the sequence 5,3, and 2.

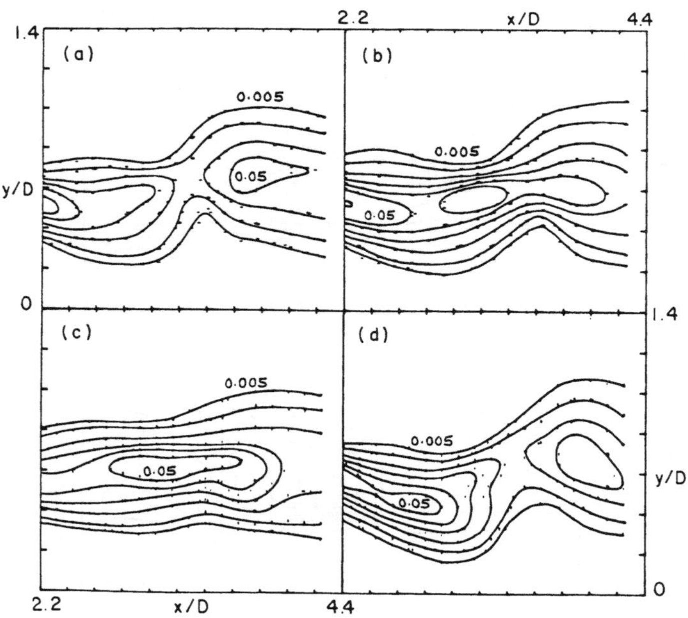

Fig. 7 The phase-average production $<P>/(f_p U_e^2)$ for the four cases in Fig. 3. Unmarked contour levels are in the sequence 0.04,0.03, 0.02, and 0.01.

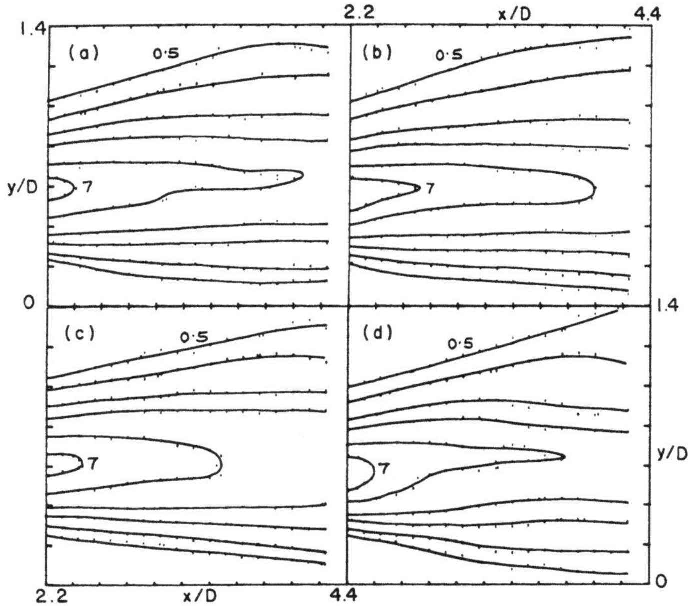

Fig. 8 The time-average strain rate $\bar{S}/f_p$ for the four cases of Fig. 3. Unmarked contour levels are in the sequence 5,3,2, and 1.

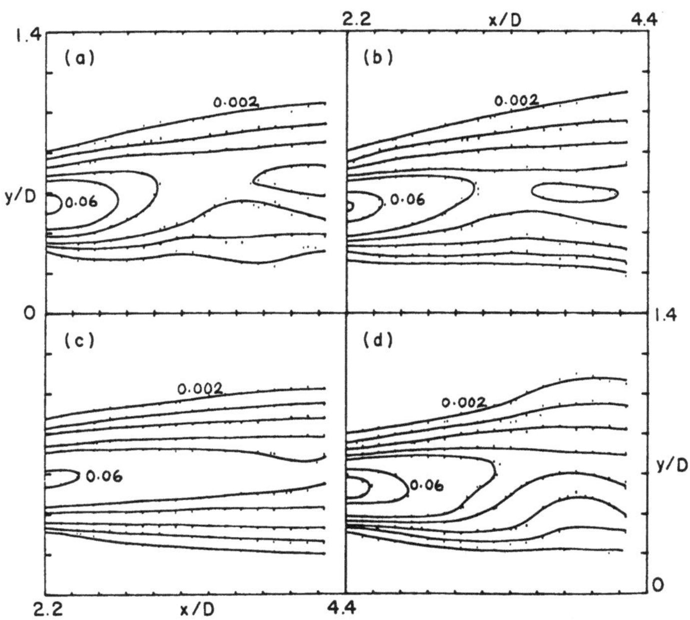

Fig. 9 The time-average production $\bar{P}/(f_p U_e^2)$ for the four cases of Fig. 3. Unmarked contour levels are in the sequence 0.08, 0.03,0.02,0.01 and 0.005.

The Effect of Forcing on the Mixing-Layer Region of a Round Jet

W. C. Reynolds and E. E. Bouchard*
Department of Mechanical Engineering
Stanford University

Summary

The ring vortex structure in the mixing-layer region of a round jet can be regulated by forcing the jet with periodic axial disturbances. Visualizations show that the initial roll-up of the laminar shear layer into ring vortices can be locked to monochromatic disturbances as weak as 0.15%. These vortices pair as they progress downstream, and this pairing results in entrainment into the jet. By forcing the jet at a sufficient amplitude and at a well-chosen frequency, the pairing, and hence the entrainment, can be suppressed in the region of the jet. At other frequencies and amplitudes only one pairing occurs, and the entrainment is related to this pairing. These observations and some simple analytical arguments indicated that entrainment into the jet occurs in two processes: first, when the initial shear layer rolls up to form discrete ring vortices, and later when (and if) the ring vortices pair.

Introduction

Large-scale, coherent structures are thought by many to be very important in governing the behavior of turbulent flows. Attention to these structures has been focused very strongly by studies of two-stream mixing layers (Brown and Roshko [1], Winant and Browand [2]), in which highly coherent vortices are observed. Since the near field of a round jet consists of a potential core surrounded by a mixing layer, it is natural to expect similar structures in the near field of a round jet. Crow and Champagne [3] observed what appeared to be ring-like vortices in the near field of a round air jet at a Reynolds number based on jet diameter and centerline discharge velocity of $Re = 20{,}000$. They also forced the jet with sinusoidal acoustic disturbances, and found that the structures and the rate of growth of the jet could be influenced by excitation. Browand and Laufer [4] made similar observations in water at lower Reynolds numbers, clearly identified the vortex rings, and found that their behavior was sensitive to the excitation conditions.

Hussain and Zaman [5,6] recently reported a comprehensive study of the vortex-pairing processes in a circular air jet under controlled, small-amplitude excitations, at $Re = 3.2 \times 10^4$. They found that the vortex-pairing process in the mixing layer is very sensitive to initial conditions and frequency, and that properly chosen excitations could produce significant increases or decreases in the turbulence intensities in the flow.

---

*Present address: Lockheed, Burbank, CA.

These and other works suggest that the entrainment processes in a jet can be controlled by properly chosen excitation. The present study deals with the nature of the vortex-pairing processes under controlled conditions, for both small and large amplitudes. The objective was to determine how the growth and entrainment in the near field of the jet are related to the pairing process, and to determine the excitation modes that can significantly increase or inhibit near-field entrainment.

This paper presents some highlights from our work. A full report on the work is now in press [7], and other, more comprehensive papers will follow.

Experimental Apparatus and Procedures

The experiments were conducted using water, which permits excellent flow visualization, in apparatus especially devised to create a very low level of background disturbances. Water is pushed upwards by a moving piston in a large cylinder, and then through a 2.54 cm diameter nozzle from which it discharges vertically as a jet into quiescent water in a large tank. The water in the supply cylinder is allowed to come to rest before each run, so the flow approaching the 36:1 contraction nozzle is virtually devoid of turbulence. This produces a very clean, thin, laminar boundary layer on the nozzle wall, and hence a very axisymmetric laminar shear layer surrounds the potential core of the jet.

The face of the piston is a diaphragm that can be excited axisymmetrically by an electromechanical system. When added to the steady motion of the piston, this produces uniform axial perturbations in the flow velocity at the nozzle exit plane; the amplitude of these oscillations can be controlled from fractions of a percent to of the order of 50% of the jet velocity. We remark that this axial excitation will produce a different sort of near-field vortical structure than will lateral excitation, which produces non-axiaymmetric structures.

Flow velocities were measured using a DISA forward-scatter, Bragg-shifted, laser anemometer operating in the tracking mode. Spectra were obtained using a spectrum analyzer with a 1-hz bandwidth. The total volume flux in the jet, Q, was determined by integrating the measured mean velocity profile:

$$Q = \int_{o}^{\infty} 2\pi r\overline{u}(r)\, dr \qquad (1)$$

This was done extremely carefully by fitting the mean profile $\overline{u}$ to segmented polynomials. An analytical "tail" was used to extrapolate the profiles to $r \to \infty$.

Visualization was accomplished by injecting neutral-density dye through a thin annular slot in the wall of the nozzle, upstream of a final contraction. The external flow was visualized by injection of dye through a small feeder tube. Vortex-formation frequencies and vortex trajectories were determined by analysis of motion pictures of the flow. For cases reported here, the jet centerline velocity was 0.226 m/s, and the slope thickness of the initial laminar shear layer was 0.86 mm. The Reynolds number of the jet was approximately 5,700.

Fourteen combinations of amplitudes and frequencies were investigated. In order to illustrate the striking changes that are possible, three cases will be described here:

Case 1: Nominally unforced

Case 2: Forcing to suppress pairing

Case 3: Forcing to permit only one pairing

This study was conducted at a fairly low Reynolds number in order to obtain very sharply defined ring vortex structures. Based on the observations in jets and mixing layer at higher Re, we believe that the work does provide insight into the near-field processes in jets at much higher Re. The recent work of Hussain and Clark [8] seems to support this view.

Case 1: Nominally Unforced

Even with no deliberate forcing, the inevitable low-level residual vibrations in the apparatus produce some repeatable disturbances. In our case the dominant disturbance was an 18 hz disturbance, apparently associated with a gear-tooth frequency, that produced a weak oscillation in the jet velocity at the nozzle exit.

Figure 1 shows a visualization of the flow in this "unforced" situation. Figure 2 shows the spectrum of axial fluctuations on the jet centerline at the nozzle exit for this nominally unforced case. Because the disturbances are monochromatic, the height of the spectral peaks gives the amplitudes of the disturbance components at the peak frequencies. The 0.5% disturbance at 36 hz should not affect the flow, because it lies outside the band of frequencies amplified by the shear layer. Indeed, no evidence of 36 hz amplification is seen in either the visualizations or in spectra taken downstream. Instead, the initial vortex-formation frequency is determined by the 0.15% peak at 18 hz, which is within the band of amplified frequencies. However, the 18 hz disturbance is weak, and this results in some jitter in the vortex-formation frequency and hence some irregularity in the spacing of the initial vortices. This irregularity is probably important in permitting vortex pairing to occur. The Strouhal number, based on jet exit velocity and diameter, is $St = 2.02$ at 18 hz.

Figure 3 shows the trajectories of the vortices for this case. Note that the first pairing consistently occurs at $x/D \simeq 1.0$. The second is more erratic, occurring in the range $1.5 < x/D < 2.5$. Beyond x/D of 4, the ring vortices develop strong azimuthal instabilities and break down into turbulent structures, which appear to retain something like a ring vortex structure.

In Fig. 1 the vortex rings rolling up from the laminar shear layer are clearly seen. The two vortices just downstream of the external dye injector are about to pair, as are the two upstream of the injector. Note the sweeping of external fluid, marked by the external dye injector, into the rings and the consolidation of this fluid in the paired structure downstream.

Case 2: Forcing to Suppress Pairing

With the proper choice of frequency and sufficient amplitude, pairing can be suppressed virtually completely. Figure 4 shows a visualization of an example situation of this type. Note that some weak interstitial vortices remain. The larger the amplitude, the weaker these interstitial vortices.

Figure 5 shows the excitation for this case. Note that it consists of a 32% amplitude oscillation at 3 hz, corresponding to $St = 0.34$, plus a number of much weaker harmonics. The vortex trajectories for this case are shown in Fig. 6. The dashed lines there show the trajectories of the weak interstitial vortices.

Case 3: Forcing To Permit Only One Pairing

Forcing at other frequencies can set up a situation in which only one pairing is observed. By forcing at 6 hz, i.e., twice the frequency of Case 2, at an amplitude of only 17%, one pairing is observed. The flow for this case is shown in Fig. 7. The excitation spectrum for this case is shown in Fig. 8, and the vortex trajectories are shown in Fig. 9. Note that the pairing occurs at $x/D \simeq 2.2$ and that the vortices orbit each other before the merging is complete at $x/D \simeq 4$. For this case, $St = 0.67$.

Effect on Entrainment and Jet Growth

The axial volume flux for these three cases is shown in Fig. 10. The regions in which pairing occur for each case are marked on the figure. Note that the volume flux for Case 1 rises in the region where the first pairing occurs, remains constant for a short way, and then rises again in the region where the second pairing occurs. For Case 2, in which there is no pairing, the volume flux is constant except in the region very near the exit, where the shear layer rolls up to form the ring vortices. For Case 3 the volume flux rises in the region where the vortices are pairing and remains constant in the region where they are orbiting. In all cases, the breakdown of the ring vortices by aximuthal disturbances produces additional growth in the jet.

These measurements suggest three mechanisms for growth and entrainment in the jet:

1. Entrainment as the shear layer rolls up to form discrete vortices;

2. Entrainment as discrete vortices merge to form stronger and more widely separated vortices;

3. Entrainment due to breakdown of the ring vortices by azimuthal disturbances.

A motion picture, presented at the conference, clearly shows the importance of the vortex-pairing mechanism. In the film, red dye is injected through an external injector, as in Figs. 1, 4, and 7. Green dye is injected in the nozzle wall boundary layer and is rolled up into green vortex rings. Under unforced conditions (Fig. 1), the red dye is entrained as it is caught up by the vortices and entwined within them as they pair. Then the forcing of Case 2 is applied, producing a flow like that of Fig. 4. The red dye begins to collect in the vicinity of the external injector, indicating that the entrainment of external flow has been reduced. Finally, the excitation is turned off and the flow returns to its "unforced" situation. Gradually, the red dye that has collected by the injector is entrained into the jet, and the area around the injector is once again clear of red dye. Clearly, pairing produces entrainment.

A simple analytical argument also suggests that pairing is responsible for entrainment. Consider an infinite row of potential ring vortices and their images behind the floor of the tank, as shown in Fig. 11. These produce a radial in-wash between the floor and the first vortex, and it is this entrainment that is responsible for the growth in volume flux immediatley downstream of the jet. Far downstream, the net radial inflow becomes zero and the average velocity profile (averaged over an x span of h) will take the shape shown in Fig. 11. The width of this shape will scale on the vortex separation h, but will be independent of the vortex strength. The only way to achieve a broader distribution is to increase the vortex separation, and this requires vortex pairing. Hence, to the degree that a round turbulent jet can be represented as the superposition of inviscid ring vortices, these vortices must be allowed to pair if the jet is to broaden.

Summary

We have shown that it is possible to control the vortex structure and merging processes in the near field of a round jet through application of appropriate disturbances, and that this effect can be used to increase or reduce the entrainment in this region. In a paper to follow, we shall present the complete data from this work and explain the relationships between the forcing parameters and the resulting structure of the near field.

Acknowledgment

This work has been supported by the Engineering Division of the National Science Foundation, as part of the program in basic research on fluid mechanics. The authors wish to express their appreciation to Dr. George K. Lea for the Foundation's support of this program.

References

1. Brown, G. L., and Roshko, A., J. Fluid Mech., 64, 775 (1974).

2. Winant, C. D., and Browand, F. K., J. Fluid Mech., 63, 237 (1974).

3. Crow, S. C., and Champagne, F. H., J. Fluid Mech., 48, 547 (1971).

4. Browand, F. K., and Laufer, J., Turb. Liquids, 5, 333, Univ. of Missouri-Rolla, (1975).

5. Zaman, K.B.M.Q., and Hussain, A.K.M.F., J. Fluid Mech., 101, 449 (1980).

6. Hussain, A.K.M.F., and Zaman, K.B.M.Q., J. Fluid Mech, 101, 493 (1980).

7. Bouchard, E.E., "The Growth and Structure of the Mixing Layer Region of a Forced Jet," Ph.D. dissertation, Dept. of Mechanical Engineering, Stanford University, June 1981.

8. Hussain, A.K.M.F., and Clark, A.R., J. Fluid Mech., 104, 203 (1981).

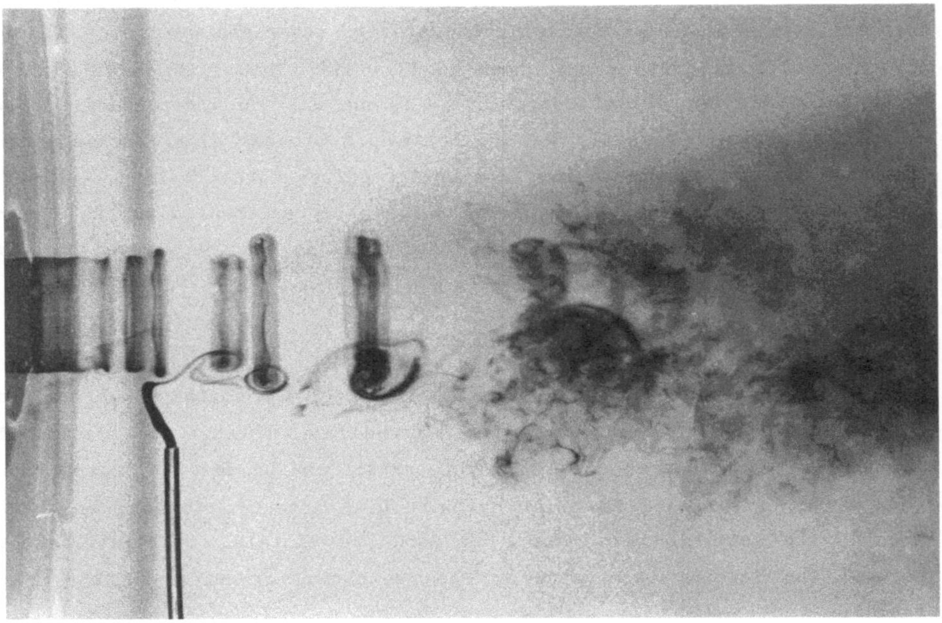

Fig. 1. Visualization without forcing; steady flow emerges from the jet at the left. Re = 5700.

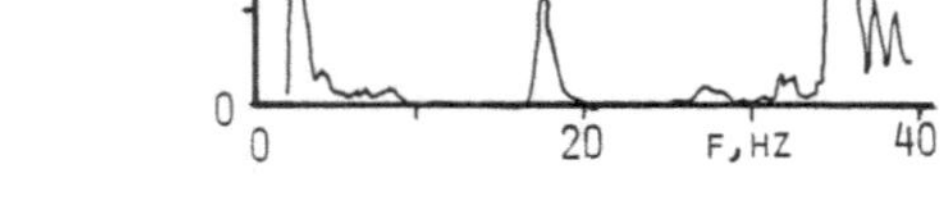

Fig. 2. Disturbance spectrum for the unforced case.

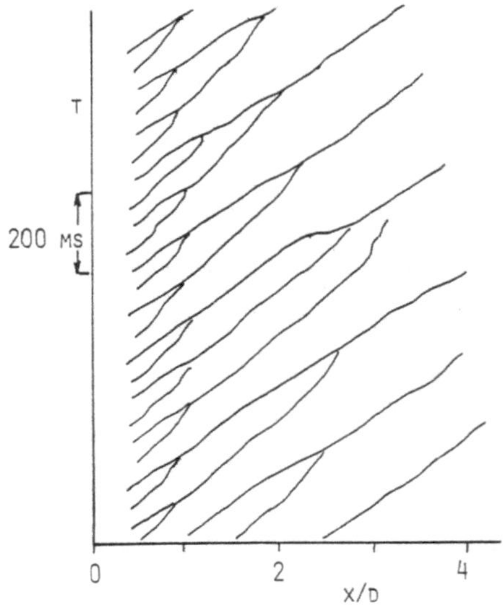

Fig. 3. Vortex trajectories for the unforced case.

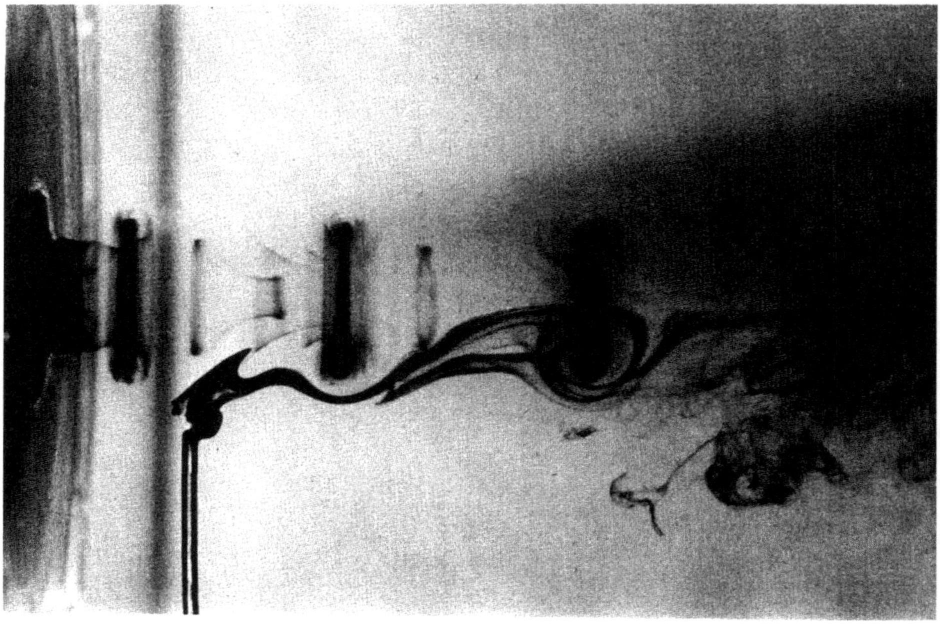

Fig. 4. Visualization of pairing-suppressed flow excitation at St = 0.34, amplitude = 32%, Re = 5700.

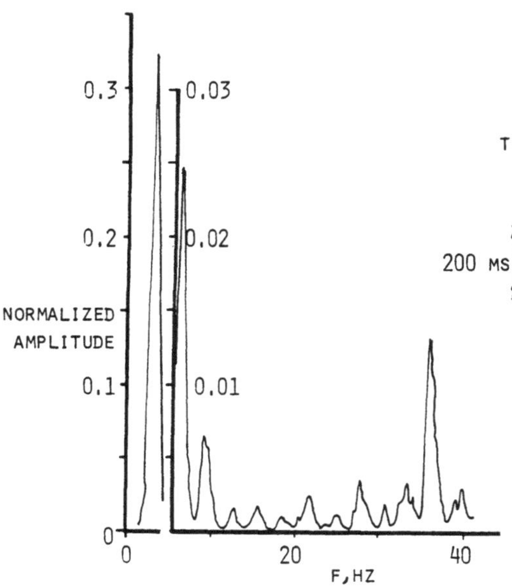

Fig. 5. Disturbance spectrum for the flow of Fig. 4.

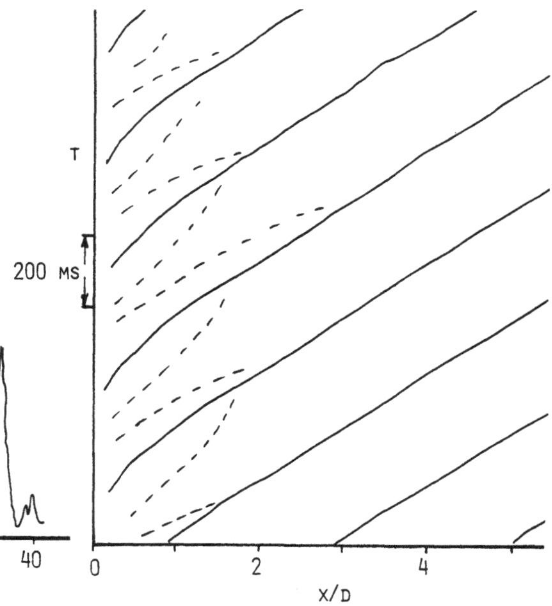

Fig. 6. Vortex trajectories for the flow of Fig. 4.

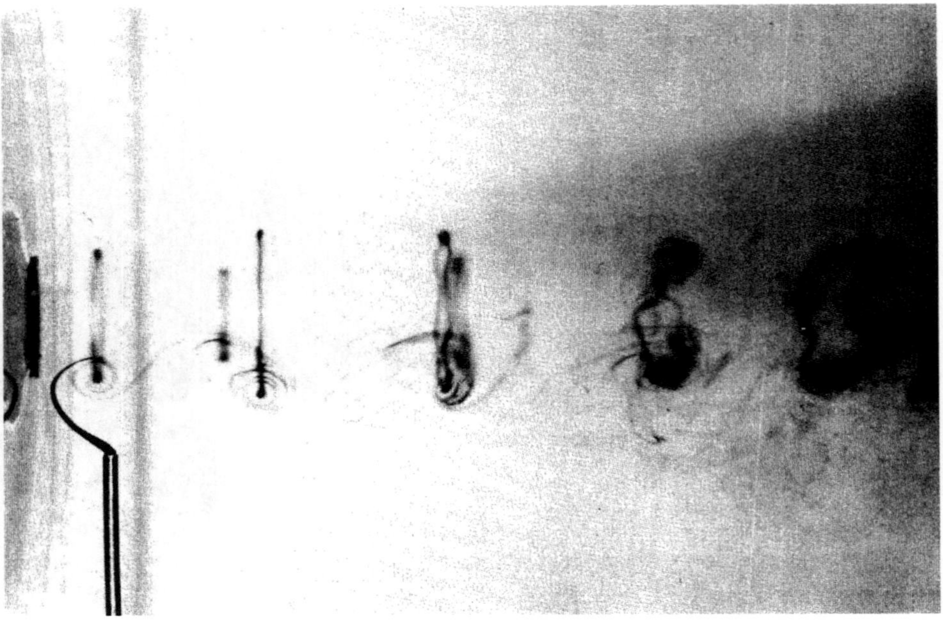

Fig. 7 Visualization of single-pairing flow; excitation at St = 0.67, amplitude = 17%, Re = 5700.

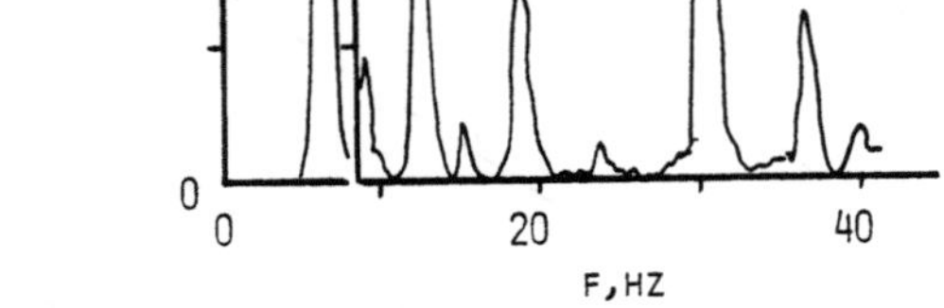

Fig. 8. Disturbance spectrum for the flow of Fig. 7.

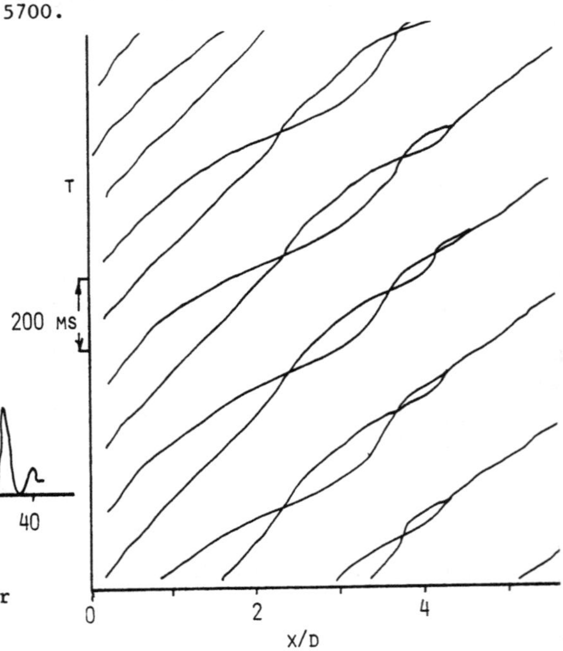

Fig. 9. Vortex trajectories for the flow of Fig. 7.

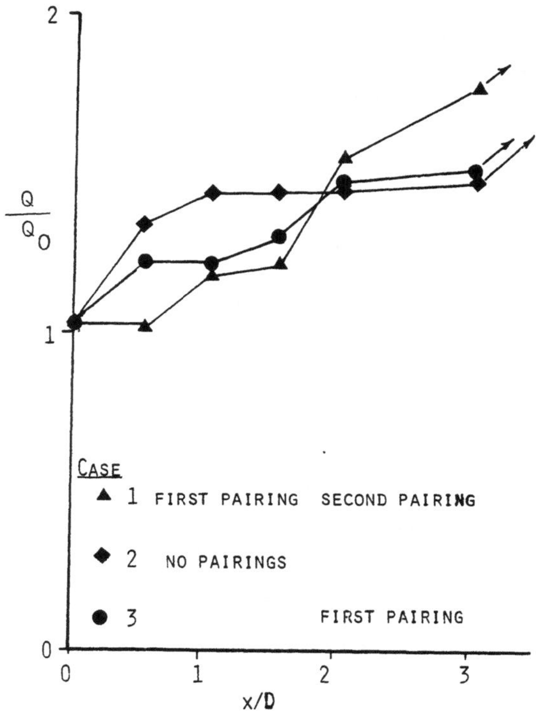

Fig. 10. Volume flow in the jet for the three cases.

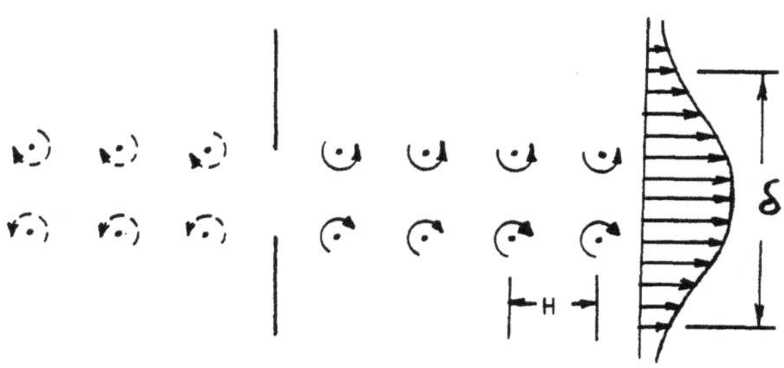

Fig. 11. Discrete ring vortex model of the round jet flow.

The Flapping Motion of a Turbulent Plane Jet: A Workable Relationship to Wave-Guide Theory

J. G. CERVANTES

School of Engineering
National University of Mexico
Mexico City, Mexico

SUMMARY

Measurements conducted in order to characterize the flapping motion of a turbulent plane jet are summarized. The flapping is defined as the lateral pseudoperiodic motion of the flow field of the jet. The technique employed in these measurements was to compute crosscorrelation functions between the velocity fluctuations (one delayed in time) in the longitudinal direction, at two points on opposite sides of the jet. Standard hot-wire anemometry and on-line digital processing instrumentation were used. It is suggested that a wave-guide representation could be used in modeling some of the experimental observations herein reported and a review of the main features of such a theory in the light of the experiments is presented.

1. Introduction

Measurements have been conducted in order to characterize the apparent flapping motion of a turbulent plane jet, Goldschmidt and Bradshaw (1), Cervantes and Goldschmidt (2). The flapping is defined as the (apparently natural) lateral pseudoperiodic motion (in the average) of the flow field of the jet. The technique employed in these measurements was to compute crosscorrelation functions between the velocity fluctuations (one delayed in time) in the longitudinal direction, at two points on opposite sides of the jet. Standard hot-wire anemometry and on-line digital processing instrumentation were used. Two types of measurements are reported in this work: (a) with the probes symmetrically positioned with respect to the centerline of the jet; and (b) with the probes at points of different longitudinal coordinate.

The existence of an apparent local flapping was indicated by a distinctive negative value at zero time delay and a quasiperiodic nature of the measured crosscorrelation functions. These were obtained at longitudinal stations ranging from 10 to 100 times the slot width. The frequency characterizing the flapping motion showed approximate self-preservation (when scaled by the half-width and the mean velocity at the centerline of the jet) for stations with longitudinal coordinate larger than 30 times the slot width. It was found that $fb/U_m \approx 0.11$. The flapping frequency did not show dependence on the lateral coordinate. The amplitude of flapping was estimated to about 20% of the jet half-width. The flapping behavior seems to travel in the downstream direction with a velocity smaller than the convective velocity of the turbulent structure.

As an attempt to put the experimental results in a proper theoretical framework, a consideration was made of the current efforts in the stability theory of turbulent shear flows, although no mathematical model is proposed here but instead a summary of the main features of such theory in the light of the reported experiments.

2. Experimental Procedures and Results

Performed Measurements

The use of correlation techniques in this investigation is circumscribed by the following normalized crosscorrelation function

$$\underline{R}\ (\tau) = \frac{\overline{u(y,t)u(-y,t+\tau)}}{\sqrt{\overline{u^2(y)}}\ \sqrt{\overline{u^2(-y)}}} \qquad (1)$$

that is, a time correlation function between axial velocity fluctuations at opposite sides of the jet, (Fig. 1a). A positive value of $\underline{R}$ for zero time delay between signals ($\tau=0$), would indicate that both velocity fluctuations change in the same direction (in the average), whereas a negative value would indicate that when one fluctuation increases, the other one generally decreases.

The experimental facility consisted of a vertical rectangular nozzle of 0.63x30.48cm (aspect ratio 48) supplied with air by a 1/4 HP squirrel-cage blower. A flexible duct connector between the blower outlet and the plenum chamber was employed in order to avoid the transmission of vibrations from the blower. The plenum chamber of dimensions 10.25cm width, 30.48cm height and 67.5cm long, was provided with a series of honeycombs and screens to ensure uniform flow. The test section downstrean of the nozzle exit was confined between two 90.2x122cm plywood horizontal plates. In this way, a turbulent jet with Reynolds number based on the slot width (0.635cm) and exit velocity (2.43×10^3cm/sec) of aproximately 10^4, was obtained. Full documentation of the mean flow can be found in Cervantes (3).

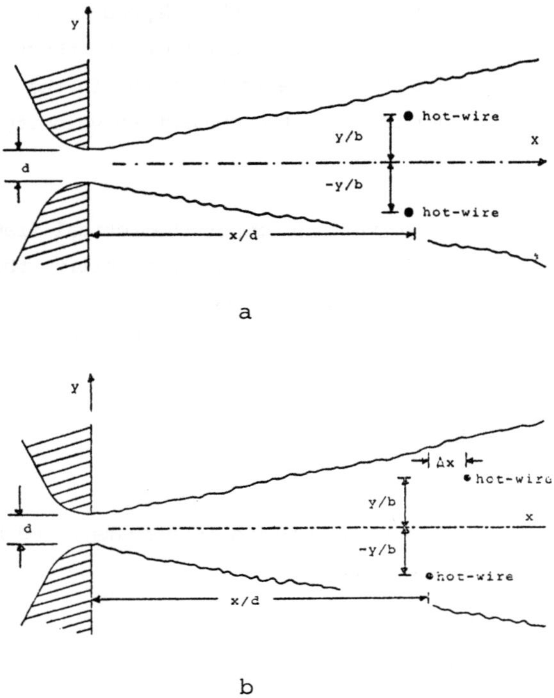

Fig. 1 Probes configurations

Two DISA 55A01 constant-temperature anemometers with DISA 55A22 and DISA 55F11 probe supports and "home-made" plated tunsten sensing elements (0.0002cm diameter), were used to obtain the velocity signals at any two points in the flow. In order to measure cross-correlation functions, a Hewlett-Packard 5452A-2114B Fourier Analyzer System was used. This on line digital system includes the Fast Fourier Transform algorithm among its software, and can be programmed to take any number of samples for a particular situation, thus obtaining a better estimate of the correlation function.

After having carefully positioned and aligned the probes, and properly programmed the analyzer, up to four estimates of the correlation function were determined for each run at each couple of points of interest. Each estimate included 250 samples.

Crosscorrelation measurements between x-component velocity signals obtained from probes set at two points symmetrically apart with respect to the centerline of the jet, were performed at x/d stations of 10,20,30,40,60,80 and 100; the probes were positioned at y/b values of 0.5, 0.75, 1.0 and 1.25 approximately, for each x/d station (Fig. 1a). Modification of the electrical motor and pulleys driving the blower, allowed runs at Reynolds numbers of 7900 and 15100. These included measurement of crosscorrelations at x/d stations of 40, 60 and 80.

A second type of crosscorrelation measurements was obtained by holding one probe at a given x/d station and placing the second probe at the opposite side of the jet but at a different x/d position (Fig. 1b). The longitudinal separation between probes Δx, was varied from 0 to 5.1cm in steps of 1.27cm. Three x/d stations (20,40 and 60) for the fixed probe were investigated.

Frequency of Flapping

All measured crosscorrelation functions between x-components of the velocities at two points on opposite sides of the centerline of the jet, are consistent and repeatable. The two most important characteristics, indicating the existence of an oscillating (in the average) lateral motion or flapping of the jet, are: an invariably negative value of the correlation function at zero time delay and an alternate succession of positive and negative values as the time delay between the two signals increases. An example of these correlations (as obtained directly from the Fourier analyzer through an x-y plotter), is given in Fig. 2.

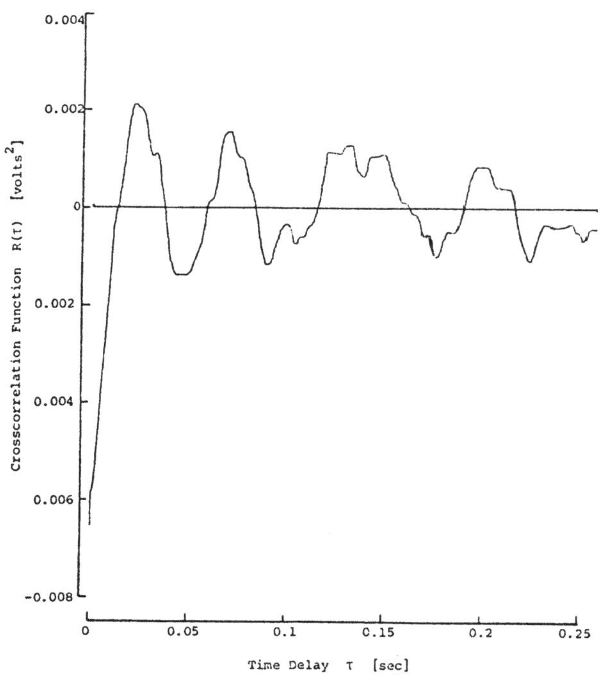

Fig. 2 An example of crosscorrelation function, x/d=40, y/b=$\pm$0.5.

The correlations exhibit a certain periodicity. This is related to and gives a measure of the flapping frequency. (Simultaneously, the magnitude of the correlation at τ=0 would give a measure of the amplitude of the oscillatory-like flapping motion). The times between local maxima or local minima, are generally not equal. However, the time to the first maximum is clearly measurable and it suggests the definition of a frequency of flapping.

The resulting frequencies of flapping for the complete survey along the x-direction, are plotted as a Strouhal number fd/U_o in Fig. 3. As noted, the flapping frequency decreases as x/d increases. On the other hand, this frecuency does not seem to depend on the lateral coordinate. This is noted by the collapse on Fig. 3, of the data points taken at four different lateral positions. This is so at least for x/d>30.

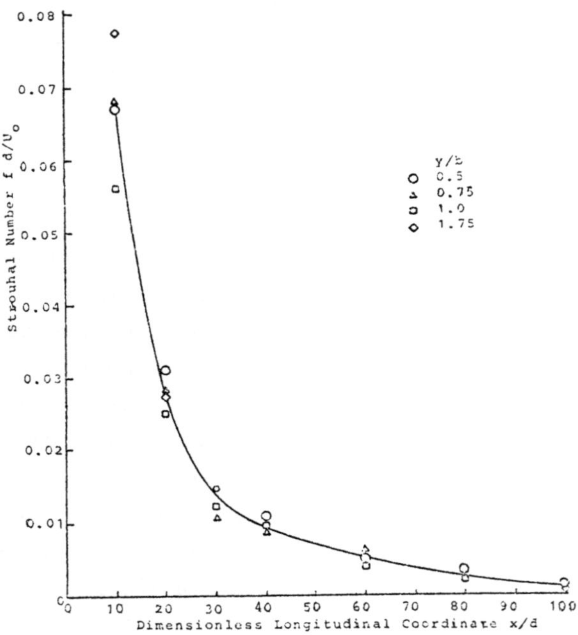

Fig. 3 Strouhal number fd/U_o vs x/d

Several scaling parameters for f were considered. The intent was to exhibit any possible similarity and self-preservation for the flapping frequency. The various jet coordinates, lenght and velocity scales (x,y,d,b,U,U_m)were combined is this search. The best results are shown in Fig. 4 and 5. A fit of the type $fy/U_m \simeq 0.11\ y/b$ in Fig. 5 seems reasonable and actually corresponds to $fb/U_m \simeq 0.11$. The flapping frecuency then, attains approximate self-preservation if scaled with the jet half-width b and the mean velocity at the centerline.

Using this results, a replotting of the correlation functions was done through a computer Calcomp system. The abscissa τ was modified to a dimensionless time delay $\tau U_m/b$ and the ordinate was divided by the product of the r.m.s values of the signals (to have a normalized crosscorrelation function). Fig. 6 is an example. It is interesting to note that not only does the flapping frequency exhibit approximate self preservation and similarity for x/d>30,

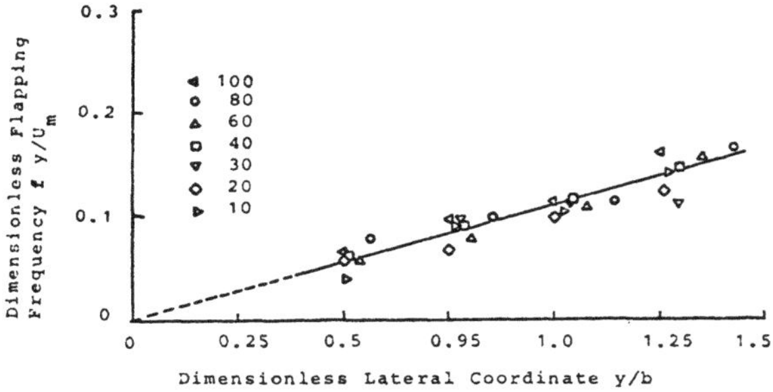

Fig. 4 Dimensionless flapping frequency fy/U_m vs x/d.

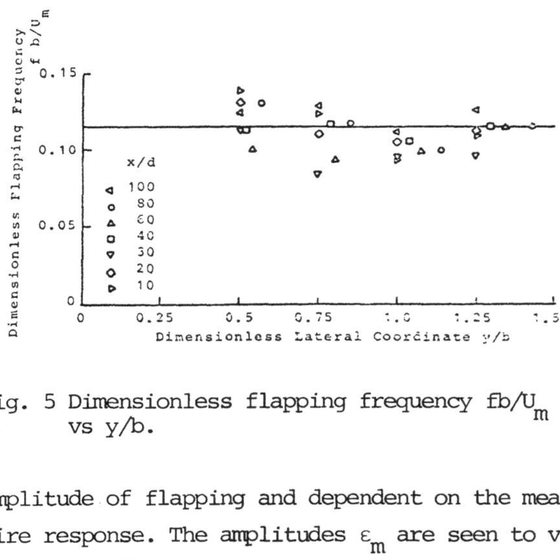

Fig. 5 Dimensionless flapping frequency fb/U_m vs y/b.

but so does the magnitude of the crosscorrelation function (at least for the first half-period).

Amplitude of Flapping

An estimate of the amplitude of flapping can be done by assuming a sinusoidal local displacement of the mean velocity profile of the jet from its mean location. In effect, the instantaneous displacement of the mean velocity profile may be assumed as $\varepsilon=\varepsilon_m \sin 2\pi ft$ where f is the frequency of flapping and ε_m is the amplitude to be estimated. As proposed in Goldschmidt and Bradshaw (1), the amplitude of the crosscorrelation at zero time delay is related to the amplitude of flapping and dependent on the mean velocity gradient and hot-wire response. The amplitudes ε_m are seen to vary from 0.15b to 0.23b as noted in Table 1.

Table 1 Amplitude of Flapping

x/d	y/b	ε_m/b
80	1.13	0.18
	0.85	0.21
	0.56	0.18
60	1.07	0.17
	0.80	0.19
	0.53	0.23
40	1.03	0.15
	0.77	0.16

Convection of the Flapping Behavior

The second kind of correlation measurements mentioned before were done with the hot-wire probes at different longitudinal coordinates on opposite sides of the jet. The original intent was to further investigate the uniformity of the apparent flapping motion. The data also gives insight on how the flapping behavior

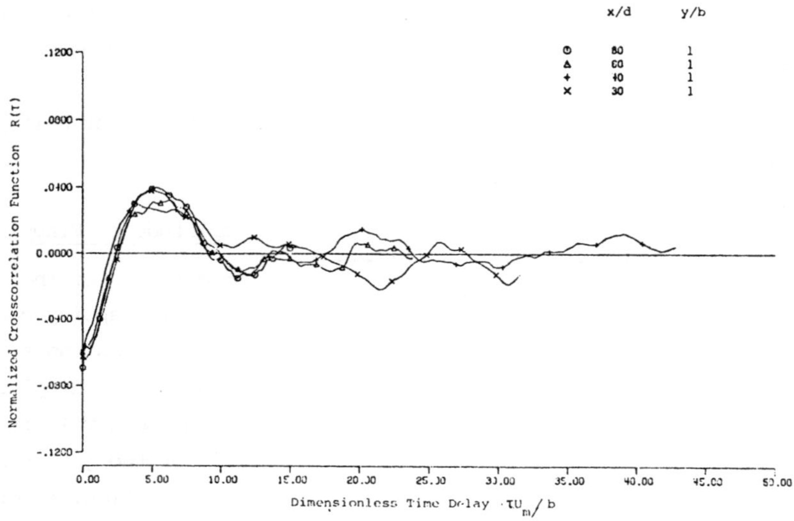

Fig. 6 Normalized crosscorrelation functions, y/b=1

is convected downstream of a given x/d position. The crosscorrelation functions were measured with one probe fixed at a certain x/d and the other was successively positioned in steps of 2d downstream. Fig. 7 show an example of a sequence of such correlation function for increasing downstream separation. It exhibits a convected pattern where the time delay for a maximum negative correlation shifts by a time $\Delta\tau$ as the probes are Δx apart. This indicates that the flapping behavior as detected at a given longitudinal station can still be sensed with considerable coherence some time later at a downstream location.

A convection velocity for the apparent flapping motion can be defined in the limit,

$$U_{cf} = \left.\frac{\Delta x}{\Delta\tau}\right|_{\Delta x \to 0} \tag{2}$$

Δx and $\Delta\tau$ can be plotted for each set of curves; an extrapolation to a zero separation distance would permit an estimate of U_{cf}. Table 2 presents the results and compares them to the data of Young (4) who, in the same setup, measured the turbulent convective velocity U_c (defining the apparent motion of the turbulent structure). These two velocities are noted to be different, with a lower value for the flapping motion.

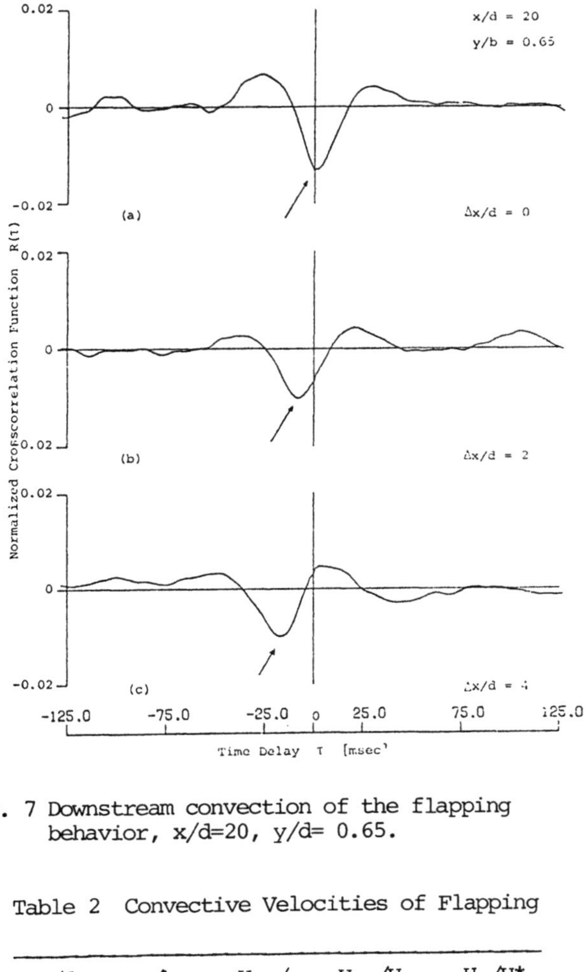

Fig. 7 Downstream convection of the flapping behavior, x/d=20, y/d= 0.65.

Table 2 Convective Velocities of Flapping

x/d	y/b	U_{cf}/u	U_{cf}/U_m	U_c/U*
10	1	1.16	0.58	
20	1	1.28	0.64	1.68
30	1	1.15	0.57	1.43
40	1	1.23	0.61	1.58
20	0.65	0.99	0.74	1.18
30	0.65	0.94	0.70	1.1
40	0.65	1.01	0.75	1.18

* Measurements by Young (4)

3. Summary of Effects

The following conclusions can be established, based on the experimental results:

1) The flapping motion of a turbulent plane jet is a distinctive and measurable natural phenomenon. This was confirmed by spot-checks at two other jet setups and by a previous preliminary investigation (Goldschmidt and Bradshaw (1)).
2) The frequency of flapping decreases in the longitudinal direction and remains unchanged in the lateral direction.
3) Approximate self-preservation is obtained for the flapping frequency (for x/d>30), if scaled with the centerline mean velocity and the half-width of the jet, giving $fb/U_m \approx 0.11$.
4) An estimate of the amplitude of flapping gave values

in the order of 20% of the jet half-width.

5) The flapping behavior travels in the downstream direction with a velocity about 25% smaller than the turbulent structure convective velocity.

4. Wave-Guide Representation of a Turbulent Shear Flow

In trying to put the summarized results in a proper theoretical framework, one is tempted to consider the current efforts in the stability theory of turbulent shear flows. Is should be stated however that no mathematical modeling is intended here, but instead a review of the main features of such theory in the light of the experiments reported in this work.

Considerable interest has developed in modeling the large-scale structure of turbulent shear flows through stability analyses. This parallel approach has been mainly motivated by the jet-noise generation problem. The theory of hydrodynamic stability has been applied with considerable success to some simple laminar flows in the verge of becoming turbulent, and the similarity between the turbulent coherent structures and the events leading to transition in laminar flows, have motivated some investigators to study the former through stability theory.

The works by Landhal (5) and by Corcos (6) are examples of this postulated relationship between the turbulent structure and the events related to transition. In attempting to relate the statistical properties of the wall pressure in a turbulent boundary layer to the characteristics of the mean flow, the first author formulated a wave-guide model (i.e., the admission of wave propagation modes) for the turbulent velocity fluctuations: considerable agreement with the experimental results reported by the second author was obtained for the streamwise decay of the fluctuations.

One method of getting an insight into shear flow structure and experimentally testing wave-guide models is to introduce a periodic disturbance of known frequency and amplitude into the flow and to determine the response of the flow to it. Such a procedure was followed by Crow and Champagne (7) for circular jet. According to their results, the preferred wave attained its maximum amplitude through the combined effects of linear amplification and nonlinear saturation at a certain axial position and then gradually decayed downstream.

In getting apropriate data for the development of the wave-guide approach of shear flow turbulence, Hussain and Reynolds (8, 9) performed experiments on a channel flow with artificially introduced wave disturbances. Using a Phase averaging technique with selective sampling through the cyclic position of the wave-maker, they surveyed downstream of the initial disturbance and extracted the organized wave motion from a hot-wire signal. Realizing that the experimental results suggested that more than one wave mode was excited by the initial disturbance, they performed an approximate single-mode linear analysis of the propagation characteristics. The dynamical equations were further derived for the small amplitude wave disturbances and closure schemes were proposed.

The wave-guide representation of turbulent shear flows has some inherent difficulties. For instance, the case of interest in a physical problem is included in spatial stability theory. However, this formulation predicts an exponential amplification of the wave in the downstream direction and the original small amplitude assumption becomes inconsistent, unless some kind of limitation is imposed on that amplification.

In any case, a realistic wave-guide model should account for a spatially growing wave, which attains a maximum growth through some kind of saturation effect after non-linear interactions. Thereafter it might begin to decay, as the experiments of Crow and Champagne (7) suggested for the circular jet.

One way of studying turbulent shear flows through a wave-guide representation, is by separating the flow field into three components: the mean flow, the instability wave and the small-scale turbulent fluctuations. Once the fluctuating quantities have been decomposed in the above three terms, they are then introduced into the basic equations. An ordinary time average is taken first, followed by a phase averaging. This leads to dynamical equations for each one of the components. The phase averaging is defined as the average over a large number of periods of the wave at a given position and is physically realizable for instance, by considering an ensemble of points having the same phase with respect to a reference oscillator, when a disturbance is artificially introduced (Hussain and Reynolds (8)).

Reynolds and Hussain (10) pointed out a serious closure problem in the equations for the organized wave as the result of the outlined averaging procedures. They further considered the necessity to take into account the interaction of the organized waves and the small scale fluctuations, perhaps through the use of an eddy viscosity model, as Landahl (5) had previously speculated.

Following the same approach of separating the flow field into three components, Liu (11) formulated a model for a compressible mixing layer and later extended it to a two-dimensional jet (Merkine and Liu (12)). The organized wave component was assumed as the product of an amplitude function, determined from the kinetic energy integral equation, and a shape function obtained from the linear eigenvalue problem corresponding to the local mean flow. The various flow quantities (assumed to include such wave component) were introduced into the basic equation of motion and time-and phase-averages were perfomed. An integral equation was in turn obtained for the kinetic energy of the instability wave and numerical solutions were sought after considered proper boundary conditions

The various energy exchange mechanisms in the equation for the kinetic energy of the instability wave determine the development of the amplitude of the organized wave. In the first stages, the amplitude increases rapidly as the result of larger production as compared with dissipation. As the development continues, the dissipation term dominates over the production one, causing the amplitude of the wave to attain its maximum value and then to decay.

The above described process is retarded for the low frequency components; in other words, the lower the frequency of the wave, the further downstream its maximum amplitude is attained.

Finally, the role of the varicose and sinuous modes (although not much differentiated at the early stages) is such that the sinuous waves have larger amplification rates and show more delayed saturation values in the downstream direction.

6. Discussion and Concluding Remarks

The existence of some kind of organization in the large-scale structure of the initial regions of turbulent shear flows, is nowadays an accepted fact, as reviewed, for instance, by Davies and Yule (13). Depending on the particular shear flow, these coherent structures are viewed as randomly spaced vortex-like patterns moving and increasing is size in the mean flow direction, and interacting with each other through a coalescence process.

Two comments however, should be made in relation to the above description. In the first place, the state of affairs so described refers to the near-field region of turbulent shear flows and the effect of this coherence in the far-field self-preserving region, if any, has not been fully studied. On the other hand, the reported investigations for the case of a plane turbulent jet are rather scarce, even with respect to the initial regions.

It may be argued in relation to the second comment, that a plane jet might essentially behave as a double mixing layer, (with each one of its components the mirror image of the other). Although this conception is geometrically true for the initial region, it is a rather simplistic view of the plane jet in its fully developed region. However, the interaction of the two layers, both as they influence the potential core in between, and as they come together in the fully marging region, could account for some of the distinctive characteristics of the plane jet even farther downstream.

In any case, it may be questioned if hidden in the general turbulent structure which characterized the self-preserving region of a plane jet, there may be any kind of manifestation of the series of events occurring in the initial region. The flapping behavior of the jet may be one of these manifestations.

A wave-quide representation could be used in modeling some of the experimental observations abridged in this work. For instance, the sinuous modes (the most amplifield ones) predicted by stability theory might account for the pseudo-periodic u-components of the velocity (in antiphase) characterizing the flapping motion. On the other hand, the theoretical result that the lower frequency wave components have the tendency to reach their maximum growth further downstream, might be related to the realtively low and decreasing frequencies associated with the flapping behavior.

In concluding this review, a perspective of the plane turbulent jet, might be as follows. In the initial region, the mixing layers, one at each side of the potencial core, seem to be characterized by the presence of large scale coherent structures. As the two layers merge together, the coherent structures interact through a yet process. As a result, the possible existence of organized waves as predicted by stability theory, might be manifested in the far downstream self-preserving region in terms of distintive phenomena, among them the flapping motion.

The objetive of this experimental investigation was to charactize the flapping behavior of a plane turbulent jet hoping to contribute to a better insight of the structural properties of turbulence, particularly in the context of the new trends in turbulence research, where a more deterministic point of view is employed.

Acknowledgments

Financial support for this investigation was provided by Consejo Nacional de Ciencia y Tecnología, of Mexico; by the National Science Foundation grant ENG 74-20780, and partially by the Office of Naval Research N00014-67-0226-0025.

References

1. Goldschmidt, V.W., and Bradshaw, P., The Physics of Fluids, 16, 354, 1973.
2. Cervantes, J.G. and Goldschmidt, V.W., Trans. ASME, J. of Fluids Engineering, (in press), 1981.
3. Cervantes, J.G., PH. D. Thesis, Purdue University, 1978.
4. Young, M.F., M.S.M.E., Thesis, Purdue University, 1973.
5. Landahl, M.T., J. Fluid Mech., 29, 441, 1967.
6. Corcos, G.M., J. Fluid Mech., 18, 353, 1964.
7. Crow, S.C. and Champagne, F.H., J. Fluid Mech., 48, 547, 1971.
8. Hussain, A.K.M.F. and Reynolds, W.C., J. Fluid Mech., 41, 241, 1970.
9. Hussain, A.K.M.F. and Reynolds, W.C., J. Fluid Mech., 54, 241, 1972.
10. Reynolds, W.C. and Hussain A.K.M.F., J. Fluid Mech., 54 263, 1972.
11. Liu, J.T.C., J. Fluid Mech., 62, 437, 1974.
12. Merkine, L. and Liu, J.T.C., J. Fluid Mech., 70, 353, 1975.
13. Davies, P.O.A.L. and Yule, A.J., J. Fluid Mech., 69, 513, 1975.